Lecture Notes in Computer Science 6656

Commenced Publication in 1973
Founding and Former Series Editors:
Gerhard Goos, Juris Hartmanis, and Jan van Leeuwen

John Domingue   Alex Galis
Anastasius Gavras   Theodore Zahariadis
Dave Lambert   Frances Cleary
Petros Daras   Srdjan Krco
Henning Müller   Man-Sze Li
Hans Schaffers   Volkmar Lotz
Federico Alvarez   Burkhard Stiller
Stamatis Karnouskos   Susanna Avessta
Michael Nilsson (Eds.)

The Future Internet

Future Internet Assembly 2011:
Achievements and Technological Promises

 Springer

Volume Editors

John Domingue
Alex Galis
Anastasius Gavras
Theodore Zahariadis
Dave Lambert
Frances Cleary

Petros Daras
Srdjan Krco
Henning Müller
Man-Sze Li
Hans Schaffers
Volkmar Lotz

Federico Alvarez
Burkhard Stiller
Stamatis Karnouskos
Susanna Avessta
Michael Nilsson

Acknowledgement and Disclaimer
The work published in this book is partly funded by the European Union under the
Seventh Framework Programme. The book reflects only the authors' views. The Union
is not liable for any use that may be made of the information contained therein.

ISSN 0302-9743
ISBN 978-3-642-20897-3
DOI 10.1007/978-3-642-20898-0
Springer Heidelberg Dordrecht London New York

e-ISSN 1611-3349
e-ISBN 978-3-642-20898-0

Library of Congress Control Number: 2011926529

CR Subject Classification (1998): C.2, H.3.5-7, H.4.3, H.5.1, K.4

LNCS Sublibrary: SL 5 – Computer Communication Networks and Telecommunications

Typesetting: Camera-ready by author, data conversion by Markus Richter, Heidelberg

Printed on acid-free paper

Springer is part of Springer Science+Business Media (www.springer.com)

List of Editors

John Domingue
Knowledge Media Institute, The Open University, STI International, Milton Keynes, UK
and STI International, Vienna, Austria
j.b.domingue@open.ac.uk

Alex Galis
Department of Electronic and Electrical Engineering, University College London, UK
a.galis@ee.ucl.ac.uk

Anastasius Gavras
Eurescom GmbH, Heidelberg, Germany
gavras@eurescom.eu

Theodore Zahariadis
Synelixis/TEI of Chalkida, Greece
zahariad@synelixis.com

Dave Lambert
Knowledge Media Institute, The Open University, Milton Keynes, UK
d.j.lambert@gmail.com

Frances Cleary
Waterford Institute of Technology – TSSG, Waterford, Ireland
fcleary@tssg.org

Petros Daras
CERTH-ITI, Thessaloniki, Greece
daras@iti.gr

Srdjan Krco
Ericsson Serbia, Belgrade, Serbia
srdjan.krco@ericsson.com

Henning Müller
Business Information Systems, University of Applied Sciences Western Switzerland,
Sierre, Switzerland
henning.mueller@hevs.ch

Man-Sze Li
IC Focus, London, UK
msli@icfocus.co.uk

Hans Schaffers
ESoCE Net, Dialogic, Aalto University School of Economics (CKIR), Aalto, Finland
hschaffers@esoce.net

Volkmar Lotz
SAP Research, Sophia Antipolis, France
volkmar.lotz@sap.com

Federico Alvarez
Universidad Politécnica de Madrid, Spain
fag@gatv.ssr.upm.es

Burkhard Stiller
University of Zürich, Switzerland
stiller@ifi.uzh.ch

Stamatis Karnouskos
SAP Research, Karlsruhe, Germany
stamatis.karnouskos@sap.com

Susanna Avéssta
Université Pierre et Marie Curie (UPMC), Paris 6, France
susanna.avessta@lip6.fr

Michael Nilsson
Centre for Distance-Spanning Technology, Luleå University of Technology, Sweden
michael.nilsson@cdt.ltu.se

Foreword

The Internet will be a catalyst for much of our innovation and prosperity in the future. It has enormous potential to underpin the smart, sustainable and inclusive growth objectives of the EU2020 policy framework and is the linchpin of the Digital Agenda for Europe. A competitive Europe will require Internet connectivity and services beyond the capabilities offered by current technologies. Future Internet research is therefore a must.

Since the signing of the Bled declaration in 2008, European research projects are developing new technologies that can be used for the Internet of the Future. At the moment around 128 ongoing projects are being conducted in the field of networks, trustworthy ICT, Future Internet research and experimentation, services and cloud computing, networked media and Internet of things. In total they represent an investment in research of almost 870 million euro, of which the European Commission funds 570 million euro.

This large-scale research undertaking involves around 690 different organizations from all over Europe, with a well-balanced blend of 50% private industries (SMEs and big companies with equal share), and 50% academic partners or research institutes. It is worth noting that it is a well-coordinated initiative, as these projects meet twice a year during the Future Internet Assembly, where they discuss research issues covering several of the domains mentioned above, in order to get a multidisciplinary viewpoint on proposed solutions.

Apart from the Future Internet Assembly, the European Commission has also launched a Public Private Partnership program on the Future Internet. This 300-million-euro program is focused on short- to middle-term research and runs from 2011 to 2014. The core of this program will be a platform that implements and integrates new generic but fundamental capabilities of the Future Internet, such as interactions with the real world through sensor/actuator networks, network virtualization and cloud computing, enhanced privacy and security features and advanced multimedia capabilities. This core platform will be based on integration of already existing research results developed over the past few years, and will be tested on large-scale use cases. The use cases that are part of the Public Private Partnership all have the potential to optimize large-scale business processes, using the properties of the core Future Internet platform. Examples of these use cases are a smarter electricity grid, a more efficient international logistics chain, a more intelligent food value chain, smart mobility, safer and smarter cities and a smarter content creation system for professional and non-professional users.

Future Internet research is an important cornerstone for a competitive Europe. We believe that all these efforts will help European organizations to be in the driving seat of many developments of the Future Internet. This book, already the third in this series, presents some of the results of this endeavor. The uniqueness of this book lies in the breadth of the topics, all of them of crucial importance for the Future Internet.

We sincerely hope that reading it will provide you with a broader view on the Future Internet efforts and achievements in Europe!

Budapest, May 2011

Luis Rodríguez-Roselló
Mário Campolargo

Preface

1 The Internet Today

Whether we use economic or societal metrics, the Internet is one of the most important technical infrastructures in existence today. One easy measure of the Internet's
impact and importance is the number of Internet users which as of June 2010 was 2
billion[1]. But of course, this does not give one the full picture. From an economic
viewpoint, in 2010 the revenue of Internet companies in the US alone was over $70
billion[2]. In Europe, IDC estimated that in 2009 the broader Internet revenues (taking
business usage into account) amounted to €159 billion and that this is projected to
grow to €229 billion by 2014[3].

The recent political protests in Egypt give us an indication of the impact the Internet has in societal terms. At the start of the demonstrations in Egypt the Internet was
closed down by the ruling government to hinder the activities of opposition groups.
Later, as the protests were having an effect, a picture emerged in the world's media of
a protester holding up a placard saying in Arabic "Thank You Facebook[4]." Protesters
in Egypt used social media to support communication and the associated Facebook
page had over 80,000 followers at its peak. It is interesting to note that here we are
talking about the power of the Internet in a country where currently Internet penetration is 21%[5] compared to say 79% for Germany[6].

2 Current Issues

The Internet has recently been in the news with stories covering two main issues
which are commonly known in the Internet research community. Firstly, recent stories
have highlighted the issue of the lack of address space associated with IPV4, which
can cater for 4 billion IP addresses[7]. Some headlines claim that the IPV4 address
space has already run out[8]. Technically, the issue has been solved through IPV6 although there is still the matter of encouraging take up.

---

[1] http://www.internetworldstats.com/stats.htm
[2] http://money.cnn.com/magazines/fortune/fortune500/2010/industries/225/index.html
[3] http://www.fi3p.eu
[4] http://www.mediaite.com/tv/picture-of-the-day-cairo-protester-holds-sign-that-says-
     thank-you-facebook/
[5] http://www.internetworldstats.com/africa.htm#eg
[6] http://www.internetworldstats.com/europa.htm#de
[7] http://www.bbc.co.uk/news/10105978
[8] http://www.ndtv.com/article/technology/internet-will-run-out-of-ip-addresses-by-friday-
     83244

A second major news item has been on net neutrality, specifically, on legislation on net neutrality in the US and UK, which take differing views. At the time of writing the US House of Representatives voted to block a proposal from the Federal Communications Commission to partially enforce net neutrality[9]. In the UK at the end of 2010 the Culture Minister, Ed Vaizey, backed a proposal to allow ISPs to manage traffic, which advocates of net neutrality argued would lead to a "two-speed Internet[10]." Vint Cerf, Sir Tim Berners-Lee and Steve Wozniak (one of the founders of Apple) have argued in favor of retaining net neutrality[11].

These two problems have gained prominence in the world's media since they are most directly linked to the political and regulatory spheres. Other issues are centered on the fact that the Internet was originally designed in a very different context and for different purposes than it is used today. Of the changes that have occurred in the decades since the Internet's inception, the main alterations which are of concern are:

- **Volume and nature of data** – the sheer volume of Internet traffic and the change from simple text characters to audio and video and also the demand for very immediate responses. For example, Cisco's latest forecast predicts that global data traffic on the Internet will exceed 767 Exabytes by 2014. Online video and high-definition TV services are expected to dominate this growth. Cisco state that the average monthly traffic in 2014 will be equivalent to 32 million people continuously streaming the 2009 Avatar film in 3D[12].
- **Mobile devices** – the Internet can now be accessed from a wide variety of mobile devices including smart phones, Internet radios, and vehicle navigation systems, which is a radically different environment from the initial Internet based on physical links. Data traffic for mobile broadband will double every year until 2014, increasing 39 times between 2009 and 2014[13].
- **Physical objects on the net** – small devices enable the emergence of the "Internet of Things" where practically any physical object can now be on the net sending location and local context data when requested.
- **Commercial services** – as mentioned above the Internet is now a conduit for a wide variety of commercial services. These business services rely on platforms which can support a wide variety of business transactions and business processes.
- **Societal expectations** – in moving from an obscure technology to a fundamental part of human communication, societal expectations have grown. The general population demand that the Internet is at least: secure, trustworthy, ubiquitous, robust, responsive and also upholds privacy.

---

[9] http://online.wsj.com/article/BT-CO-20110217-718244.html

[10] http://www.bbc.co.uk/news/uk-politics-11773574

[11] See http://googleblog.blogspot.com/2005/11/vint-cerf-speaks-out-on-net-neutrality.html, http://www.scientificamerican.com/article.cfm?id=long-live-the-web, http://www.theatlantic.com/technology/archive/2010/12/steve-wozniak-to-the-fcc-keep-the-internet-free/68294/

[12] http://www.ispreview.co.uk/story/2010/06/10/cisco-forecasts-quadruple-jump-in-global-internet-traffic-by-2014.html

[13] http://www.ispreview.co.uk/story/2010/06/10/cisco-forecasts-quadruple-jump-in-global-internet-traffic-by-2014.html

3   FIA Overview

This book is based on the research that is carried out within the Future Internet Assembly (FIA). FIA is part of the European response to the problems outlined above. In short, FIAs bring together over 150 research projects that are part of the FP7 Challenge 1 ICT Programme to strengthen Europe's Future Internet research activities and also to maintain the EU's global competitiveness in the space. The projects are situated within established units which cover the following areas:

- The network of the future
- Cloud computing, Internet of services and advanced software engineering
- Internet-connected objects
- Trustworthy ICT
- Networked media and search systems
- Socio-economic considerations for the Future Internet
- Application domains for the Future Internet
- Future Internet research and experimentation (FIRE)

Researchers and practitioners associated with the Future Internet gather at the FIAs every six months for a dialogue and interaction on topics which cross the above areas. In conjunction with the meetings the FIA Working Groups sustain activity throughout the year working toward a common vision for the Future Internet based on scenarios and roadmaps. Since the opening FIA in the spring of 2008, we have now held FIAs in the following cities: Bled, Madrid, Prague, Stockholm, Valencia and Ghent, with the next meetings scheduled for Budapest and Poznan. An overview of FIAs and the FIA working groups can be found at the EU Future Internet portal: http://www.future-internet.eu/.

4   Book Overview

This book, the third in the series, contains a sample of the results from the recent FIAs. Our goal throughout the series has been to support the dissemination of results to all researchers as widely as possible. Therefore, as with the previous two books, the content is freely available online as well as in print form[14].

The selection process for the chapters in this text was as follows. In the middle of 2010 a call was issued for abstracts of up to 2 pages covering a relevant Future Internet topic. Accompanying this was a description of the authors indicating their experience and expertise related to FIA and Challenge 1 projects. Of the 67 abstracts submitted a subset were selected after each was reviewed by at least two editors, and the authors were then asked to produce a full chapter. A second reviewing process on the

---

[14] The previous two FIA books can be found online at http://www.booksonline.iospress.nl/Content/View.aspx?piid=12006 and http://www.booksonline.iospress.nl/Content/View.aspx?piid=16465.

full papers, where each chapter was subjected to at least two reviews, resulted in a final set of 32 chapters being selected.

The book is structured into the following sections each of which is preceded by a short introduction.

- Foundations
  - Architectural Issues
  - Socio-economic Issues
  - Security and Trust
  - Experiments and Experimental Design
- Future Internet Areas
  - Networks
  - Services
  - Content
- Applications

FIA Budapest will be the seventh FIA since the kickoff in Bled and in that time a community has emerged which continues to collaborate across specific topic areas with the common goal of investigating the issues related to the creation of a new global communications platform within a European context. This text holds a sample of the latest results of these endeavors. We hope that you find the contents valuable.

Budapest, May 2011

John Domingue
Alex Galis
Anastasius Gavras
Theodore Zahariadis
Dave Lambert
Frances Cleary
Petros Daras
Srdjan Krco
Henning Müller
Man-Sze Li
Hans Schaffers
Volkmar Lotz
Federico Alvarez
Burkhard Stiller
Stamatis Karnouskos
Susanna Avéssta
Michael Nilsson

Table of Contents

Part I: Future Internet Foundations: Architectural Issues

Part II: Future Internet Foundations: Socio-economic Issues

Part III: Future Internet Foundations: Security and Trust

Part IV: Future Internet Foundations: Experiments and Experimental Design

Part V: Future Internet Areas: Network

Part VI: Future Internet Areas: Services

Part VII: Future Internet Areas: Content

Part VIII: Future Internet Applications

Part I:

Future Internet Foundations: Architectural Issues

Introduction

The Internet has evolved from a slow, person-to-machine, communication channel to the most important medium for information exchange. Billions of people all over the world use the Internet for finding, accessing and exchanging information, enjoying multimedia communications, taking advantage of advanced software services, buying and selling, keeping in touch with family and friends, to name a few. The success of the Internet has created even higher hopes and expectations for new applications and services, which the current Internet may not be able to support to a sufficient level. On one hand, the increased reliability, availability and interoperability requirements of the new networked services, and on the other hand the extremely high volumes of multimedia content challenge the today's Internet. As a result, the "Future Internet" research and development threads have been gaining momentum all over the world and as such the international race to create a new generation Internet is in full swing.

The current Internet has been founded on a basic architectural premise, that is: a simple network service can be used as a universal means to interconnect both dumb and intelligent end systems. The simplicity of the current Internet has pushed complexity into the endpoints, and has allowed impressive scale in terms of interconnected devices. However, while the scale has not yet reached its limits, the growth of functionality and the growth of size have both slowed down and may soon reach both its architectural capability and capacity limits. The current Internet capability limit will be stressed further by the expected growth, in the next years, in order of magnitude of more Internet services, the likely increase in the interconnection of smart objects and items (Internet of Things) and its integration with enterprise applications.

Although the current Internet, as a ubiquitous and universal means for communication and computation, has been extraordinarily successful, there are still many unsolved problems and challenges some of which have basic aspects. Many of these aspects could not have been foreseen when the first parts of the Internet were built, but these do need to be addressed now. The very success of the Internet is now creating obstacles to the future innovation of both the networking technology that lies at the Internet's core and the services that use it.

We are faced with an Internet that is good at delivering packets, but shows a level of inflexibility at the network and service layers and a lack of built-in facilities to support any non-basic functionality.

In order to move forward new architectures that can meet the research and societal challenges and opportunities of Digital Society are needed. Incremental changes to existing architectures, which are enhancing the existing Internet, are also of significant importance. Such new architectures, enhancements related artefacts would be based on:

- Emerging promising concepts, which have the potential reach beyond current Internet core networking and servicing protocols, components, mechanisms and requirements.
- Integration models enabling better incorporation and usage of the communication-centric, information-centric, resource-centric, content-centric, service/computation-centric, context-centric faces and internet of things-centric facets.

- Structures and infrastructures for control, configuration, integration, composition, organisation and federation.
- Unification and higher degree of integration of the communication, storage, content and computation as the means of enabling change from capacity concerns towards increased and flexible capability with operation control.
- Higher degree of virtualisation for all systems: applications, services, networks, storage, content, resources and smart objects.
- Fusion of diverse design requirements, which include openness, economic viability, fairness, scalability, manageability, evolvability and programmability, autonomicity, mobility, ubiquitous access, usage, security including trust and privacy.

The content of this area includes eight chapters covering some of the above architectural research in Future Internet.

The "*Towards a Future Internet Architecture*" chapter identifies the fundamental limitations of Internet, which are not isolated but strongly dependent on each other. Increasing the bandwidth would significantly help to address or mitigate some of these problems, but would not solve their root cause. Other problems would nevertheless remain unaddressed. The transmission can be improved by utilising better data processing & handling and better data storage, while the overall Internet performance would be significantly improved by control & self-* functions. As an overall result this chapter proposes the following: extensions, enhancements and re-engineering of today's Internet protocols may solve several challenging limitations. Yet, addressing the fundamental limitations of the Internet architecture is a multi-dimensional problem. Improvements in each dimension combined with a holistic approach of the problem space are needed.

The "*Towards In-Network Clouds in Future Internet*" chapter explores the architectural co-existence of new and legacy services and networks, via virtualisation of connectivity and computation resources and self-management capabilities, by fully integrating networking with cloud computing in order to create In-Network Clouds. It also presents the designs and experiments with a number of In-Network Clouds platforms, which have the aim to create a flexible environment for autonomic deployment and management of virtual networks and services as experimented with and validated on large-scale testbeds.

The "*Flat Architectures: Towards Scalable Future Internet Mobility*" chapter provides a comprehensive overview and review of the scalability problems of mobile Internet nowadays and to show how the concept of flat and ultra flat architectures emerges due to its suitability and applicability for the future Internet. It also aims to introduce the basic ideas and the main paradigms behind the different flat networking approaches trying to cope with the continuously growing traffic demands. The analysis of these areas guides the readers from the basics of flat mobile Internet architectures to the paradigm's complex feature set and power creating a novel Internet architecture for future mobile communications.

The "*Review and Designs of Federated Management in Future Internet Architectures*" chapter analyses issues about federated management targeting information sharing capabilities for heterogeneous infrastructure. An inter-operable, extensible, reusable and manageable new Internet reference model is critical for Future Internet realisation and deployment. The reference model must rely on the fact that high-level applications make use of diverse infrastructure representations and not use of resources directly. So when resources are not being required to support or deploy services they can be used in other tasks or services. As an implementation challenge for controlling and harmonising these entire resource management requirements, the federation paradigm emerges as a tentative approach and potentially optimal solution. This chapter provides, in a form of realistic implementations, research results and solutions addressing the rationale for federation, and all these activities are developed under the umbrella of the federated management work in the Future Internet.

The *"An Architectural Blueprint for a Real-World Internet"* chapter reviews a number of architectures developed in projects in the area of Real-World Internet (RWI), Internet of Things (IoT), and Internet Connected Objects. All of these systems are faced with very similar problems in their design with very limited interoperability among these systems. To address these issues and to speed up development and deployment while at the same time reduce development and maintenance costs, reference architectures are an appropriate tool. As reference architectures require agreement among all stakeholders, they are usually developed through an incremental process. This chapter presents the current status of the development of a reference architecture for the RWI as an architectural blueprint.

The *"Towards a RESTful Architecture for Managing a Global Distributed Interlinked Data-Content-Information Space"* chapter analyses the concept of "Content-Centric" architecture, lying between the Web of Documents and the generalized Web of Data, in which explicit data are embedded in structured documents enabling consistent support for the direct manipulation of information fragments. It presents the InterDataNet (IDN) infrastructure technology designed to allow the RESTful management of interlinked information resources structured around documents. IDN deals with globally identified, addressable and reusable information fragments; it adopts an URI-based addressing scheme; it provides a simple, uniform Web-based interface to distributed heterogeneous information management; it endows information fragments with collaboration-oriented properties, namely: privacy, licensing, security, provenance, consistency, versioning and availability; it glues together reusable information fragments into meaningful structured and integrated documents without the need of a predefined schema.

The *"A Cognitive Future Internet Architecture"* chapter proposes a novel Cognitive Framework as a reference architecture for the Future Internet (FI), which is based on so-called Cognitive Managers. The objective of the proposed architecture is twofold. On one hand, it aims at achieving a full interoperation among the different entities constituting the ICT environment, by means of the introduction of Semantic Virtualization Enablers. On the other hand, it aims at achieving an internetwork and inter-layer cross-optimization by means of a set of Cognitive Enablers, which are in charge of taking consistent and coordinated decisions according to a fully cognitive approach, availing of information coming from both the transport and the service/content layers of all networks. Preliminary test studies, realized in a home environment, confirm the potentialities of the proposed solution.

The "*Title Model Ontology for Future Internet Networks*" chapter contributes to the use of ontologies in the Future Internet, with the proposal of semantic formalization of the Entity Title Model. It is also suggested the use of semantic representation languages in place of protocols.

Alex Galis and Theodore Zahariadis

Towards a Future Internet Architecture

Theodore Zahariadis[1], Dimitri Papadimitriou[2], Hannes Tschofenig[3], Stephan Haller[4], Petros Daras[5], George D. Stamoulis[6], and Manfred Hauswirth[7]

[1] Synelixis Solutions Ltd/TEI of Chalkida, Greece
zahariad@{synelixis.com, teihal.gr}
[2] Alcatel-Lucent, Belgium
dimitri.papadimitriou@alcatel-lucent.com
[3] Nokia Siemens Networks, Germany
hannes.tschofenig@nsn.com
[4] SAP, Germany
stephan.haller@sap.com
[5] Center of Research and Technology Hellas/ITI, Greece
daras@iti.gr
[6] Athens University of Economics and Business, Greece
gstamoul@aueb.gr
[7] Digital Enterprise Research Institute, Ireland
manfred.hauswirth@deri.org

Abstract. In the near future, the high volume of content together with new emerging and mission critical applications is expected to stress the Internet to such a degree that it will possibly not be able to respond adequately to its new role. This challenge has motivated many groups and research initiatives worldwide to search for structural modifications to the Internet architecture in order to be able to face the new requirements. This paper is based on the results of the Future Internet Architecture (FIArch) group organized and coordinated by the European Commission (EC) and aims to capture the group's view on the Future Internet Architecture issue.

Keywords: Internet Architecture, Limitations, Processing, Handling, Storage, Transmission, Control, Design Objectives, EC FIArch group.

1   Introduction

The Internet has evolved from a remote access to mainframe computers and slow communication channel among scientists to the most important medium for information exchange and the dominant communication environment for business relations and social interactions. Billions of people all over the world use the Internet for finding, accessing and exchanging information, enjoying multimedia communications, taking advantage of advanced software services, buying and selling, keeping in touch with family and friends, to name a few. The success of the Internet has created even higher hopes and expectations for new applications and services, which the current Internet may not be able to support to a sufficient level. It is expected that the number

J. Domingue et al. (Eds.): Future Internet Assembly, LNCS 6656, pp. 7–18, 2011.

of nodes (computers, terminals mobile devices, sensors, etc.) of the Internet will soon grow to more than 100 billion [1]. Reliability, availability, and interoperability required by new networked services, and this trend will escalate in the future. Therefore, the requirement of increased robustness, survivability, and collaborative properties is imposed to the Internet architecture. In parallel, the advances in video capturing and content/media generation have led to very large amounts of multimedia content and applications offering immersive experiences(e.g., 3D videos, interactive environments, network gaming, virtual worlds, etc.) compared to the quantity and type of data currently exchanged over the Internet. Based on [2], out of the 42 Exabytes (10^{18}) of consumer Internet traffic likely to be generated every month in 2014, 56% will be due to Internet video, while the average monthly consumer Internet traffic will be equivalent to 32 million people streaming Avatar in 3D, continuously, for the entire month.

All these applications create new demands and requirements, which to a certain extent can be addressed by means of "over-dimensioning" combined with the enhancement of certain Internet capabilities over time. While this can be a satisfactory (although sometimes temporary) solution in some cases, analyses have shown [3],[4] that increasing the bandwidth on the backbone network will not suffice due to new qualitative requirements concerning, for example, highly critical services such as e-health applications, clouds of services and clouds of sensors, new social network applications like collaborative 3D immersive environments, new commercial and transactional applications, new location-based services and so on.

In other words, the question is to determine if the architecture and its properties might become the limiting factor of Internet growth and of the deployment of new applications. For instance, as stated in [5] *"the end-to-end arguments are insufficiently compelling to outweigh other criteria for certain functions such as routing and congestion control"*. On the other hand, the evolution of the Internet architecture is carried out by means of incremental and reactive additions [6], rather than by major and proactive modifications. Moreover, studies on the impact of research results have shown that better performance or richer functionality implying an architectural change define necessary but not sufficient conditions for such change in the Internet architecture and/or its components. Indeed, the Internet architecture has shown since so far the capability to overcome such limits without requiring radical architectural transformation. Hence, before proposing or designing a new Internet Architecture (if a new one is needed), it is necessary to demonstrate the fundamental limits of the current architecture [7]. Thus, scientists and researchers from both the industry and academia worldwide are working towards understanding these architectural limits so as to progressively determine the principles that will drive the Future Internet architecture that will adequately meet at least the abovementioned challenges [EIFFEL], [4WARD], [COAST].

The Future Internet as a global and common communication and distributed information system may be considered from various interrelated perspectives: the networks and shared infrastructure perspective, the services and application perspective as well as the media and content perspective. Significant efforts world-wide have already been devoted to investigate some of its pillars [8] [9] [10] [11] [12] [13]. In

Europe, a significant part of the Information and Communication Technology (ICT) of the Framework Program 7 is devoted to the Future Internet [14]. Though many proposals for a Future Internet Architecture have already been developed, no specific methodology to evaluate the efficiency (and the need) for such architecture proposals exist. The purpose of this paper is to capture the view of the Future Internet Architecture (FIArch) group organized and coordinated by the European Commission.

Since so far, the FIArch group has identified and reached some understanding and agreement on the different types of limitations of the Internet and its architecture. Interested readers may also refer to [15] for more information[1].

2   Definitions

Before describing the approach followed by the FIArch Group, we define the terms used in our work. Based on [16], we define as *"architecture"* a set of functions, states, and objects/information together with their behavior, structure, composition, relationships and spatio-temporal distribution. The specification of the associated functional, object/ informational and state models leads to an architectural model comprising a set of components (i.e. procedures, data structures, state machines) and the characterization of their interactions (i.e. messages, calls, events, etc.).

We also qualify as a *"fundamental limitation"* of the Internet architecture a functional, structural, or performance restriction or constraint that cannot be effectively resolved with current or clearly foreseen "architectural paradigms" as far as our understanding/knowledge goes. On the other hand, we define as *"challenging limitation"* a functional, structural, or performance restriction or constraint that could be resolved as far as our understanding/knowledge goes by replacing and/or adding/removing a component of the architecture so that this would in turn change the global properties of the Internet architecture (e.g. separation of the locator and identifier role of IP addresses).

In the following, we use the term *"data"* to refer to any organized group of bits a.k.a. data packets, data traffic, information, content (audio, video, multimedia), etc. and the term *"service"* to refer to any action performed on data or other services and the related Application Programming Interface (API).[2] Note however that this document does not take position on the localization and distribution of these APIs.

3   Analysis Approach

Since its creation, the Internet is driven by a small set of fundamental design principles rather than a formal architecture that is created on a whiteboard by a standardization or research group. Moreover, the necessity for backwards compatibility and the trade-off between Internet redesign and proposing extensions, enhancements and re-engineering of today's Internet protocols are heavily debated.

---

[1]   Interested readers may also search for updated versions at the FIArch site:
   http://ec.europa.eu/information_society/activities/foi/research/fiarch/index_en.htm
[2]   The definition of service does not include the services offered by humans using the Internet

The emergence of new needs at both functional and performance levels, the cost and complexity of Internet growth, the existing and foreseen functional and performance limitations of the Internet's architectural principles and design model put the following elementary functionalities under pressure:

- *Processing/handling of "data"*: refers to forwarders (e.g. routers, switches, etc.), computers (e.g., terminals, servers, etc.), CPUs, etc. and handlers (software programs/routines) that generate and treat as well as query and access data.
- *Storage of "data"*: refers to memory, buffers, caches, disks, etc., and associated logical data structures.
- *Transmission of "data"*: refers to physical and logical transferring/exchange of data.
- *Control of processing, storage, transmission of systems and functions*: refers to the action of observation (input), analysis, and decision (output) whose execution affects the running conditions of these systems and functions. Note that by using these base functions, the data communication function can be defined as the combination of processing, storage, transmission and control functions applied to "data". The term control is used here to refer to control functionality but also management functionality, e.g. systems, networks, services, etc.

For each of the above functionalities, the FIArch group has tried to identify and analyze the presumed problems and limitations of the Internet. This work was carried out by identifying an extensive list of limitations and potentially problematic issues or missing functionalities, and then selecting the ones that comply with the aforementioned definition of a fundamental limitation.

3.1   Processing and Handling Limitations

The fundamental limitations that have been identified in this category are:

i.   The Internet does not allow hosts to diagnose potential problems and the network offers little feedback for hosts to perform root cause discovery and analysis. In today's Internet, when a failure occurs it is often impossible for hosts to describe the failure (what happened?) and determine the cause of the failure (why it happened?), and which actions to take to actually correct it. The misbehavior that may be driven by pure malice or selfish interests is detrimental to the cooperation between Internet users and providers. Non-intrusive and non-discriminatory means to detect misbehavior and mitigate their effects while keeping open and broad accessibility to the Internet is a limitation that is crucial to overcome [16].

ii. *Lack of data identity is damaging the utility of the communication system.* As a result, data, as an 'economic object', traverses the communication infrastructure multiple times, limiting its scaling, while lack of content 'property rights' (not only author- but also usage-rights) leads to the absence of a fair charging model.

iii. Lack of methods for dependable, trustworthy processing and handling of network and systems infrastructure and essential services in many critical environments, such as healthcare, transportation, compliance with legal regulations, etc.

iv. *Real-time processing.* Though this is not directly related to the Internet Architecture itself, the limited capability for processing data on a real-time basis poses limitations in terms of the applications that can be deployed over the Internet. On the other hand, many application areas (e.g. sensor networks) require real-time Internet processing at the edges nodes of the network.

3.2 Storage Limitations

The fundamental restrictions that have been identified in this category are:

i. *Lack of context/content aware storage management:* Data are not inherently associated with knowledge of their context. This information may be available at the communication end-points (applications) but not when data are in transit. So, it is not feasible to make efficient storage decisions that guarantee fast storage management, fast data mining and retrieval, refreshing and removal optimized for different types of data [18].

ii. *Lack of inherited user and data privacy:* In case data protection/ encryption methods are employed (even using asymmetric encryption and public key methods), data cannot be efficiently stored/handled. On the other hand, lack of encryption, violates the user and data privacy. More investigations into the larger privacy and data-protection ecosystem are required to overcome current limits of how current information systems deal with privacy and protection of information of users, and develop ways to better respect the needs and expectations [30], [31], [32]

iii. *Lack of data integrity, reliability and trust,* targeting the security and protection of data; this issue covers both unintended disclosure and damage to integrity from defects or failures, and vulnerabilities to malicious attacks.

iv. *Lack of efficient caching & mirroring:* There is no inherited method for on-path caching along the communication path and mirroring of content compared to off-path caching that is currently widely used (involving e.g. connection redirection). Such methods could deal with issues like flash crowding, as the onset of the phenomenon will still cause thousands of cache servers to request the same documents from the original site of publication.

3.3 Transmission Limitations

The fundamental restrictions that have been identified in this category are:

i. *Lack of efficient transmission of content-oriented traffic:* Multimedia content-oriented traffic comprises much larger volumes of data as compared to any other information flow, while its inefficient handling results in retransmission of the same data multiple times. Content Delivery Networks (CDN) and more generally architectures using distributed caching alleviate the problem under certain conditions but can't extend to meet the Internet scale [19]. Transmission from centralized locations creates unnecessary overheads and can be far from optimal when massive amounts of data are exchanged.

ii. *Lack of integration of devices with limited resources to the Internet as autonomous addressable entities.* Devices in environments such as sensor networks or even nano-networks/smart dust as well as in machine-to-machine (M2M) environments operate with such limited processing, storage and transmission capacity that only partly run the protocols necessary in order to be integrated in the Internet as *autonomous addressable entities.*

iii. *Security requirements of the transmission links:* Communications privacy does not only mean protecting/encrypting the exchanged data but also not disclosing that communication took place. It is not sufficient to just protect/encrypt the data (including encryption of protocols/information/content, tamper-proof applications etc) *but also protect the communication* itself, including the relation/interaction between (business or private) parties.

3.4   Control Limitations

The fundamental limitations that have been identified in this category are:

i. *Lack of flexibility and adaptive control[3][4].* In the current Internet model, design of IP (and more generally communication) control components have so far being driven exclusively by i) cost/performance ratio considerations and ii) pre-defined, static, and open loop control processes. The first limits the capacity of the system to adapt/react in a timely and cost-effective manner when internal or external events occur that affect its value delivery; this property is referred to as flexibility [20][21]. Moreover, the current trend in unstructured addition of ad-hoc functionality to partly mitigate this lack of flexibility has resulted in increased complexity and (operational and system) cost of the Internet. Further, to maintain/sustain or even increase its value delivery over time, the Internet will have to provide flexibility in its functional organization, adaptation, and distribution. Flexibility at run time is essential to cope with the increasing uncertainty (unattended and unexpected events) as well as breadth of expected events/ running conditions for which it has been initially designed. The latter results in such a complexity that leaves no possibility for individual systems to adapt their control decisions and tune their execution at running time by taking into account their internal state, its activity/behavior as well as the environment/external conditions.

ii. *Improper segmentation of data and control.* The current Internet model segments (horizontally) data and control, whereas from its inception the control functionality has a transversal component. Thus, on one hand, the IP functionality isn't limited anymore to the "network layer", and on the other, IP is not totally decoupled from the underlying "layers" anymore (by the fact IP/MPLS and underlying layers

---

3   Some may claim that this limitation is "very important" or "very challenging" but not a "fundamental" one. As we consider it significant anyway, we include it here for the sake of completeness.

4   This limitation is often named by the potential approach aimed to address it, including autonomic networking, self-mamagenent, etc. However, none of them has shown ability to support flexibility at run time to cope with increasing uncertainty (since the control processes they accommodate are still those pre-determined at design time).

share the same control instance). Hence, the hour-glass model of the Internet does not account for this evolution of the control functionality when considered as part of the design model.

iii. *Lack of reference architecture of the IP control plane.* The IP data plane is itself relatively simple but its associated control components are numerous and sometimes overlapping, as a result of the incremental addition of ad-hoc control components over time, and thus their interactions are becoming more and more complex. This leads to detrimental effects for the controlled entities, e.g., failures, instability, inconsistency between routing and forwarding (leading to e.g. loops) [22][23].

iv. *Lack of efficient congestion control.* Congestion control cannot be realized as a pure end-to-end function: congestion is an inherent network phenomenon that can only be resolved efficiently by some cooperation of end-systems and the network, since it is a shared communication infrastructure. Hence, substantial benefit could be expected by further assistance from the network, but, on the other hand, such network support could lead to duplication of functions, which may harmfully interact with end-to-end principle and resulting protocol mechanisms. Addressing effectively the trade-off of network support without decreasing its scaling properties by requiring maintenance of per-flow state is one of the Internet's main challenges [16].

3.5 Limitations That May Fall in More than One Category

Certain fundamental limitations of current Internet may fall in more than one category. Examples of such limitations include:

i. *Traffic growth vs heterogeneity in capacity distribution:* Hosts connected to the Internet do not have the possibility to enforce the path followed by their traffic. Hence, even if multiple alternatives to reach a given destination would be offered to the host, they are unable to enforce their decision across the network. On the other hand, as the Internet enables any-to-any connectivity, there is no effective means to predict the spatial distribution of the traffic within a timescale that would allow providers to install needed capacity when required or at least expected to prevent overload of certain network segments. This results into serious capacity shortage (and thus congestion) over certain segments of the network. Especially, the traffic exchange points (as well as certain international and the transatlantic links) are in many cases significantly overloaded. In some cases, building out more capacity to handle this new congestion may be infeasible or unwarranted. Two main types of limitations are seen in this respect: i) not known scalable means to overcome the result of network infrastructure abstraction, and ii) those related to congestion and diagnosability. These are related to at least the base functions of control and processing/handling.

ii. *The current inter-domain routing system is reaching fundamental limits* in terms of routing table scalability but also adaptation to topology and policy dynamics (perform efficiently under dynamic network conditions) that in turn impact its convergence, and robustness/stability properties. Both dimensions increase memory requirements but also the processing capacity of routing engines [23][7] Related projects: [EULER] [ResumeNet].

iii. *Scaling to deal with flash crowding.* The huge number of (mobile) terminals combined with a sudden peak in demand for a particular piece of data may result in phenomena that cannot be handled; such phenomena can be related at to all the base functions.

iv. The amount of foreseen data and information[5] requires significant *processing power / storage / bandwidth for indexing / crawling and (distributed) querying* and also solutions for large *scale / real-time data mining / social network analysis,* so as to achieve successful retrieval and integration of information from an extremely high numer of sources across the network. All the aforementioned issues imply the need for addressing new architectural challenges capable to cope with the fast and scalable identification and discovery of and access to data. *The exponential growth of information makes it increasingly harder to identify relevant information ("drowning in information while starving for knowledge"). This information overload becomes more and more acute and existing search and recommendation tools are not filtering and ranking the information adequately and lack the required granularity (document-level vs. individual information item).*

v. *Security of the whole Internet Architecture.* The Internet architecture is not intrinsically secure and is based on add-ons to, e.g. protocols, to secure itself. The consequence is that protocols may be secure but the overall architecture is not self-protected against malicious attacks.

vi. *Support of mobility* when using IP address as both network and host identifier but also TCP connection identifier results in Transmission Control Protocol (TCP) connection continuity problem. Its resolution requires decoupling between the identifier of the position of the mobile host in the network graph (network address) from the identifier used for the purpose of TCP connection identification. Moreover, when mobility is enabled by wireless networks, packets can be dropped because of corruption loss (when the wireless link cannot be conditioned to properly control its error rate or due to transient wireless link interruption in areas of poor coverage), rendering the typical reaction of congestion control mechanism of TCP inappropriate. As a result, non-congestive loss may be more prevalent in these networks due to corruption loss. This limitation results from the existence of heterogeneous links, both wired and wireless, yielding a different trade-off between performance, efficiency and cost, and affecting several base functions again.

4   Design Objectives

The purpose of this section is to document the design objectives that should be met by the Internet architecture. We distinguish between "high-level" and "low-level" design objectives. High-level objectives refer to the cultural, ethical, socio-economic, but also technological expectations to be met by the Internet as global and common information communication system. High-level objectives are documented in [15]. By low-level design objectives, we mean here the functional and performance properties as well as the structural and quality properties that the architecture of this global and

---

5   Eric Schmidt, the CEO of Google, the world's largest index of the Internet, estimated the size at around 5 million terabytes of data (2005). Eric commented that Google has indexed roughly 200 terabytes of that is 0,004% of the total size.

common information communication system is expected to meet. From the previous sections, some of low-level objectives are met and others are not by the (present) architecture of the Internet. We also emphasize here that these objectives are commonly shared by the Internet community at large

The remaining part of this Section translates a first analysis of the properties that should be met by the Internet architecture starting from the initial of objectives as enumerated in various references (see [27], [28], [29]). One of the key challenges is thus to determine the necessary addition/improvement of current architecture principles and the improvement (or even removal of architectural components needed to eliminate or at least tangibly mitigate/avoid the known effects of the fundamental limitations. It is to be emphasized that a great part of research activities in this domain consists in identifying hidden relationships and effects.

As explained in [27], the Internet architecture has been structured around eight foundational objectives: i) to connect existing networks, ii) survivability, iii) to support multiple types of services, iv) to accommodate a variety of physical networks, v) to allow distributed management, vi) to be cost effective, vii) to allow host attachment with a low level of effort and, viii) to allow resource accountability. Moreover, RFC 1287, published in 1991 by the IAB [36], underlines that the Internet architecture needs to be able to scale to 10^9 IP networks recognizing the need to add scalability as a design objective. In this context, the followed approach consists of starting from the existing Internet design objectives compared to the approach that would consist of applying a tabula rasa approach, i.e., completely redefine from scratch the entire set of Internet design objectives.

Based on previous sections, the present section describes the design objectives that are currently met, partly met or not met at all by the current architecture. In particular, the low-level design objectives of the architecture are to provide:

- *Accessibility* (open and by means of various/heterogeneous wireless/radio and wired interfaces) to the communication network but also to heterogeneous data, applications, and services, nomadicity, and mobility (while providing means to maintain continuity of application communication exchanges when needed). Accessibility and nomadicity are currently addressed by current Internet architecture. On the other hand, mobility is still realized in most cases by means of dedicated/separated architectural components instead of Mobile IP. see Subsection 3.5. Point 6

- *Accountability* of resource usage and security without impeding user privacy, utility and self-arbitration: see Subsection.3.1.Point.2

- *Manageability*, implying distributed, organic, automated, and autonomic/self-adaptive operations: see Subsection 3.5 and *Diagnosability* (i.e. root cause detection and analysis): see Subsection.3.1.Point.1

- *Transparency*, i.e. the terminal/host is only concerned with the end-to-end service; in the current Internet this service is the connectivity even if the notion of "service" is not embedded in the architectural model of the Internet: initially addressed but loosing ground.

- *Distribution of processing, storage, and control functionality and autonomy* (organic deployment): addressed by current architecture; concerning storage and processing, several architectural enhancements might be required, e.g. for the integration of distributed but heterogeneous data and processes.

- *Scalability*, including routing and addressing system in terms of number of hosts/terminals, number of shared infrastructure nodes, etc. and management system: - see Subsection.3.5.Point.2
- *Reliability*, referring here to the capacity of the Internet to perform in accordance to what it is expected to deliver to the end-user/hosts while coping with a growing number of users with increasing heterogeneity in applicative communication needs.
- *Robustness/stability*, *resiliency*, and *survivability:* see Subsection.3.5.Point.2
- *Security*: see Subsection.3.5 point 5, Subsection 3.1.Point.2 and 3.
- *Generality* e.g. support of plurality of applications and associated data traffic such as non/real-time streams, messages, etc., independently of the shared infrastructure partitioning/divisions, and independently of the host/terminal: addressed and to be reinforced (migration of mobile network to IPv6 Internet, IPTV moving to Internet TV, etc.) otherwise leading to segmentation and specialization per application/service.
- *Flexibility*, i.e. capability to adapt/react in a timely and cost-effective manner upon occurrence of internal or external events that affect its value delivery, and *Evolutivity* (of time variant components): not addressed see Subsection 3.4.Point.1.
- *Simplicity and cost-effectiveness*: deeper analysis is needed but simplicity seems to be progressively decreasing see Section 3.4 Point 3. Note that simplicity is explicitly added as a design objective to -at least- prevent further deterioration of the complexity of current architecture (following the "Occam's razor principle"). Indeed, lowering complexity for the same level of performance and functionality at a given cost is a key objective.
- *Ability to offer information-aware transmission and distribution*: Subsection 3.3, Point 1, and Subsection 3.5, Point 4.

5   Conclusions

In this article we have identified fundamental limitations of Internet architecture following a systematic investigation thereof from a variety of different viewpoints. Many of the identified fundamental limitations are not isolated but strongly dependent on each other. Increasing the bandwidth would significantly help to address or mitigate some of these problems, but would not solve their root cause. Other problems would nevertheless remain unaddressed. The **transmission** can be improved by utilizing better **data processing and handling** (e.g. network coding, data compression, intelligent routing) and better **data storage** (e.g. network/terminals caches, data centers/mirrors etc.), while the overall Internet performance would be significantly improved by **control and self-*** functions. As an overall finding we may conclude the following: **Extensions, enhancements and re-engineering of today's Internet protocols may solve several challenging limitations. Yet, addressing the fundamental limitations of the Internet architecture is a multi-dimensional and challenging research topic. While improvements are needed in each dimension, these should be combined by undertaking a holistic approach of the problem space.**

Acknowledgements. This article is the based on the work that has been carried out by the EC Future Internet Architecture (FIArch) group (to which the authors belong), which is coordinated by the EC FP7 Coordination and Support Actions (CSA) projects in the area of Future Internet: NextMedia, IOT-I, SOFI, EFFECTS+, EIFFEL, Cho-rus+, SESERV and Paradiso 2, and supported by the EC Units D1: Future Networks, D2: Networked Media Systems, D3: Software & Service Architectures & Infrastruc-tures, D4: Networked Enterprise & Radio Frequency Identification (RFID) and F5: Trust and Security. The authors would like to acknowledge and thank all members of the group for their significant input and the EC Scientific Officers Isidro Laso Balles-teros, Jacques Babot, Paulo De Sousa, Peter Friess, Mario Scillia, Arian Zwegers for coordinating the activities.

The authors would like also to acknowledge the FI architectural work performed under the project FP7 COAST ICT-248036 [COAST].

References

[1] AKARI Project: New Generation Network Architecture AKARI Conceptual Design (ver1.1). AKARI Architecture Design Project, Original Publish (Japanese) June 2008, English Translation October 2008, Copyright © 2007-2008 NICT (2008)

[2] Medeiros, F.: ICT 2010: Digitally Driven, Brussels, 29 September 2010, Source Cisco VNL (2010)

[3] Mahonen, P. (ed.), Trossen, D., Papadimitrou, D., Polyzos, G., Kennedy, D.: Future Net-worked Society., EIFFEL whitepaper (Dec. 2006)

[4] Jacobson, V., Smetters, D., Thornton, J., Plass, M., Briggs, N., Braynard, R.: Networking Named Content. Proceeding of ACM CoNEXT 2009. Rome, Italy (December 2009)

[5] Moors, T.: A critical review of "End-to-end arguments in system design". In: Proceedings of IEEE International Conference on Communications (ICC) 2002, New-York City (New Jersey), USA (April/May 2002)

[6] RFC 1958: The Internet and its architecture have grown in evolutionary fashion from modest beginnings, rather than from a Grand Plan

[7] Li, T. (ed.): Design Goals for Scalable Internet Routing. Work in progress, draft-irtf-rrg-design-goals-02 (Sep. 2010)

[8] http://www.nsf.gov/pubs/2010/nsf10528/nsf10528.htm

[9] http://www.nsf.gov/funding/pgm_summ.jsp?pims_id=503325

[10] http://www.nets-find.net

[11] http://www.geni.net/?p=1339

[12] http://akari-project.nict.go.jp/eng/overview.htm

[13] http://mmlab.snu.ac.kr/fiw2007/presentations/architecture_tschoi.pdf

[14] http://www.future-internet.eu/

[15] FIArch Group: Fundamental Limitations of Current Internet and the path to Future Inter-net (December 2010)

[16] Perry, D., Wolf, A.: Foundations for the Study of Software Architecture. ACM SIGSOFT Software Engineering Notes 17, 4 (1992)

[17] Papadimitriou, D., et al. (eds.): Open Research Issues in Internet Congestion Control. Internet Research Task Force (IRTF), RFC 6077 (February 2011)

[18] Akhlaghi, S., Kiani, A., Reza Ghanavati, M.: Cost-bandwidth tradeoff in distributed storage systems (published on-line). ACM Computer Communications 33(17), 2105–2115 (2010)

[19] Freedman, M.: Experiences with CoralCDN: A Five-Year Operational View. In: Proc. 7th USENIX/ACM Symposium on Networked Systems Design and Implementation (NSDI '10) San Jose, CA (May 2010)

[20] Dobson, S., et al.: A survey of autonomic communications. ACM Transactions on Autonomous and Adaptive Systems (TAAS) 1(2), 223–259 (2006)

[21] Gelenbe, E.: Steps toward self-aware networks. ACM Communications 52(7), 66–75 (2009)

[22] Evolving the Internet, Presentation to the OECD (March 2006), http://www.cs.ucl.ac.uk/staff/m.handley/slides/

[23] Meyer, D., et al.: Report from the IAB Workshop on Routing and Addressing, IETF, RFC 4984 (Sep. 2007)

[24] Mahonen, P. (ed.), Trossen, D., Papadimitrou, D., Polyzos, G., Kennedy, D.: Future Networked Society. EIFFEL whitepaper (Dec. 2006)

[25] Trosse, D.: Invigorating the Future Internet Debate. ACM SIGCOMM Computer Communication Review 39(5) (2009)

[26] Eggert, L.: Quality-of-Service: An End System Perspective. In: MIT Communications Futures Program – Workshop on Internet Congestion Management, QoS, and Interconnection, Cambridge, MA, USA, October 21-22 (2008)

[27] Ratnasamy, S., Shenker, S., McCanne, S.: Towards an evolvable internet architecture. SIGCOMM Comput. Commun. Rev. 35(4), 313–324 (2005)

[28] Cross-ETP Vision Document, http://www.future-internet.eu/fileadmin/documents/reports/Cross-ETPs_FI_Vision_Document_v1_0.pdf

[29] Clark, D.D.: The Design Philosophy of the DARPA Internet Protocols, Proc SIGCOMM 88 (reprinted in ACM CCR 25(1), 102-111, 1995). ACM CCR 18(4), 106–114 (1988)

[30] Saltzer, J.H., Reed, D.P., Clark, D.D.: End-To-End Arguments in System Design. ACM TOCS 2(4), 277–288 (1984)

[31] Carpenter, B.: Architectural Principles of the Internet, Internet Engineering Task Force (IETF), RFC 1958 (July 1996)

[32] Krishnamurthy, B.: I know what you will do next summer., ACM SIGCOMM Computer Communication Review (Oct. 2010), http://www2.research.att.com/~bala/papers/ccr10-priv.pdf

[33] W3C Workshop on Privacy for Advanced Web APIs 12/13 July 2010, London (2010), http://www.w3.org/2010/api-privacy-ws/report.html

[34] Workshop on Internet Privacy, co-organized by the IAB, W3C, MIT, and ISOC, 8 and 9 December (2010), http://www.iab.org/about/workshops/privacy/

[35] Clark, D., et al.: Towards the Future Internet Architecture, Internet Engineering Task Force (IETF); RFC 1287 (December 1991)

[36] http://www.iso.org/iso/iso_technical_committee.html?commid=45072

[37] http://www.4ward-project.eu/

[Alicante]   http://www.ict-alicante.eu/

[ANA]   http://www.ana-project.org/

[COAST]   http://www.fp7-coast.eu/

[COMET]   http://www.comet-project.org/

[ECODE]   http://www.ecode-project.eu/

[EIFFEL]   http://www.fp7-eiffel.eu/

[EULER]   http://www.euler-project.eu/

[IoT-A]   http://www.iot-a.eu/

[nextMedia]   http://www.fi-nextmedia.eu/

[OPTIMIX]   http://www.ict-optimix.eu/

[ResumeNet]   http://www.resumenet.eu/

[SelfNet]   https://www.ict-selfnet.eu/

[TRILOGY]   http://trilogy-project.org/

[UniverSelf] http://www.univerself-project.eu

Towards In-Network Clouds in Future Internet

Alex Galis[1], Stuart Clayman[1], Laurent Lefevre[2], Andreas Fischer[3],
Hermann de Meer[3], Javier Rubio-Loyola[4], Joan Serrat[5], and Steven Davy[6]

[1] University College London, United Kingdom, {a.galis,s.clayman}@ee.ucl.ac.uk
[2] INRIA, France, laurent.lefevre@ens-lyon.fr
[3] University of Passau, Germany {andreas.fischer,hermann.demeer}@uni-passau.de
[4] CINVESTAV Tamaulipas, Mexico, jrubio@tamps.cinvestav.mx
[5] Universitat Politècnica de Catalunya, Spain, serrat@nmg.upc.edu
[6] Waterford Institute of Technology, Ireland, sdavy@tssg.org

Abstract. One of the key aspect fundamentally missing from the current Internet infrastructure is an advanced service networking platform and facilities, which take advantage of flexible sharing of available connectivity, computation, and storage resources. This paper aims to explore the architectural co-existence of new and legacy services and networks, via virtualisation of connectivity and computation resources and self-management capabilities, by fully integrating networking with cloud computing in order to create In-Network Clouds. It also presents the designs and experiments with a number of In-Network Clouds platforms, which have the aim to create a flexible environment for autonomic deployment and management of virtual networks and services as experimented with and validated on large-scale testbeds.

Keywords: In-Network Clouds, Virtualisation of Resources, Self-Management, Service plane, Orchestration plane and Knowledge plane.

1   Introduction

The current Internet has been founded on a basic architectural premise, that is: a simple network service can be used as a universal means to interconnect both dumb and intelligent end systems. The simplicity of the current Internet has pushed complexity into the endpoints, and has allowed impressive scale in terms of inter-connected devices. However, while the scale has not yet reached its limits [1, 2], the growth of functionality and the growth of size have both slowed down and may soon reach both its architectural capability and capacity limits. Internet applications increasingly require a combination of capabilities from traditionally separate technology domains to deliver the flexibility and dependability demanded by users. Internet use is expected to grow massively over the next few years with an order of magnitude more Internet services, the interconnection of smart objects from the Internet of Things, and the integration of increasingly demanding enterprise and societal applications.

The Future Internet research and development trends are covering the main focus of the current Internet, which is connectivity, routing, and naming as well as defining

J. Domingue et al. (Eds.): Future Internet Assembly, LNCS 6656, pp. 19–33, 2011.

and design of all levels of interfaces for Services and for networks' and services' resources. As such, the Future Internet covers the complete management and full lifecycle of applications, services, networks and infrastructures that are primarily constructed by recombining existing elements in new and creative ways.

The aspects which are fundamentally missing from the current Internet infrastructure, include the advanced service networking platforms and facilities, which take advantage of flexible sharing of available resources (e.g. connectivity, computation, and storage resources).

This paper aims to explore the architectural co-existence of new and legacy services and networks, via virtualisation of resources and self-management capabilities, by fully integrating networking [4, 8, 10, 15] with cloud computing [6, 7, 9] in order to produce In-Network Clouds. It also presents the designs and experiments with a number of In-Network Clouds platforms [9, 10], which have the aim to create a flexible environment for autonomic deployment and management of virtual networks and services as experimented with and validated on large-scale testbeds [3].

2   Designs for In-Network Clouds

Due to the existence of multiple stakeholders with conflicting goals and policies, modifications to the existing Internet are now limited to simple incremental updates and deployment of new technology is next to impossible and very costly. In-Network clouds have been proposed to bypass this ossification as a diversifying attribute of the future inter-networking and inter-servicing paradigm. By allowing multiple heterogeneous network and service architectures to cohabit on a shared physical substrate, In-Network virtualisation provides flexibility, promotes diversity, and promises security and increased manageability.

We define In-Network clouds as an integral part of the differentiated Future Internet architecture, which supports multiple computing clouds from different service providers operating on coexisting heterogeneous virtual networks and sharing a common physical substrate of communication nodes and servers managed by multiple infrastructure providers. By decoupling service providers from infrastructure providers and by integrating computing clouds with virtual networks the In-Network clouds introduce flexibility for change.

In-Network Network and Service Clouds can be represented by a number of distributed management systems described with the help of five abstractions: Virtualisation Plane (VP), Management Plane (MP), Knowledge Plane (KP), Service Plane (SP), and Orchestration Plane (OP) as depicted in Fig. 1.

These planes are new higher-level artefacts, used to make the Future Internet of Services more intelligent, with embedded management functionality. At a logical level, the VMKSO planes gather observations, constraints and assertions, and apply rules to these in order to initiate proper reactions and responses. At the physical level, they are embedded and execute on network hosts, devices, attachments, and servers

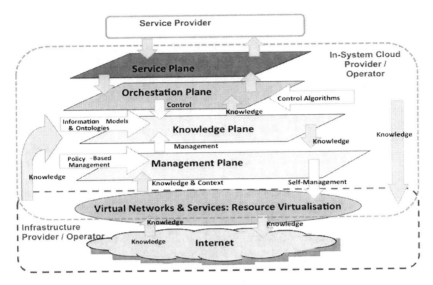

Fig. 1. In-Network Cloud Resources

within the network. Together these distributed systems form a software-driven network control infrastructure that will run on top of all current networks (i.e. fixed,
wireless, and mobile networks) and service physical infrastructures in order to provide an autonomic virtual resource overlay.

2.1 Service Plane Overview

The Service Plane (SP) consists of functions for the automatic (re-) deployment of
new management services, protocols, as well as resource-facing and end-user facing
services. It includes the enablers that allow code to be executed on the network entities. The safe and controlled deployment of new code enables new services to be
activated on-demand. This approach has the following advantages:

- Service deployment takes place automatically and allows a significant number of
  new services to be offered on demand;
- It offers new, flexible ways to configure network entities that are not based on
  strict configuration sets;
- Services that are not used can be automatically disabled. These services can be
  enabled again on-demand, in case they are needed;
- It eases the deployment of network-wide protocol stacks and management services;
- It enables secure but controlled execution environments;
- It allows an infrastructure that is aware of the impact on the existing services of a
  new deployment;
- It allows optimal resource utilization for the new services and the system.

2.2   Orchestration Plane Overview

The purpose of the Orchestration Plane is to coordinate the actions of multiple auto-nomic management systems in order to ensure their convergence to fulfil applicable business goals and policies. It supervises and it integrates all other planes' behaviour ensuring integrity of the Future Internet management operations. The Orchestration Plane can be thought of as a control framework into which any number of compo-nents can be plugged into, in order to achieve the required functionality. These com-ponents could have direct interworking with control algorithms, situated in the control plane of the Internet (i.e. to provide real time reaction), and interworking with other management functions (i.e. to provide near real time reaction).

The Orchestration Plane is made up of one or more Autonomic Management Sys-tems (AMS), one or more Distributed Orchestration Components (DOC), and a dy-namic knowledge base consisting of a set of information models and ontologies and appropriate mapping logic and buses. Each AMS represents an administrative and/or organisational boundary that is responsible for managing a set of devices, subnet-works, or networks using a common set of policies and knowledge. The Orchestration Plane acts as control workflow for all AMS ensuring bootstrapping, initialisation, dynamic reconfiguration, adaptation and contextualisation, optimisation, organisation, and closing down of an AMS. It also controls the sequence and conditions in which one AMS invokes other AMS in order to realize some useful function (i.e., an orches-tration is the pattern of interactions between AMS). An AMS collects appropriate monitoring information from the virtual and non-virtual devices and services that it is managing, and makes appropriate decisions for the resources and services that it gov-erns, either by itself (if its governance mode is individual) or in collaboration with other AMS (if its governance mode is distributed or collaborative), as explained in the next section. The OP is build on the concepts identified in [13], however it differs in several essential ways:

- Virtual resources and services are used.
- Service Lifecycle management is introduced.
- The traditional management plane is augmented with a narrow knowledge plane, consisting of models and ontologies, to provide increased analysis and inference capabilities.
- Federation, negotiation, distribution, and other key framework services are pack-aged in a distributed component that simplifies and directs the application of those framework services to the system.

The Distributed Orchestration Component (DOC) provides a set of framework net-work services. Framework services provide a common infrastructure that enables all components in the system under the scope of the Orchestration Plane to have plug_and_play and unplug_and_play behaviour. Applications compliant with these framework services share common security, metadata, administration, and manage-ment services. The DOC enables the following functions across the orchestration plane: federation, negotiation, distribution and governance. The federation functional-ity of the OP is represented by the composition/decomposition of networks & services under different domains. Since each domain may have different SLAs, security and administrative policies, a federation function would trigger a negotiation between domains and the re-deployment of service components in the case that the new policies and high level goals of the domain are not compatible with some of the deployed services. The negotiation functionality of the OP enables separate domains to reach composition/ decomposition agreements and to form SLAs for deployable services. The distribution functionality of the OP provides communication and control services that enable management tasks to be split into parts that run on multiple AMSs within the same domain. The distribution function controls the deployment of AMSs and their components. The governance functionality of the OP monitors the consistency of the AMSs' actions, it enforces the high level policies and SLAs defined by the DOCs and it triggers for federation, negotiation and distribution tasks upon noncompliance.

The OP is also supervising the optimisation and the distribution of knowledge within the Knowledge Plane to ensure that the required knowledge is available in the proper place at the proper time. This implies that the Orchestration Plane may use very local knowledge to deserve a real time control as well as a more global knowledge to manage some long-term processes like planning.

2.3   Virtualisation Plane Overview

Virtualisation hides the physical characteristics [14, 16] of the computing and networking resources being used, from its applications and users. This paper uses system virtualisation to provide virtual services and resources. System virtualisation separates an operating system from its underlying hardware resources; resource virtualisation abstracts physical resources into manageable units of functionality. For example, a single physical resource can appear as multiple virtual resources (e.g., the concept of a virtual router, where a single physical router can support multiple independent routing processes by assigning different internal resources to each routing process); alternatively, multiple physical resources can appear as a single physical resource (e.g., when multiple switches are "stacked" so that the number of switch ports increases, but the set of stacked switches appears as a single virtual switch that is managed as a single unit). Virtualisation enables optimisation of resource utilisation. However, this optimisation is confined to inflexible configurations within a single administrative domain. This paper extends contemporary virtualisation approaches and aims at building an infrastructure in which virtual machines can be dynamically relocated to any physical node or server regardless of location, network, and storage configurations and of administrative domain.

The virtualisation plane consists of software mechanisms to abstract physical resources into appropriate sets of virtual resources that can be organised by the Orchestration Plane to form components (e.g., increased storage or memory), devices (e.g., a switch with more ports), or even networks. The organisation is done in order to realise a certain business goal or service requirement. Two dedicated interfaces are needed: the vSPI and the vCPI (Virtualisation System Programming Interface and Virtualisation Component Programming Interface, respectively). A set of control loops is formed using the vSPI and the vCPI, as shown in Figure 2.

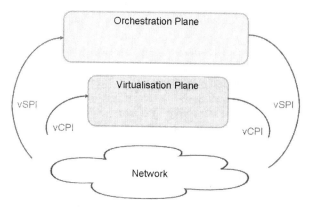

Fig. 2. Virtualisation Control Loop

Virtualisation System Programmability Interface (vSPI). The vSPI is used to enable the Orchestration Plane (and implicitly the AMS and DOC that are part of a given Orchestration Plane) to govern virtual resources, and to construct virtual services and networks that meet stated business goals having specified service requirements. The vSPI contains the "macro-view" of the virtual resources that a particular Orchestration Plane governs, and is responsible for orchestrating groups of virtual resources in response to changing user needs, business requirements, and environmental conditions. The low-level configuration (i.e., the "micro-view") of a virtual resource is provided by the vCPI. For example, the vSPI is responsible for informing the AMS that a particular virtual resource is ready for use, whereas the vCPI is responsible for informing the AMS that a particular virtual resource has been successfully reconfigured. The governance is performed by the set of AMS that are responsible for managing each component or set of components; each AMS uses the vSPI to express its needs and usage of the set of virtual resources to which it has access. The vSPI is responsible for determining what portion of a component (i.e., set of virtual resources) is allocated to a given task. This means that all or part of a virtual resource can be used for each task, providing an optimised partitioning of virtual resources according to business need, priority and other requirements. Composite virtual services can thus be constructed using all or part of the virtual resources provided by each physical resource.

Virtualisation Component Programming Interface (vCPI). Each physical resource has an associated and distinct vCPI. The vCPI is fulfilling two main functions: monitoring and management. The management functionality enables the AMS to manage the physical resource, and to request virtual resources to be constructed from that physical resource by the vCPI of the Virtualisation Plane. The AMS sends abstract (i.e., device-independent) commands via the vCPI, which are translated into device- and vendor-specific commands that reconfigure the physical resource (if necessary) and manage the virtual resources provided by that physical resource. The vCPI also provides monitoring information from the virtual resources back to the AMS that controls that physical resource. Note that the AMS is responsible for obtaining management data describing the physical resource. The vCPI is responsible for providing dynamic management data to its governing AMS that states how many virtual resources are currently instantiated, and how many additional virtual resources of what type can be supported.

2.4   Knowledge Plane Overview

The Knowledge Plane was proposed by Clark et al. [1] as a new dimension to a network architecture, contrasting with the data and control planes; its purpose is to provide knowledge and expertise to enable the network to be self-monitoring, self-analysing, self-diagnosing and self-maintaining. A narrow functionality Knowledge Plane (KP), consisting of context data structured in information models and ontologies, which provide increased analysis and inference capabilities is the basis for this paper. The KP brings together widely distributed data collection, wide availability of that data, and sophisticated and adaptive processing or KP functions, within a unifying structure. Knowledge extracted from information/data models forms facts. Knowledge extracted from ontologies is used to augment the facts, so that they can be reasoned about. Hence, the combination of model and ontology knowledge forms a universal lexicon, which is then used to transform received data into a common form that enables it to be managed. The KP provides information and context services as follows:

- information-life cycle management, which includes storage, aggregation, transformations, updates, distribution of information;
- triggers for the purpose of contextualisation of management systems (supported by the context model of the information model);
- support for robustness enabling the KP to continue to function as best possible, even under incorrect or incomplete behaviour of the network itself;
- support of virtual networks and virtual system resources in their needs for local control, while enabling them to cooperate for mutual benefit in more effective network management.

The goal of making the control functions of Networks context-aware is therefore essential in guaranteeing both a degree of self-management and adaptation as well as supporting context-aware communications that efficiently exploit the available network resources. Furthermore, context-aware networking enables new types of applications and services in the Future Internet.

Context Information Services. The Context Information Service Platform (CISP), within the KP, has the role of managing the context information, including its distribution to context clients/consumers. Context clients are context-aware services, either user-facing services or network management services, which make use of or/and adapt themselves to context information. Network services are described as the services provided by a number of functional entities (FEs), and one of the objectives of this description is to investigate how the different FEs can be made context-aware, i.e. act as context clients. The presence of CISP functionality helps to make the interactions between the different context sources and context clients simpler and more efficient. It acts as a mediating unit and reduces the numbers of interactions and the overhead control traffic. CISP is realised by four basic context-specific functional entities: (i) the Context Executive (CE) Module which interfaces with other entities/context clients, (ii) the Context Processing (CP) Module which implements the core internal operations related to the context processing, (iii) the Context Information Base (CIB) which acts as a context repository, and (iv) the Context Flow Controller (CFC) which performs context flow optimization activities (see Fig. 3).

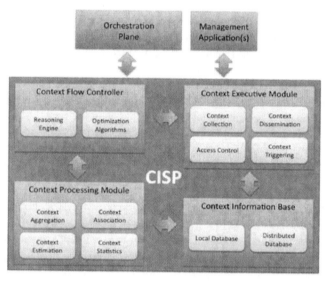

Fig. 3. Context Information Service Platform

The Context Executive Module (CE) is introduced to meet the requirements of creating a gateway into the CISP architecture and deals with indexing, registering, authorising and resolving context names into context or location addresses. Furthermore, the CE meets the requirements of context collection, context dissemination, interfaces with the Context Information Base and supports for access control. The Context Processing Module (CP) is responsible for the context management, including context aggregation, estimation and creation of appropriate context associations between clients and sources. The context association allows the CISP to decide where a specific context should be stored. Furthermore, the CP collects statistics about context usage. We note that these context statistics should be optimised in terms of memory usage for scalability purposes. In practice, the CP creates meta-context from context using mechanisms that exploit the business requirements, other forms of context and context usage statistics. The meta-context carries information that supports better the self-management functionalities of the context-aware applications. In general, the CE module is responsible for the communication of the CISP with the other management applications/components and the CP module for the optimisation of the context information. The Context Information Base (CIB) provides flexible storage capabilities, in support of the Context Executive and Context Processor modules. Context is distributed and replicated within the domain in order to improve the performance of context lookups. The CIB stores information according to a common ontology. The Context Flow Controller configures the Context Processing and Context Executive Modules based on the requirements of the Management Application and the general guidelines from the Orchestration Plane. These configuration settings are enabling certain behaviours in terms of context flow optimization with respect to these guidelines.

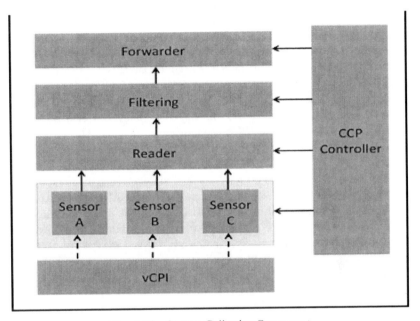

Fig. 4. The Context Collection Component

Context Collection Points. The Context Collection Points (CCP) act as sources of information: they monitor hardware and software for their state, present their capabilities, or collect configuration parameters. A monitoring mechanism and framework was developed to gather measurements from relevant physical and virtual resources and CCPs for use within the CISP. It also offers control mechanisms of the relevant probes and it also controls the context aggregation points (CAP). Such a monitoring framework has to have a minimal runtime foot–print, avoiding to be intrusive, so as not to adversely affect the performance of the network itself or the running management elements. The CISP Monitoring System supports three types of monitoring queries to an CCP: (i) 1-time queries, which collect information that can be considered static, e.g., the number of CPUs, (ii) N-time queries, which collect information periodically, and (iii) continuous queries that monitor information in an on-going manner. CCPs should be located near the corresponding sources of information in order to reduce management overhead. Filtering rules based on accuracy objectives should be applied at the CCPs, especially for the N-time and continuous queries, for the same reason. Furthermore, the CCPs should not be many hops away from the corresponding context aggregation point (CAP). Fig. 4 shows the structure of a CCP, which we have designed and implemented, consisting of 5 main components: the sensors, a reader, a filter, a forwarder and a CCP controller. These are described below.

The *sensors* can retrieve any information required. This can include common operations such as getting the state of a server with its CPU or memory usage, getting the state of a network interface by collecting the number of packets and number of bytes coming in and out, or getting the state of disks on a system presenting the total volume, free space, and used space. In our implementation, each sensor runs in its own thread allowing each one to collect data at different rates and also having the ability to turn them on and off if they are not needed. We note that the monitoring information retrieval is handled by the Virtualisation Plane.

The *reader* collects the raw measurement data from all of the sensors of a CCP. The collection can be done at a regular interval or as an event from the sensor itself. The reader collects data from many sensors and converts the raw data into a common measurement object used in the CISP Monitoring framework. The format contains meta-data about the sensor and the time of day, and it contains the retrieved data from the sensor.

The *filter* takes measurements from the reader and can filter them out before they are sent on to the forwarder. Using this mechanism it is possible to reduce the volume of measurements from the CCP by only sending values that are significantly different from previous measurements. For example, if a 5% filter is set, then only measurements that differ from the previous measurement by more than 5% will be passed on. By using filtering in this way, the CCP reduces the load on the network. In our case, the filtering percentage matches the accuracy objective of the management application requesting the information.

The *forwarder* sends the measurements onto the network. The common measurement object is encoded into a network amenable measurement format.

The CCP Controller controls and manages the other CCP components. It controls (i) the lifecycle of the sensors, being able to turn them on and off, and to set the rate at which they collect data; (ii) the filtering process, by changing the filter or adapting an existing filter; (iii) the forwarder, by changing the attributes of the network (such as IP address and port) that the ICP is connected to.

The vCPI supports the extension with additional functions, implicitly allowing the creation of other types of sensors, and thus helping the CCP to get more information. Also various sensors, which can measure attributes from CPU, memory, and network components of a server host, were created. We can also measure the same attributes of virtualised hosts by interacting with a hypervisor to collect these values. Finally, there are sensors that can send emulated measurements. These are useful for testing and evaluation purposes, with one example being an emulated response time, which we use in our experiments.

2.5 Management Plane Overview

The Management Plane is a basic building block of the infrastructure, which governs the physical and virtual resources, is responsible for the optimal placement and continuous migration of virtual routers into hosts (i.e., physical nodes and servers) subject to constraints determined by the Orchestration Plane. The Management Plane is designed to meet the following functionality:

- Embedded (Inside) Network functions: The majority of management functionality should be embedded in the network and it is abstracted from the human activities. As such the Management Plane components will run on execution environments supported by the virtual networks and systems, which run on top of all current networks (i.e. fixed, wireless and mobile networks) and service physical infrastructures.
- Aware and Self-aware functions: It monitors the network and operational context as well as internal operational network state in order to assess if the network current behaviour serve its service purposes.
- Adaptive and Self-adaptive functions: It triggers changes in network operations (state, configurations, functions) as a result of the changes in network and service context.
- Automatic self-functions: It enables self-control (i.e. self-FCAPS, self-*) of its internal network operations, functions and state. It also bootstraps itself and it operates without manual external intervention. Only manual/external input is provided in the setting-up of the business goals.
- Extensibility functions: It adds new functions without disturbing the rest of the system (Plug-and-Play / Unplug_and_Play / Dynamic programmability of management functions & services).
- System functions: Minimise life-cycle network operations' costs and minimise energy footprint.

In addition the Management Plane, as it governs all virtual resources, is responsible for the optimal placement and continuous migration of virtual routers into hosts (i.e. physical nodes and servers) subject to constraints determined by the Orchestration Plane.

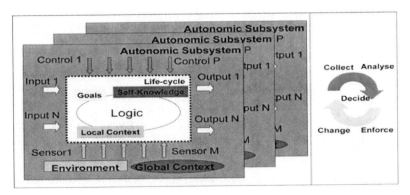

Fig. 5. Autonomic Control Loops

The Management Plane consists of Autonomic Management Systems (AMS). AMS is an infrastructure that manages a particular network domain, which may be an Autonomous System (AS). An AMS implements its own control loops, consisting of context collection, analysis, decision-making and decision enforcement. Each AMS includes interfaces to a dedicated set of models and ontologies and interfaces to one or more Distributed Orchestration Components (DOC), which manage the interoperation of two or more AMS domains. Mapping logic enables the data stored in models to be transformed into knowledge and combined with knowledge stored in ontologies to provide a context-sensitive assessment of the operation of one or more virtual resources. The AMS communicate through sets of interfaces that: (i) enable management service deployment and configuration (i.e., the ANPI and vSPI interfaces), (ii) manipulate physical and virtual resources (i.e., the vCPI interface).

The AMS are design to follow the autonomic control loops depicted in Fig. 5. The AMS is designed to be federated, enabling different AMS that are dedicated to govern different types of devices, resources, and services, to be combined. In order to support this, each AMS uses the models and ontologies to provide a standard set of capabilities that can be advertised and used by other AMS. The capabilities of an AMS can be offered for use to other AMS through intelligent negotiations (e.g., pre-defined agreements, auctioning, bargaining and other mechanisms). An AMS collects appropriate monitoring information from the virtual resources that is managing and makes appropriate decisions for the resources and management services that it governs, either by itself (if its governance mode is individual) or in collaboration with other AMS (if its governance mode is distributed or collaborative).

Since the AMS implement their own control loops, they can have their own goals. However, their goals should be harmonised to the high-level goals coming from the DOC that is responsible for each particular AMS. Each DOC is responsible for a set of AMS that form a network domain, called Orchestrated Domain (OD). An OD may belong to a single organisation that has the same high-level goals. We note that the entry point for the high-level goals is the Orchestration Plane. For example, a set of AMS may re-establish a local service in case of failure without interacting with the Orchestration Plane. However, this new establishment should follow the same guidelines that this local service used to follow. So, there is a significant difference between management and orchestration. Orchestration, actually, harmonises the different management components to one or more common goals.

3   Realisation: In-Network Cloud Functionality

A set of integrated service-centric platforms and supporting systems have been developed and issued as open source [10], which aims to create a highly open and flexible environment for In-Network Clouds in Future Internet. They are briefly described herewith. Full design and implementation of all software platforms are presented in [10].

- vCPI (Virtual Component Programming Interface is the VP's main component dealing with the heterogeneity of virtual resources and enabling programmability of network elements In each physical node there is an embedded vCPI, which is aware of the structure of the virtual resources, which are hosted in the physical node.

- CISP (Context Information Service Platform) is the KP's main component supported by a distributed monitoring platform for resources & components. CISP has the role of managing the context information, including its distribution to context clients/consumers.
- ANPI (Autonomic Network Programming Interface) is the SP's main component that enables large-scale autonomic services deployment on virtual networks.
- MBT (Model-Based Translator) platform, part of the KP, which takes configuration files compliant with an Information Model and translates them to device specific commands.
- LATTICE Monitoring Framework, also part of the KP, provides functionality to add powerful and flexible monitoring facilities to system clouds (virtualisation of networks and services). Lattice has a minimal runtime footprint and is not intrusive, so as not to adversely affect the performance of the system itself or any running applications. The monitoring functionality can be built up of various components provided by the framework, so creating a bespoke monitoring sub-system. The framework provides data sources, data consumers, and a control strategy. In a large distributed system there may be hundreds or thousands of measurement probes, which can generate data.
- APE (Autonomic Policy-based Engine), a component of the MP, supports context-aware policy-driven decisions for management and orchestration activities.
- XINA is a modular scalable platform that belong to the CISP and enables the deployment, control and management of programmable or active sessions over virtual entities, such as servers and routers.
- RNM (Reasoning and Negotiation Module), a core element of the KP, which mediates and negotiates between separate federated domains.

These In-Network Cloud platforms were integrated and validated on 2 testbeds enabling experimentation with thousands of virtual machines: V^3 – UCL's Experimental Testbed located in London consisting of 80 cores with a dedicated 10 Gbits/s infrastructure and Grid5000 - an Experimental testbed located in France consisting of 5000 cores and linked by a dedicated 10 Gbits/s infrastructure. Validation and performance analysis are fully described in [13]. Demonstrations are available at: http://clayfour. ee.ucl.ac.uk/demos/ and they are used for:

- Autonomic deployment of large-scale virtual networks (In-Network Cloud Provisioning);
- Self – management of virtual networks (In-Network Cloud Management);
- Autonomic service provisioning on In-Network Clouds (Service Computing Clouds).

4   Conclusion

This work has presented the design of an open software networked infrastructure (In-Network Cloud) that enables the composition of fast and guaranteed services in an efficient manner, and the execution of these services in an adaptive way taking into account better shared network and service resources provided by an virtualisation environment. We have also described the management architectural and system model for our Future Internet, which were described with the help of five abstractions and distributed systems – the OSKMV planes: Virtualisation Plane (VP), Management Plane (MP), Knowledge Plane (KP), Service Plane (SP) and Orchestration Plane (OP). The resulting software-driven control network infrastructure was fully exercised and relevant analysis on network virtualisation and service deployments were carried out on a large-scale testbed.

Virtualising physical network and server resources has served two purposes: Managing the heterogeneity through introduction of homogeneous virtual resources and enabling programmability of the network elements. The flexibility gained through this approach helps to adapt the network dynamically to both unforeseen and predictable changes in the network. A vital component of such a virtualisation approach is a common management and monitoring interface of virtualised resources. Such an interface has exported management and monitoring functions that allow management components to control the virtual resources in a very fine-grained way through a single, well defined interface. By enabling such fine-grained control, this interface can then form the basis for new types of applications and services in the Future Internet.

Acknowledgments. This work was partially undertaken in the context of the FP7-EU Autonomic Internet [10] and the RESERVOIR [9] research projects, which were funded by the Commission of the European Union. We also acknowledge the support of the Spanish Ministerio de Innovación grant TEC2009-14598-C02-02.

References

1. D. Clark, et al., "NewArch: Future Generation Internet Architecture",
   http://www.isi.edu/newarch/
2. Clark, D., et al.: NewArch: Future Generation Internet Architecture,
   http://www.isi.edu/newarch/
3. Galis, A., et al.: Management and Service-aware Networking Architectures (MANA) for Future Internet Position Paper: System Functions, Capabilities and Requirements. Invited paper IEEE ChinaCom09 26-28, Xi'an, China (August 2009),
   http://www.chinacom.org/2009/index.html
4. Rubio-Loyola, J., et al.: Platforms and Software Systems for an Autonomic Internet. IEEE Globecom 2010; 6-10 Dec., Miami, USA (2010)
5. Galis, A., et al.: Management Architecture and Systems for Future Internet Networks. In: Towards the Future Internet, IOS Press, Amsterdam (2009)
6. Chapman, C., et al.: Software Architecture Definition for On-demand Cloud Provisioning. ACM HPDC, 21-25, Chicago hpdc2010.eecs.northwestern.edu (June 2010)
7. Rochwerger, B., et al.: An Architecture for Federated Cloud Computing. In: Cloud Computing, Wiley, Chichester (2010)

8. Chapman, C., et al.: Elastic Service Management in Computational Clouds. 12th IEEE/IFIP NOMS2010/CloudMan 2010 19-23 April, Osaka (2010) http://cloudman2010.lncc.br/

9. Clayman, S., et al.: Monitoring Virtual Networks with Lattice. NOMS2010/ManFI 2010-Management of Future Internet 2010; 19-23 April, Osaka, Japan (2010), http://www.manfi.org/2010/

10. RESERVOIR project, http://www.reservoir-fp7.eu

11. AutoI project http://ist-autoi.eu

12. Clark, D., Partridge, C., Ramming, J.C.: and, J. T. Wroclawski "A Knowledge Plane for the internet". In: Proceedings of the 2003 Conference on Applications, Technologies, Architectures, and Protocols For Computer Communications (Karlsruhe, Germany, SIGCOMM '03, Karlsruhe, Germany, August 25–29, 2003, pp. 3–10. ACM, New York (2003)

13. Jennings, B., Van Der Meer, S., Balasubramaniam, S., Botvich, D., Foghlu, M., Donnelly, W., Strassner, J.: Towards autonomic management of communications networks. IEEE Communications Magazine 45(10), 112–121 (2007)

14. Deliverable D6.3 Final Results AutoI Approach http://ist-autoi.eu/

15. Mosharaf, N.M., Chowdhury, K., Boutaba, R., Cheriton, D.R.: A Survey of Network Virtualization. Journal Computer Networks: The International Journal of Computer and Telecommunications Networking 54(5) (2010)

16. Galis, A., Denazis, S., Bassi, A., Berl, A., Fischer, A., de Meer, H., Strassner, J., Davy, S., Macedo, D., Pujolle, G., Loyola, J.R., Serrat, J., Lefevre, L., Cheniour, A.: Management Architecture and Systems for Future Internet Networks. In: Towards the Future Internet – A European Research Perspective, p. 350. IOS Press, Amsterdam (2009), http://www.iospress.nl/

17. Berl, A., Fischer, A., De Meer, H.: Using System Virtualization to Create Virtualized Networks. Electronic Communications of the EASST 17, 1–12 (2009), http://journal.ub.tu-berl.asst/article/view/218/219

Flat Architectures: Towards Scalable Future Internet Mobility

László Bokor, Zoltán Faigl, and Sándor Imre

Budapest University of Technology and Economics, Department of Telecommunications
Mobile Communication and Computing Laboratory – Mobile Innovation Centre
Magyar Tudosok krt. 2, H-1117, Budapest Hungary
{goodzi, szlaj, imre}@mcl.hu

Abstract. This chapter is committed to give a comprehensive overview of the scalability problems of mobile Internet nowadays and to show how the concept of flat and ultra flat architectures emerges due to its suitability and applicability for the future Internet. It also aims to introduce the basic ideas and the main paradigms behind the different flat networking approaches trying to cope with the continuously growing traffic demands. The discussion of the above areas will guide the readers from the basics of flat mobile Internet architectures to the paradigm's complex feature set and power creating a novel Internet architecture for future mobile communications.

Keywords: mobile traffic evolution, network scalability, flat architectures, mobile Internet, IP mobility, distributed and dynamic mobility management

1 Introduction

Mobile Internet has recently started to become a reality for both users and operators thanks to the success of novel, extremely practical smartphones, portable computers with easy-to-use 3G USB modems and attractive business models. Based on the current trends in telecommunications, vendors prognosticate that mobile networks will suffer an immense traffic explosion in the packet switched domain up to year 2020 [1–4]. In order to accommodate the future Internet to the anticipated traffic demands, technologies applied in the radio access and core networks must become scalable to advanced future use cases.

There are many existing solutions aiming to handle the capacity problems of current mobile Internet architectures caused by the mobile traffic data evolution. Reserving additional spectrum resources is the most straightforward approach for increasing the throughput of the radio access, and also spectrum efficiency can be enhanced thanks to new wireless techniques (e.g., High Speed Packet Access, and Long Term Evolution). Heterogeneous systems providing densification and offload of the macrocellular network throughout pico, femtocells and relays or WiFi/WiMAX interfaces also extend the radio range. However, the deployment of novel technologies providing higher radio throughput (i.e., higher possible traffic rates) easily generates new

J. Domingue et al. (Eds.): Future Internet Assembly, LNCS 6656, pp. 35–50, 2011.

usages and the traffic increase may still accelerate. Since today's mobile Internet architectures have been originally designed for voice services and later extended to support packet switched services only in a very centralized manner, the management of this ever growing traffic demand is quite hard task to deal with. The challenge is even harder if we consider fixed/mobile convergent architectures managing mobile customers by balancing user traffic between a large variety of access networks. Scalability of traffic, network and mobility management functions has become one of the most important questions of the future Internet.

The growing number of mobile users, the increasing traffic volume, the complexity of mobility scenarios, and the development of new and innovative IP-based applications require network architectures able to deliver all kind of traffic demands seamlessly assuring high end-to-end quality of service. However, the strongly centralized nature of current and planned mobile Internet standards (e.g., the ones maintained by the IETF or by the collaboration of 3GPP) prevents cost effective system scaling for the novel traffic demands. Aiming to solve the burning problems of scalability from an architectural point of view, flat and fully distributed mobile architectures are gaining more and more attention today.

The goal of this chapter is to provide a detailed introduction to the nowadays emerging scalability problems of the mobile Internet and also to present a state of the art overview of the evolution of flat and ultra flat mobile communication systems. In order to achieve this we first introduce the issues relating to the continuously growing traffic load inside the networks of mobile Internet providers in Section 2. Then, in Section 3 we present the main evolutionary steps of flat architectures by bringing forward the most important schemes, methods, techniques and developments available in the literature. This is followed, in Section 4, by an introduction of distributed mobility management schemes which can be considered as the most essential building block of flat mobile communications. As a conclusion we summarize the benefits and challenges concerning flat and distributed architectures in Section 5.

2   Traffic Evolution Characteristics and Scalability Problems of the Mobile Internet

2.1   Traffic Evolution Characteristics of the Mobile Internet

One of the most important reasons of the traffic volume increase in mobile telecommunications is demographical. According to the current courses, world's population is growing at a rate of 1.2 % annually, and the total population is expected to be 7.6 billion in year 2020. This trend also implies a net addition of 77 million new inhabitants per year [5]. Today, over 25% of the global population – this means about two billion people – are using the Internet. Over 60% of the global population – now we are talking about five billion people – are subscribers of some mobile communication service [1][6]. Additionally, the number of wireless broadband subscriptions is about to exceed the total amount of fixed broadband subscriptions and this development becomes even more significant considering that the volume of fixed broadband sub-scriptions is gathering much slower.

The expansion of wireless broadband subscribers not only inflates the volume of mobile traffic directly, but also facilitates the growth in broadband wireless enabled terminals. However, more and more devices enable mobile access to the Internet, only a limited part of users is attracted or open to pay for the wireless Internet services meaning that voice communication will remain the dominant mobile application also in the future. Despite this and the assumption of [5] implying that the increase in the number of people potentially using mobile Internet services will likely saturate after 2015 in industrialized countries, the mobile Internet subscription growth potential will be kept high globally by two main factors. On one hand the growth of subscribers continues unbrokenly in the developing markets: mobile broadband access through basic handhelds will be the only access to the Internet for many people in Asia/Pacific. On the other hand access device, application and service evolution is also expected to sustain the capability of subscriber growth.

The most prominent effect of services and application evolution is the increase of video traffic: it is foreseen that due to the development of data-hungry entertainment services like television/radio broadcasting and VoD, 66% of mobile traffic will be video by 2014 [2]. A significant amount of this data volume will be produced by mobile Web-browsing which is expected to become the biggest source of mobile video traffic (e.g., YouTube). Cisco also forecasts that the total volume of video (including IPTV, VoD, P2P streaming, interactive video, etc.) will reach almost 90 percent of all consumer traffic (fixed and mobile) by the year 2012, producing a substantial increase of the overall mobile traffic of more than 200% each year [7]. Video traffic is also anticipated to grow so drastically in the forthcoming years that it could overstep Peer-to-Peer (P2P) traffic [4]. Emerging web technologies (such as HTML5), the increasing video quality requirements (HDTV, 3D, SHV) and special application areas (virtual reality experience sharing and gaming) will further boost this process and set new challenges to mobile networks. Since video and related entertainment services seems to become dominant in terms of bandwidth usage, special optimization mechanisms focusing on content delivery will also appear in the near future. The supposed evolution of Content Delivery Networking (CDN) and smart data caching technologies might have further impact on the traffic characteristics and obviously on mobile architectures.

Another important segment of mobile application and service evolution is social networking. As devices, networks and modes of communications evolve, users will choose from a growing scale of services to communicate (e.g., e-mail, Instant Messaging, blogging, micro-blogging, VoIP and video transmissions, etc.). In the future, social networking might evolve even further, like to cover broader areas of personal communication in a more integrated way, or to put online gaming on the next level deeply impregnated with social networking and virtual reality.

Even though video seems to be a major force behind the current traffic growth of the mobile Internet, there is another emerging form of communications called M2M (Machine-to-Machine) which has the potential to become the leading traffic contributor in the future. M2M sessions accommodate end-to-end communicating devices without human intervention for remote controlling, monitoring and measuring, road safety, security/identity checking, video surveillance, etc. Predictions state that there will be 225 million cellular M2M devices by 2014 with little traffic per node but resulting significant growth in total, mostly in uplink direction [3]. The huge number of sessions with tiny packets creates a big challenge for the operators. Central network functions may not be as scalable as needed by the increasing number of sessions in the packet-switched domain.

As a summary we can state that the inevitable mobile traffic evolution is foreseen thanks to the following main factors: growth of the mobile subscriptions, evolution of mobile networks, devices, applications and services, and significant device increase potential resulted by the tremendous number of novel subscriptions for Machine-to-Machine communications.

2.2 Scalability Problems of the Mobile Internet

Existing wireless telecommunication infrastructures are not prepared to handle this traffic increase, current mobile Internet was not designed with such requirements in mind: mobile architectures under standardization (e.g., 3GPP, 3GPP2, WiMAX Forum) follow a centralized approach which cannot scale well to the changing traffic conditions.

On one hand user plane scalability issues are foreseen for anchor-based mobile Internet architectures, where mechanisms of IP address allocation and tunnel establishment for end devices are managed by high level network elements, called anchor points (GGSN in 3GPP UMTS, PDN GW in SAE, and CSN for WiMAX networks). Each anchor point maintains special units of information called contexts, containing binding identity, tunnel identifier, required QoS, etc. on a per mobile node basis. These contexts are continuously updated and used to filter and route user traffic by the anchor point(s) towards the end terminals and vice versa. However, network elements (hence anchor points too) are limited in terms of simultaneous active contexts. Therefore, in case of traffic increase new equipments should be installed or existing ones should be upgraded with more capacity.

On the other hand, scalability issues are also foreseen on the control plane. The well established approach of separating service layer and access layer provides easy service convergence in current mobile Internet architectures but introduces additional complexity regarding session establishment procedures. Since service and access network levels are decomposed, special schemes have been introduced (e.g., Policy and Charging Control architecture by 3GPP) to achieve interaction between the two levels during session establishment, modification and release routines. PCC and similar schemes ensure that the bearer established on the access network uses the resources corresponding to the session negotiated at the service level and allowed by the operator policy and user subscription. Due to the number of standardized interfaces (e.g., towards IP Multimedia Subsystem for delivering IP multimedia services), the interoperability between the service and the access layer can easily cause scalability and QoS issues even in the control plane.

As a consequence, architectural changes are required for dealing with the ongoing traffic evolution: future mobile networks must specify architecture optimized to maximize the end-user experience, minimize CAPEX/OPEX, energy efficiency, network performance, and to ensure mobile networks sustainability.

3   Evolution of Flat Architectures

3.1   Evolution of the Architecture of 3GPP Mobile Networks

Fixed networks were firstly subject to similar scalability problems. The evolution of DSL access architecture has shown in the past that pushing IP routing and other functions from the core to the edge of the network results in sustainable network infrastructure. The same evolution was started to happen within the wireless telecommunication and mobile Internet era.

The 3GPP network architecture specifications having the numbers 03.02 [8] and 23.002 [9] show the evolution of the 3GPP network from GSM Phase 1 published in 1995 until the Evolved Packet System (EPS) specified in Release 8 in 2010. The core part of EPS called Evolved Packet Core (EPC) is continuously extended with new features in Release 10 and 11. The main steps of the architecture evolution are summarized in the followings. Fig. 1 illustrates the evolution steps of the packet-switched domain, including the main user plane anchors in the RAN and the CN.

In Phase 1 (1995) the basic elements of the GSM architecture have been defined. The reasons behind the hierarchization and centralization of the GSM architecture were both technical and economical. Primarily it offloaded the switching equipments (cross-bar switch or MSC). In parallel, existing ISDN switches could be re-used as MSCs only if special voice encoding entities were introduced below the MSCs, hence further strengthening the hierarchical structure of the network. However, with the introduction of the packet-switched domain (PS) and the expansion of the PS traffic the drawbacks of this paradigm started to appear very early.

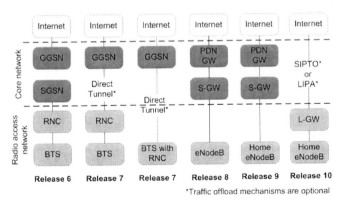

Fig. 1. The evolution of the packet-switched domain of the 3GPP architecture, including the main user plane anchors in the RAN and the CN.

The main driver to introduce packet-switching was that it allowed multiplexing hence resources could be utilized in a greater extent. In Phase 2+ (1997) the PS domain is described, hence centralized General Packet Radio Service (GPRS) support nodes are added to the network. Release 1999 (2002) describes the well known UMTS architecture clearly separating the CS and PS domains. Seeing that UMTS was designed to be the successor of GSM, it is not strange that the central anchors remained in place in 3G and beyond.

Progress of mobile and wireless communication systems introduced some fundamental changes. The most drastic among them is that IP has become the unique access protocol for data networks and the continuously increasing future wireless traffic is also based on packet data (i.e., Internet communication). Due to the collateral effects of this change a convergence procedure started to introduce IP-based transport technology in the core and backhaul network: Release 4 (2003) specified the Media gateway function, Release 5 (2003) introduced the IP Multimedia Subsystem (IMS) core network functions for provision of IP services over the PS domain, while Release 6 standardized WLAN interworking and Multimedia Broadcast Multicast Service (MBMS).

With the increasing IP-based data traffic flattening hierarchical and centralized functions became the main driving force in the evolution of 3GPP network architectures. Release 7 (also called Internet HSPA, 2008) supports the integration of the RNC with the NodeB providing a one node based radio access network. Another architectural enhancement of this release is the elaboration of Direct Tunnel service [10][11]. Direct Tunnel allows to offload user traffic from SGSN by bypassing it. The Direct Tunnel enabled SGSNs can initiate the reactivation of the PDP context to tunnel user traffic directly from the RNC to the GGSN or to the Serving GW introduced in Release 8. This mechanism tries to reduce the number of user-plane traffic anchors. However it also adds complexity in charging inter-PS traffic because SGSNs can not account the traffic passing in direct tunnels. When Direct Tunnel is enabled, SGSNs still handle signaling traffic, i.e., keep track of the location of mobile devices and participate in GTP signaling between the GGSN and RNC.

Release 8 (2010) introduces a new PS domain, i.e., the Evolved Packet Core (EPC). Compared to four main GPRS PS domain entities of Release 6, i.e. the base station (called NodeB), RNC, SGSN and GGSN, this architecture has one integrated radio access node, containing the precious base station and the radio network control functions, and three main functional entities in the core, i.e. the Mobility Management Entity (MME), the Serving GW (S-GW) and the Packet data Network GW (PDN GW).

Release 9 (2010) introduces the definition of Home (e)NodeB Subsystem. These systems allow unmanaged deployment of femtocells at indoor sites, providing almost perfect broadband radio coverage in residential and working areas, and offloading the managed, pre-panned macro-cell network [14].

In Release 10 (2010) Selective IP Traffic Offload (SIPTO) and Local IP Access (LIPA) services have been published [15]. These enable local breakout of certain IP traffic from the macro-cellular network or the H(e)NodeB subsystems, in order to offload the network elements in the PS and EPC PS domain. The LIPA function enables an IP capable UE connected via Home(e)NodeB to access other IP capable entities in the same residential/enterprise IP network without the user plane traversing the core network entities. SIPTO enables per APN and/or per IP flow class based traffic offload towards a defined IP network close to the UE's point of attachment to the access network. In order to avoid SGSN/S-GW from the path, Direct Tunnel mode should be used.

The above evolutionary steps resulted in that radio access networks of 3GPP became flattened to one single serving node (i.e., the eNodeB), and helped the distribution of previous centralized RNC functions. However, the flat nature of LTE and LTE-A architectures concerns only the control plane but not the user plane: LTE is linked to the Evolved Packet Core (EPC) in the 3GPP system evolution, and in EPC, the main packet switched core network functional entities are still remaining centralized, keeping user IP traffic anchored. There are several schemes to eliminate the residual centralization and further extend 3GPP.

3.2   Ultra Flat Architecture

One of the most important schemes aiming to further extend 3GPP standards is the Ultra Flat Architecture (UFA) [16–20]. Authors present and evaluate an almost green field approach which is a flat and distributed convergent architecture, with the exception of certain control functions still provided by the core. UFA represents the ultimate step toward flattening IP-based core networks, e.g., the EPC in 3GPP. The objective of UFA design is to distribute core functions into single nodes at the edge of the network, e.g., the base stations. The intelligent nodes at the edge of the network are called UFA gateways. Fig. 2 illustrates the UFA with HIP and PMIP-based mobility control.

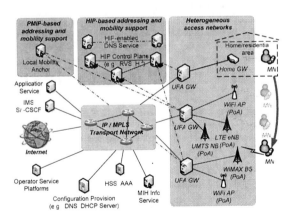

Fig. 2. The Ultra Flat Architecture with HIP and PMIP-based mobility control

Since mobility introduces frequent IP-level handovers a Session Initialization Protocol (SIP) based handover procedure has been described in [16]. It has been shown by a numerical analysis, and in a later publication with measurements on a testbed [17] that seamless handovers can be guaranteed for SIP-based applications. SIP Back-to-

Back User Agents (B2BUAs) in UFA GWs can prepare for fast handovers by communicating the necessary contexts, e.g., the new IP address before physical handover. This scheme supports both mobile node (MN) and network decided handovers.

In the PS domain, IP multimedia services require a two-level session establishment procedure. First, the MN and the correspondent node (CN) negotiate the session parameters using SIP on the service level, then the Policy and Charging Control (PCRF), ensures that the bearer established in the access layer uses the resources corresponding to the negotiated session. The problem is that service level is not directly notified about access layer resource problems, and, e.g., it is difficult to adapt different application session components of the same service to the available resources in the access layer. In order to solve this problem, a novel SIP-based session establishment and session update procedure is introduced in [16] for the UFA.

Interworking with Internet applications based on non SIP control protocol is a technical challenge for mobile operators. One of their aims is to provide seamless handovers for any application. IP-mobility control can be provided by protocols below the application layer. A Mobile IPv6 and a Host Identity Protocol (HIP) based signaling scheme alternative has been introduced for UFA by Z. Faigl et al. [18]. L. Bokor et al. describe a new HIP extension service which enables signaling delegation [19]. This service is applied in HIP-based handover and session establishment procedures of UFA, to reduce the number of HIP Base Exchanges in the access and core network, and to enable delegation of HIP-level signaling of the MN by the UFA GWs. Moreover, a new cross-layer access authorization mechanism for L2 and HIP has been introduced, to replace certificate-based access authorization with a more lightweight access authorization. In [20] authors clearly define the terminal attachment, session establishment and handover procedures, further enhance the original idea by providing two integrated UFA schemes (i.e., SIP–IEEE 802.21–HIP and SIP–IEEE 802.21–PMIP) and analyze the suitability of the two solutions using the Multiplicative Analytic Hierarchy Process.

4   Distributed Mobility Management in Flat Architectures

4.1   Motivations for Distributing Mobility Functions

Flat mobile networks not only require novel architectural design paradigms, special network nodes and proprietary elements with peculiar functions, but also demand certain, distinctive mobility management schemes sufficiently adapted to the distributed architecture. In fact the distributed mobility management mechanisms and the relating decision methods, information, command and event services form the key routines of the future mobile Internet designs. The importance of this research area is also emphasized by the creation of a new IETF non-working group called Distributed Mobility Management (DMM) in August 2010, aiming to extend current IP mobility solutions for flat network architectures.

Current mobility management solutions rely on hierarchical and centralized architectures which employ anchor nodes for mobility signaling and user traffic forwarding. In 3G UMTS architectures centralized and hierarchical mobility anchors are implemented by the RNC, SGSN and GGSN nodes that handle traffic forwarding tasks using the apparatus of GPRS Tunneling Protocol (GTP). The similar centralization is noticeable in Mobile IP (MIP) [21] where the Home Agent –an anchor node for both signaling and user plane traffic– administers mobile terminals' location information, and tunnels user traffic towards the mobile's current locations and vice versa. Several enhancements and extensions such as Fast Handoffs for Mobile IPv6 (FMIP) [22], Hierarchical Mobile IPv6 (HMIP) [23], Multiple Care-of Addresses Registration [24], Network Mobility (NEMO) Basic Support [25], Dual-Stack Mobile IPv6 [26], and Proxy Mobile IPv6 (PMIP) [27], were proposed to optimize the performance and broaden the capabilities of Mobile IP, but all of them preserve the centralized and anchoring nature of the original scheme.

There are also alternate schemes in the literature aiming to integrate IP-based mobility protocols into cellular architectures and to effectively manage heterogeneous networks with special mobility scenarios. Cellular IP [28] introduces a gateway router dealing with local mobility management while also supporting a number of handoff techniques and paging. A similar approach is the handoff-aware wireless access Internet infrastructure (HAWAII) [29], which is a separate routing protocol to handle micromobility. Terminal Independent Mobility for IP [30] combines some advantages from Cellular IP and HAWAII, where terminals with legacy IP stacks have the same degree of mobility as terminals with mobility-aware IP stacks. Authors of [31] present a framework that integrates 802.21 Media Independent Handover [32] and Mobile IP for network driven mobility. However, these proposals are also based on centralized functions and generally rely on MIP or similar anchoring schemes.

Some of the above solutions are already standardized [12][13][33] for 3G and beyond 3G architectures where the introduced architectural evolution is in progress: E-UTRAN (Evolved Universal Terrestrial Radio Access Network) or LTE (Long Term Evolution) base stations (eNodeBs) became distributed in a flatter scheme allowing almost complete distribution of radio and handover control mechanisms together with direct logical interfaces for inter-eNodeB communications. Here, traffic forwarding between neighboring eNodeBs is temporarily allowed during handover events providing intra-domain mobility. However, traffic forwarding and inter-gateway mobility operations remain centralized thanks to S-GW, PDN-GW, Local Mobility Anchor and Home Agent, responsible for maintaining and switching centralized, hierarchical and overlapping system of tunnels towards mobile nodes. Also, offloading with LIPTO and SIPA extensions cannot completely solve this issue: mobility management mechanisms in current wireless and mobile networks anchor the user traffic relatively far from users' location. This results in centralized, unscalable data plane and control plane with non-optimal routes, overhead and high end-to-end packet delay even in case of motionless users, centralized context maintenance and single point of failures. Anchor-based traffic forwarding and mobility management solutions also cause deployment issues for caching contents near the user..

To solve all these problems and questions novel, distributed and dynamic mobility management approaches must be envisaged, applicable to intra- and inter-technology mobility cases as well.

4.2   Application Scenarios for DMM Schemes

The basic idea is that anchor nodes and mobility management functions of wireless and mobile systems could be distributed to multiple locations in different network segments, hence mobile nodes located in any of these locations could be served by a close entity.

A first alternative for achieving DMM is core-level distribution. In this case mobility anchors are topologically distributed and cover specific geographical area but still remain in the core network. A good example is the Global HA to HA protocol [34], which extends MIP and NEMO in order to remove their link layer dependencies on the Home Link and distribute the Home Agents in Layer 3, at the scale of the Internet. DIMA (Distributed IP Mobility Approach) [35] can also be considered as a core-level scheme by allowing the distribution of MIP Home Agent (the normally isolated central server) to many and less powerful interworking servers called Mobility Agents (MA). These new nodes have the combined functionality of a MIP Home Agent and HMIP/PMIP Mobility Anchor Points. The administration of the system of distributed MAs is done via a distributed Home Agent overlay table structure based on a Distributed Hash Table (DHT) [36]. It creates a virtual Home Agent cluster with distributed binding cache that maps a mobile node's permanent identifier to its temporary identifier.

A second alternative for DMM is when mobility functions and anchors are distributed in the access part of the network. For example in case of pico- and femto cellular access schemes it could be very effective to introduce Layer 3 capability in access nodes to handle IP mobility management and to provide higher level intervention and even cross-layer optimization mechanisms. The concept of UMTS Base Station Router (BSR) [37] realizes such an access-level mobility management distribution scheme where a special network element called BSR is used to build flat cellular systems. BSR merges the GGSN, SGSN, RNC and NodeB entities into a single element: while a common UMTS network is built from a plethora of network nodes and is maintained in a hierarchical and centralized fashion, the BSR integrates all radio access and core functions. Furthermore, the BSR can be considered a special wireless edge router that bridges between mobile/wireless and IP communication. In order to achieve this, mobility support in the BSR is handled at three layers: RF channel mobility, Layer 2 anchor mobility, and Layer 3 IP mobility. The idea of Liu Yu et al. [38] is quite similar to the BSR concept. Here a node called Access Gateway (AGW) is introduced to implement distributed mobility management functionalities at the access level. The whole flat architecture consists of two kinds of elements, AGW on the access network side and terminals on the user side. Core network nodes are mainly simple IP routers. The scheme applies DHT and Loc/ID separation: each mobile node has a unique identifier (ID) keeping persistent, and an IP address based locator (Loc) changed by every single mobility event. The (Loc,ID) pair of each mobile is stored inside AGW nodes and organized/managed using DHTs.

A third type of DMM application scenarios is the so-called host-level or peer-to-peer distributed mobility management where once the correspondent node is found, communicating peers can directly exchange IP packets. In order to find the correspondent node, a special information server is required in the network, which can also be centralized or distributed. A good example for host-level schemes in the IP layer is MIPv6 which is able to bypass the user plane anchor (i.e., Home Agent) due to its route optimization mechanism, therefore providing a host-to-host communication method. End-to-end mobility management protocols working in higher layers of the TCP/IP stack such as Host Identity Protocol (HIP) [39], TCP-Migrate [40], MSOCKS [41], Stream Control Transmission Protocol (SCTP) [42], or Session Initiation Protocol (SIP) [43] can also be efficiently employed in such schemes.

4.3   Distribution Methods of Mobility Functions

Mobility management functions can be distributed in two main ways: partially and fully.

Partially distributed schemes can be implemented either by distinguishing signaling and user planes based on their differences in traffic volume or end-host behavior (i.e., only the user plane is distributed), or by granting mobility support only to nodes that actually need it (i.e., actually eventuate mobility event), hence achieving more advanced resource management. Note that these two approaches may also be combined.

Today's mobility management protocols (e.g., Mobile IP, NEMO BS and Proxy Mobile IP without route optimization) do not separate signaling and user planes which means that all control and data packets traverse the centralized or hierarchized mobility anchor. Since the volume of user plane traffic is much higher compared to the signaling traffic, the separation of signaling and user planes together with the distribution of the user plane but without eliminating signaling anchors can still result in effective and scalable mobility management. This is exploited by the HIP based UFA scheme [18–20] where a relatively simple inter-UFA GW protocol can be used thanks to the centralized HIP signaling plane, but the user plane is still fully distributed. Mobile IP based DMM solutions also rely on the advantages of this partial distribution concept when they implement route optimization, hence separate control packets from data messages after a short period of route optimization procedure.

The second type of partially distributed mobility management is based on the capability to turn off mobility signaling when such mechanisms are not needed. This so-called dynamic mobility management dynamically executes mobility functions only for mobile nodes that are actually subjected to handover event, and lack transport or application-layer mobility support. In such cases, thanks to the removal of unwanted mobility signaling, handover latency and control overhead can be significantly reduced. Integrating this concept with distributed anchors, the algorithms supporting dynamic mobility could also be distributed. Such integration is accomplished in [44][45] where authors introduce and evaluate a scheme to dynamically anchor mobile nodes' traffic in distributed Access Nodes (AN), depending on mobiles' actual location when sessions are getting set up. The solution's dynamic nature lies in the fact that sessions of mobile nodes are dynamically anchored on different ANs depending on the IP address used. Based on this behavior, the system is able to avoid execution of mobility management functions (e.g., traffic encapsulation) as long as a particular mobile node is not moving. The method is simultaneously dynamic and distributed, and because mobility functions are fully managed at the access level (by the ANs), it is appropriate for flat architectures. Similar considerations are applied in [46] for MIP, in [47] for HMIP and in [48] for PMIP. The MIP-based scheme introduces a special mode for the mobility usage in IP networks: for all the IP sessions opened and closed in the same IP sub-network no MIP functions will be executed even if the mobile node is away from its home network; standard MIP mechanisms will be used only for the ongoing communications while the mobile node is in motion between different IP sub-networks. The HMIP-based method proposes a strategy to evenly distribute the signaling burden and to dynamically adjust the micromobility domain (i.e., regional network) boundary according to real-time measurements of handover rates or traffic load in the networks. The PMIP-based solution discusses a possible deployment scheme of Proxy Mobile IP for flat architecture. This extension allows to dynamically distributing mobility functions among access routers: the mobility support is restricted to the access level, and adapted dynamically to the needs of mobile nodes by applying traffic redirection only to MNs' flows when an IP handover event occurs.

Fully distributed schemes bring complete distribution of mobility functions into effect (i.e., both data plane and control plane are distributed). This implies the introduction of special mechanisms in order to identify the anchor that manages mobility signaling and data forwarding of a particular mobile node, and in most cases this also requires the absolute distribution of mobility context database (e.g., for binding information) between every element of the distributed anchor system. Distributed Hash Table or anycast/broadcast/multicast communication can be used for the above purposes. In such schemes, usually all routing and signaling functions of mobility anchor nodes are integrated on the access level (like in [49]), but less flat architectures (e.g., by using Hi3 [50] for core-level distribution of HIP signaling plane) are also feasible.

5   Conclusion

Flat architectures infer high scalability because centralized anchors – the main performance bottlenecks – are removed, and traffic is forwarded in a distributed fashion. The flat nature also provides flexibility regarding the evolution of broadband access, e.g., the range extension of RANs with unmanaged micro-, pico- and femtocells, without concerns of capacity in centralized entities covering the actual area in a hierarchical structure.

In flat architectures, integrated and IP-enabled radio base station (BS) entities are directly connected to the IP core infrastructure. Therefore, they provide convenient and implicit interoperability between heterogeneous wireless technologies, and facilitate a convenient way of sharing the infrastructure for the operators. Flattening also infers the elimination of centralized components that are access technology specific. Thanks to the integrated, "single box" nature of these advanced base stations, the additional delay that user and signaling plane messages perceive over a hierarchical and multi-element access and core network (i.e., transmission and queuing delays to a central control node) are also reduced or even eliminated. This integrated design of

BS nodes also minimizes the feedback time of intermodule communication, i.e., signaling is handled as soon as it is received locally, on the edge of the operator's network, and enables to incorporate sophisticated cross-layer optimization schemes for performance improvements.

The application of general-purpose IP equipments produced in large quantities has economic advantages as well. In flat architectures the radio access network components could be much cheaper compared to HSPA and LTE devices today because of the economy of scale. Also operational costs can be reduced as a flat network has fewer integrated components, and lacks of hierarchical functions simultaneously influenced by management processes. The higher competition of network management tools due to the apparition of tools developed formerly for the Internet era may reduce the operational expenditures as well.

Failure tolerance/resistance, reliability and redundancy of networks also can be refined and strengthen by flat design schemes. Anchor and control nodes in hierarchical and centralized architectures are often single point of failures and their shortfall can easily cause serious breakdowns in large service areas. Within flat architectures no such single points of failure exist, and the impact of possible shortfalls of the distributed network elements (i.e., BSs) can smoothly narrowed to a limited, local area without complex failure recovery operations.

Another important benefit of flat architectures is the potential to prevent suboptimal routing situations and realize advanced resource efficiency. In a common hierarchical architecture, all traffic passes through the centralized anchor nodes, which likely increases the routing path and results in suboptimal traffic routing compared to the flat use-cases.

However, in order to exploit all the above benefits and advantages, some challenges that flat architectures face must be concerned.

In flat architectures, network management and configuration together with resource control must be done in a fully distributed and decentralized way. It means that self-configuration and self-optimization capabilities are to be introduced in the system. Closely related to self-optimization and self-configuration, self-diagnosis and self-healing is essential for continuous and reliable service provision in flat networking architectures. This is reasoned by the fact that IP equipments are more sensible to failures: due to lack of core controller entities base stations are no more managed centrally; hence failure diagnostics and recovery must be handled in a fully distributed and automated way. This is a great challenge but it comes with the benefits of scalability, fault tolerance and flexibility.

Optimization of handover performance is another key challenge for flat networks. Unlike in hierarchical and centralized architectures which usually provide efficient fast handover mechanisms using Layer 2 methods, in flat architectures IP-based mobility management protocol – with advanced micromobility extension – must be used. Since all the BSs are connected directly to the IP core network, hiding mobility events from the IP layer is much harder.

Last but not least Quality of Service provision is also an important challenge of flat architectures. This problem emerges because current QoS assurance mechanisms in the IP world require improvements to replace the Layer 2 QoS schemes of the traditional hierarchical and centralized mobile telecommunication architectures. The IP network that deals with the interconnection of base stations in flat networks must be able to assure different QoS levels (e.g., in means of bandwidth and delay) and manage resources for adequate application performance.

Based on the collected benefits and the actual challenges of flat architectures we can say that applying flat networking schemes together with distributed and dynamic mobility management is one of the most promising alternatives to change the current mobile Internet architecture for better adaptation to future needs.

Acknowledgments. This work was made in the frame of Mobile Innovation Centre's 'MEVICO.HU' project, supported by the National Office for Research and Technology (EUREKA_Hu_08-1-2009-0043) under the co-operation of the Celtic Call7 Project MEVICO.

References

1. UMTS Forum White Paper: Recognising the Promise of Mobile Broadband (June 2010)
2. Cisco VNI: Global Mobile Data Traffic Forecast Update, 2009-2014 (Feb. 2010)
3. Dohler, M., Watteyne, T., Alonso-Zárate, J.: Machine-to-Machine: An Emerging Communication Paradigm, Tutorial. In: GlobeCom'10 (Dec. 2010)
4. Schulze, H., Mochalski, K.: Ipoque, Internet Study 2008/2009, Ipoque (Jan. 2011)
5. UMTS Forum, REPORT NO 37, Magic Mobile Future 2010-2020 (April 2005)
6. International Telecommunication Union, Press Release: ITU sees 5 billion mobile subscriptions globally in 2010 (February 2010)
7. Cisco VNI: Hyperconnectivity and the Approaching Zettabyte Era (June 2010)
8. ETSI GTS GSM 03.02-v5.1.0: Digital cellular telecommunications system (Phase 2+) - Network architecture (GSM 03.02) (1996)
9. 3GPP TS 23.002: Network architecture, V10.1.1, Release 10 (Jan. 2011)
10. 3GPP TR 23.919: Direct Tunnel Deployment Guideline, Release 7, V1.0.0 (May 2007)
11. 3GPP TS 23.401: General Packet Radio Service (GPRS) enhancements for Evolved Universal Terrestrial Radio Access Network (E-UTRAN) access, Rel.8, V8.12 (Dec. 2010)
12. 3GPP TS 29.275, Proxy Mobile IPv6 (PMIPv6) based Mobility and Tunneling protocols, Stage 3, Release 10, V10.0.0 (Dec. 2010)
13. 3GPP TS 24.303, Mobility management based on Dual-Stack Mobile IPv6, Stage 3, Release 10, V10.1.0 Dec (2010)
14. FemtoForum: Femtocells – Natural Solution for Offload – a Femto Forum brief (June 2010)
15. 3GPP TR 23.829: Local IP Access and Selected IP Traffic Offload, Release 10, V1.3 (2010)
16. Daoud, K., Herbelin, P., Crespi, N.: UFA: Ultra Flat Architecture for high bitrate services in mobile networks. In: Proc. of PIMRC'08, Cannes, France, pp. 1–6 (2008)
17. Daoud, K., Herbelin, P., Guillouard, K., Crespi, N.: Performance and Implementation of UFA: a SIP-based Ultra Flat Mobile Network Architecture. In: Proc. of PIMRC (Sep. 2009)
18. Faigl, Z., Bokor, L., Neves, P., Pereira, R., Daoud, K., Herbelin, P.: Evaluation and comparison of signaling protocol alternatives for the Ultra Flat Architecture, ICSNC, pp. 1–9 (2010)

19. Bokor, L., Faigl, Z., Imre, S.: A Delegation-based HIP Signaling Scheme for the Ultra Flat Architecture. In: Proc. of the 2nd IWSCN, Karlstad, Sweden, pp. 9–16 (2010)

20. Faigl, Z., Bokor, L., Neves, P., Daoud, K., Herbelin, P.: Evaluation of two integrated signalling schemes for the ultra flat architecture using SIP, IEEE 802.21, and HIP/PMIP protocols. In: Journal of Computer Networks (2011), doi:10.1016/j.comnet.2011.02.005

21. Johnson, D., Perkins, C., Arkko, J.: IP Mobility Support in IPv6, IETF RFC 3775 (2004)

22. Koodli, R. (ed.): Fast Handoffs for Mobile IPv6, IETF RFC 4068 (July 2005)

23. Soliman, H., Castelluccia, C., El Malki, K., Bellier, L.: Hierarchical Mobile IPv6 Mobility Management (HMIPv6), IETF RFC 4140 (Aug. 2005)

24. Wakikawa, R. (ed.): V. Devarapalli, G. Tsirtsis, T. Ernst, K. Nagami: Multiple Care-of Addresses Registration, IETF RFC 5648 (October 2009)

25. Devarapalli, V., Wakikawa, R., Petrescu, A., Thubert, P.: Network Mobility (NEMO) Basic Support Protocol, IETF RFC 3963 (Jan. 2005)

26. Soliman, H. (ed.): Mobile IPv6 Support for Dual Stack Hosts and Routers, IETF RFC 5555 (June 2009)

27. Gundavelli, S. (ed.): K. Leung, V. Devarapalli, K. Chowdhury, B. Patil: Proxy Mobile IPv6, IETF RFC 5213 (Aug. 2008)

28. Valko: Cellular IP: A New Approach to Internet Host Mobility, ACM SIGCOMM Comp. Commun. Rev., 29 (1), 50-65 (1999)

29. Ramjee, R., Porta, T.L., Thuel, S., Varadhan, K., Wang, S.: HAWAII: A Domain-Based Approach for Supporting Mobility in Wide-area Wireless Networks. In: IEEE Int. Conf. Network Protocols (1999)

30. Grilo, A., Estrela, P., Nunes, M.: Terminal Independent Mobility for IP (TIMIP). IEEE Communications Magazine 39(12), 34–41 (2001)

31. Melia, T., de la Oliva, A., Vidal, A., Soto, I., Corujo, D., Aguiar, R.L.: Toward IP converged heterogeneous mobility: A network controlled approach. Com. Networks 51 (2007)

32. IEEE, IEEE Standard for Local and metropolitan area networks- Part 21: Media Independent Handover, IEEE Std 802.21-2008 (Jan. 2009)

33. 3GPP TS 23.402, Architecture enhancements for non-3GPP accesses, Rel.10,V10.2 (2011)

34. Thubert, P., Wakikawa, R., Devarapalli, V.: Global HA to HA protocol, IETF Internet-Draft, draft-thubert-nemo-global-haha-02.txt (Sept. 2006)

35. Fischer, M., Andersen, F.-U., Kopsel, A., Schafer, G., Schlager, M.: A Distributed IP Mobility Approach for 3G SAE. In: Proc. of 19th PIMRC, ISBN: 978-1-4244-2643-0 (Sept. 2008)

36. Farha, R., Khavari, K., Abji, N., Leon-Garcia, A.: Peer-to-peer mobility management for all-ip networks. In: Proc. of ICC '06, V. 5, pp. 1946–1952 (June 2006)

37. Bauer, M., Bosch, P., Khrais, N., Samuel, L.G., Schefczik, P.: The UMTS base station router. Bell Labs Tech. Journal, I. 11(4), 93–111 (2007)

38. Liu Yu, Zhao Zhijun, Lin Tao, Tang Hui: Distributed mobility management based on flat network architecture. In: Proc. of 5th WICON, pp. 1-5, Singapore (2010)

39. Moskowitz, R., Nikander, P., Jokela, P. (eds.): T. Henderson: Host Identity Protocol, IETF RFC 5201 (April 2008)

40. Snoeren, A.C., Balakrishnan, H.: An End-to-End Approach to Host Mobility. In: Proc. of MobiCom (Aug. 2000)

41. Maltz, D., Bhagwat, P.: MSOCKS: An Architecture for Transport Layer Mobility. In: Proc. INFOCOM, pp. 1037-1045 (Mar 1998)

42. Stewart, R. (ed.): Stream Control Transmission Protocol, IETF RFC 4960 (Sept. 2007)

43. Rosenberg, J., Schulzrinne, H., Camarillo, G., Johnston, A., Peterson, J., Sparks, R., Handley, M., Schooler, E.: SIP: Session Initiation Protocol, IETF RFC 3261 (June 2002)

44. Bertin, P., Bonjour, S., Bonnin, J.-M.: A Distributed Dynamic Mobility Management Scheme Designed for Flat IP Architectures. In: Proc. of NTMS '08, pp.1-5 (2008)
45. Bertin, P., Bonjour, S., Bonnin, J.: Distributed or centralized mobility? In: Proc. of the 28th IEEE conference on Global telecommunications (GLOBECOM'09), Honolulu, HI (2009)
46. Kassi-Lahlou, M., Jacquenet, C., Beloeil, L., Brouckaert, X.: Dynamic Mobile IP (DMI), IETF Internet-Draft, draft-kassi-mobileip-dmi-01.txt (Jan. 2003)
47. Song, M., Huang, J., Feng, R., Song, J.: A Distributed Dynamic Mobility Management Strategy for Mobile IP Networks. In: Proc. of 6th ITST, pp. 1045-1050 (June 2006)
48. Seite, P., Bertin, P.: Dynamic Mobility Anchoring, IETF Internet-Draft (May 2010)
49. Yan, Z., Lei, L., Chen, M.: WIISE - A Completely Flat and Distributed Architecture for Future Wireless Communication Systems, Wireless World Research Forum (Oct. 2008)
50. Gurtov, A., et al.: Hi3: An efficient and secure networking architecture for mobile hosts. Journal of Computer Communications 31(10), 2457–2467 (2008)

Review and Designs of Federated Management in Future Internet Architectures

Martín Serrano[1], Steven Davy[1], Martin Johnsson[1], Willie Donnelly[1], and Alex Galis[2]

[1] Waterford Institute of Technology – WIT
Telecommunications Software and Systems Group – TSSG, Co. Waterford, Ireland
{jmserrano, sdavy, mjohnsson, wdonnelly}@tssg.org
[2] University College London – UCL
Department of Electronic and Electrical Engineering, Torrington Place, London, U.K.
a.galis@ee.ucl.ac.uk

Abstract. The Future Internet as a design conception is network and service-aware addressing social and economic trends in a service oriented way. In the Future Internet, applications transcend disciplinary and technology boundaries following interoperable reference model(s). In this paper we discuss issues about federated management targeting information sharing capabilities for heterogeneous infrastructure. In Future Internet architectures, service and network requirements act as design inputs particularly on information interoperability and cross-domain information sharing. An inter-operable, extensible, reusable and manageable new Internet reference model is critical for Future Internet realisation and deployment. The reference model must rely on the fact that high-level applications make use of diverse infrastructure representations and not use of resources directly. So when resources are not being required to support or deploy services they can be used in other tasks or services. As implementation challenge for controlling and harmonising these entire resource management requirements, the federation paradigm emerges as a tentative approach and potentially optimal solution. We address challenges for a future Internet Architecture perspective using federation. We also provide, in a form of realistic implementations, research results and solutions addressing rationale for federation, all this activities are developed under the umbrella of federated management activity in the Future Internet.

Keywords: Federation, Management, Reference Model, Future Internet, Architectures and Systems, Autonomics, Service Management, Semantic Modelling and Management, Knowledge Engineering, Networking Data and Ontologies, Future Communications and Internet.

1   Introduction

In recent years convergence on Internet technologies for communication's, computation's and storage's networks and services has been a clear trend in the Information and Communications Technology (ICT) domain. Although widely discussed and

J. Domingue et al. (Eds.): Future Internet Assembly, LNCS 6656, pp. 51–66, 2011.

researched, this trend has not fully run its course in terms of implementation, due to many complex issues involving deployment of non-interoperable and management infrastructural aspects and also due to technological, social, economic restrictions and bottlenecks in the Future Internet.

In the Future Internet, services and networks follow a common goal: to provide solutions in a form of implemented interoperable mechanisms. Telecommunications networks have undergone a radical shift from a traditional circuit-switched environment with heavy/complex signalling focused on applications-oriented perspective, towards a converged service-oriented space, mostly Internet interaction by customer as end-user and network operators as service providers. The benefits of this shift reflect cost reduction and increase systems flexibility to react to user demands, by replacing a plethora of proprietary hardware and software platforms with generic solutions supporting standardised development and deployment stacks.

The Future Internet as design conception is service-aware of the network infrastructure addressing service-oriented, social trends and economic commitments. In the Future Internet trans-disciplinary solutions (applications that transcend disciplinary boundaries) following reference model(s) are crucial for a realistic integrated management realisation. Challenges in the future communications systems mainly demand, in terms of end user requirements, personalized provisioning, service-oriented performance, and service-awareness networking.

Additionally to those technology requirements, necessities to support information interoperability as result of more service-oriented demands exist. Reliable services and network performance act as technology requirements for more secure and reliable communication systems supporting end user and network requirements. Demands on data models integration are requirements to be considered during the design and implementation phases of any ICT system.

The emergence and wide-scale deployment of wireless access network technologies calls into question the viability of basing the future Internet on IP and TCP – protocols that were never intended for use across highly unreliable and volatile wireless interfaces. Some, including the GENI NSF-funded initiative [1], to rebuild the Internet, argue that the future lies in layers of overlay networks that can meet various requirements whilst keeping a very simplistic, almost unmanaged, IP for the underlying Internet. Others initiatives such as Clean Slate program [2] Stanford University, and Architecture Design Project for New Generation Network [3] argue that the importance of wireless access networks requires a more fundamental redesign of the core Internet Protocols themselves.

We argue that service agnostic network design are no longer a way to achieve interactive solutions in terms of service composition and information sharing capabilities for heterogeneous infrastructure support. A narrow focus on designing optimal networking protocols in isolation is too limited. Instead, a more holistic and long-term view is required, in which networking issues are addressed in a manner that focuses on the supporting role various protocols play in delivering communications services that meet the rapidly changing needs of the communities of users for which the hour glass architecture model become in a critical infrastructure.

In this paper service and network requirements [4][5][6][7][8][9] acts as inputs particularly on information interoperability and cross-domain information sharing controlling communication systems for the Future Internet. We support the idea of interoperable, extensible, reusable, common and manageable new Internet reference model is critical for Future Internet realization and deployment. The new Internet reference model must rely on the fact that high-level applications make use of diverse infrastructure representations and not use of resources directly. So when resources are not being required to support or deploy services they can be used in other tasks or services. As implementation challenge for controlling and harmonize this entire resource management requirements and architectural design issues the federation paradigm emerges as a tentative approach and optimal solution. We address challenges for a future Internet Architecture perspective using federation. We also provide, in a form of realistic implementations, research results and solutions addressing basics for federation.

Federated management scenarios are investigated [5][10] on what information enterprise application management systems can provide to allow the latter to more robustly and efficiently allocate network services.

This paper is organized as follows: Section II presents a brief review of the challenges about Future Internet architectures in terms of cross-domain interoperability. Section III presents the rationale about federation as crucial concept in the framework of this Future Internet research. Section IV presents a Federated Management Reference Model and its implications for networks and services. Section V describes what we consider as critical functional blocks for an Inter-disciplinary approach towards the specification of mechanisms for federated management. Section VI introduces End-to-End service management scenarios; we also investigate what information enterprise application management systems can provide to federated management systems allowing network and services allocation. Section VII presents the summary and outlook of this research. Finally some bibliography references supporting this research are included.

2   Challenges for Future Internet Architectures

This section focuses on inter-disciplinary approaches to specify data link and cross-domain interoperability to, collectively, constitute a reference model that can guide the realisation of future communications environments in the Future Internet [4][11][12][13]. The Future Internet architecture must provide societal services and, in doing so, support and sustain interactions between various communities of users in straight relation with communication infrastructure mechanisms. Service-awareness [4] has many aspects to consider as challenges, including: delivery of content and service logic with consumers' involvement and control; fulfilment of business and other service characteristics such as Quality of Service (QoS) and Service Level Agreements (SLA); optimisation of the network resources during the service delivery; composition and decomposition on demand of control and network domains; interrelation and unification of the communication, storage, content and computation substrata.

Networking-awareness [4] challenges imply the consumer-facing and the resource-facing services are aware of the properties, the requirements, and the state of the network environment, which enable services to self-adapt according the changes in the network context and environment. It also means that services are both executed and managed within network execution environments and that both the services and the network resources can be managed uniformly in an integrated way. Uniform management allows services and networks to harmonize their decisions and actions [14]. The design of both networks and services is moving forward to include higher levels of automation, and autonomicity, which includes self-management.

The optimization of resources [15][16][17] using federation in the Future Internet relies on classify and identify properly what resources need to be used, thus dynamically the service composition and service deployed can be executed by result of well known analysis on network and services.

3   Rationale for Federation in the Future Internet

Federation is relatively a new paradigm in communications, currently studied as the alternative to solve interoperability problems promoting scalability issues and exploring towards solving complexity when multiple applications/systems need to interact with a common goal. In this paper federation is handled as the mechanism used by communications management systems providing autonomic control loops.

In this section, the rationale for federated, autonomic management of communications services is addressed from the perspective of end-to-end applications and services in the Future Internet. Federation in the Future Internet envisions management systems (networks and services) made up of possibly heterogeneous components, each of which has some degree of local autonomy to realize business goals. Such business goals provide services that transcend legal and organizational boundaries in dynamic networks of consumers and providers. All the management systems with their own autonomy level contribute to satisfy more complex business goals, a single entity would not be able to achieve.

A visionary perspective for what federation can offer in communications systems and how federation contributes enabling information exchange has been described in previous works [18][19]. The intention in this paper is not to define what the Federation in future communications is, or which advantages it can offer either basics definition(s) in communications, but rather to provide a realistic approach in form of functional architecture, research results and implementation advances as well to show in kind how federation acts as feasible alternative towards solving interoperability problems in service and application management systems.

Future Internet environments consist of heterogeneous administrative domains, each providing a set of different services. In such complex environment, there is no single central authority; rather, each provider has at least one (and usually multiple) separate resources and/or services that must be shared and/or negotiated.

The term Federation in communications was discussed in a previous work [20] and currently many definitions have been proposed. We particularly follow a federated management definition as *"A federation is a set of domains that are governed by either a single central authority or a set of distributed collaborating governing authorities in which each domain has a set of limited powers regarding their own local interests*" because it fits better to management systems and due federation has two important implications nor considered in previous definitions i) federation must facilitates designing platforms without unnecessarily replicating functionalities, and ii) to achieve federation is necessary building inter-connected, inter-operating and/or inter-working platforms. Federation also implies that any virtual and /or real resource, which reside on another domain are managed correctly.

4   Federated Management Activity in the Future Internet

This section references theoretical foundation for the development of interdisciplinary Future Internet visions about a Federated Management and their implications for networks and services. These principles can be validated via direct industrial investment, and roll out real integrated test beds to trial new network and service infrastructures.

In future Internet end user, service, application and network requirements act as guidelines to identify study and clarify part of complex requirements. The relationships between Network Virtualisation and Federation [16][21][22][23] and the relationship between Service virtualisation (service clouds) and federation [17] are the support of a new world of solutions defining the Future Internet.

Next generation networks and services [3][4][24] can not be conceived without systems acting and reacting in a dynamic form to the changes in its surrounding (context-awareness, data link and information interoperability), even more the systems must be able for self-managing considering end-user requirements and acting in autonomous forms offering added value services (Autonomics) [6][7][25] where traditional definitions describing self-management emerged. However, most of them are based on very-high level human directives, rather than fully or partially automatic low-level management operations. While many aspects of the network will be self-managed based on high-level policies, the aggregation and subsequent understanding of monitoring/fault data is a problem that has not yet been completely solved here is where federation take place and acquire importance.

4.1   Federated Autonomic Management Reference Model

Federated refers to the ability of a system to enable network and service management as result of threading negotiations for evolving value chains composed of providers and/or consumers [14]. *Autonomic* reflects the ability of such systems to be aware of both themselves and their environment, so that they can self-govern their behaviour within the constraints of the business goals that they *collectively* seek to achieve. *Management* refers to the ability of such systems not just to configure, monitor and control a network element or service, but also to optimize the administration of lifecycle aspects of that resource(s) or service(s) in a programmable way. This enables end-users to take a more proactive role managing their quality of experience in a semantic way. In the depicted representation for the federated autonomic management reference model shown in the Figure 1 service and network domains must interact to exchange relevant information facilitating services and network operations. These cross-domain interactions demand certain level of abstraction to deal with mapping requirements from different information and data domains. This higher level of abstraction enables *business* and *service foundations* to be met by the network, and emphasizes offering federated services in a portable manner that is independent of the utilized networks. The objective is to effectively deliver and manage end-to-end communications services over an interconnected, but *heterogeneous infrastructure* and establishes *communication foundations*.

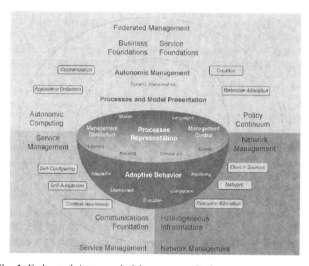

Fig. 1. Federated Autonomic Management Reference Representation

A greater degree of coordination and cooperation is required between communication resources, the software that manages them, and the actors who direct such management. In federation management end-to-end communication services involve configuring service and network resources in accordance to the policies of the actors involved in the management process. An autonomic management system provides automatic feedback to progressively improve management policies as service usage and resource utilization change. A goal of autonomic systems is to provide rich usage data to guide rapid service innovation.

Concepts related to Federation such as Management Distribution, Management Control and process representation are clear on their implications to the network management, however up to date there are no clear implications around what federation offers in communications either what federation to the next generation networks and in the Future internet design with service systems using heterogeneous network technologies imply. A clear scenario where federation is being identified as useful mechanism is the Internet service provisioning, in today's Internet it is observed the growing trend for services to be both provided and consumed by loosely coupled value networks of consumers, providers and combined consumer and providers. These consumer valued networks acting *"ideally"* as independent self-management entities must combine efforts to offer *"common"* and *"agreed"* services even with many technological restrictions and conflicts blocking such activity. A set of scenarios is introduced in a following section describing federation in more detail.

4.2   Federated Management Service Life Cycle

Management and configuration of large-scale and highly distributed and dynamic enterprise and networks applications [26] is everyday increasingly in complexity. In the current Internet typical large enterprise systems contain thousands of physically distributed software components that communicate across different networks [27] to satisfy end-to-end services client requests. Given the possibility of multiple network connection points for the components cooperating to serve a request (e.g., the components may be deployed in different data centres), and the diversity on service demand and network operating conditions, it is very difficult avoid conflicts [14][20][28] between different monitoring and management systems to provide effective end-to-end applications managing the network.

The Figure 2 depicts the federated autonomic reference model service life cycle for the Future Internet. We are exploring how the definition and contractual agreements between different enterprises (*1.Definition*) establish the process for monitoring (*2.Observation*) and also identify particular management data at application, service, middleware and hardware levels (*3.Analysis*) that can be gathered, processed, aggregated and correlated (*4.Mapping*) to provide knowledge that will support management operations of large enterprise applications (*5.Federated Agreements*) and the network services they require (*6.Federated Regulations*). We support the idea that monitoring data at the network and application level can be used to generate knowledge that can be used to support enterprise application management in a form of control loops in the information; a feature necessary in the Future Internet service provisioning process (*7.Federated Decisions*). Thus infrastructure can be re-configurable and adaptive to business goals based on information changes (*8.Action*).

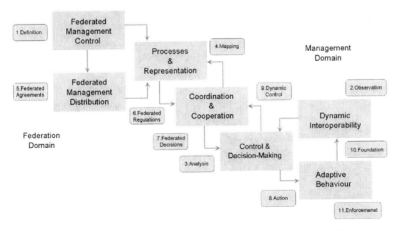

Fig. 2. Federated Management Life Cycle Reference Model

We also consider appropriate ways on how information from enterprise applications and from management systems can be provided to federate management systems allowing to more robustly and efficiently be processed to generate adaptive changes in the infrastructure (*9.Dynamic Control*). Appropriate means of normalising, interpreting, sharing and visualising this information as knowledge (*10.Foundations*) thus allocate new federated network services (*11.Enforcement*).

In a federated system the interaction between domains and the operations in between represent a form of high-level control to perform the negotiations and regulations to achieve the compromises pertaining to a federation and mainly resolve negotiations (represented as transition processes normally) not considered between individual or autonomous self-management domains. The transitions processes are not a refinement or finite process itself rather than that a transition represents information transformed into knowledge and their respectively representation. The operations and processes management in a federated architecture must support a finite goal to define, control, and coordinate service and network management conditions as much as possible from an application management high level control view, the so called federation. The federated functional architecture and its logic operations are depicted in Figure 2 and described more in detail here after.

- *Dynamic Interoperability* - Autonomic functionality, which can be result of negotiations between management components, devices or systems by using a same language or information model, non formal representation is necessary.
- *Adaptive Behaviour* - Autonomic and inherent functionality assigned to the components can be managed by federated conditions and regulations, which communicate with other non-autonomic components, thus virtual and/or real resources can be expand and contract the network infrastructure.
- *Control and Decision-Making* - The Functionality of a component(s) and system(s) to conduct its own affairs based on inputs considered as conditions and define outputs considered as actions.
- *Coordination & Cooperation* - Functionality associated to promote and resolve high level representation and mapping of data and information. Negotiations in form of data representation between different data and information models by components in the system(s) are associated to this feature.
- *Management Control* - Administration functionality for the establishment of cross-domain regulations considering service and network regulations and requirements as negotiations.
- *Management Distribution* - Organizational functionality for the adoption and enforcement of cross-domain regulations as result of service and network requirements.
- *Process & Representation* - Autonomic functionality assigned to components or systems, orchestrating the behaviour(s) in the components of managed systems.

5   Federated Management Architecture

This section describes designing principles for inter-domain federated management architectures in the Future Internet. These designs about architecture for the federated reference model by functional blocks addresses the specification of mechanisms including models, algorithms, processes, methodologies and architectures. The functional architecture collectively constitute, in terms of implementation efforts, framework(s), toolkit(s) and components that can guide the realisation of federated communications environments to effectively provide complex services (interoperable boundaries) and, in doing so, support and sustain service offering between various communities of users (heterogeneous data & infrastructure).

The federated architecture must be enabled for ensuring the information is available allowing useful transfer of knowledge (information interoperability) across multiple interfaces. It is likely that adaptive monitoring is used to optimise the efficiency of the federated process. A specific set of service applications, domain independent, and configurations for managing services and networks are used to ensure transference of results to other systems as result of sensitivity analysis. Simulation studies and analytical work is being conducted to back up further experimental results.

Designing a federated platform implies the combination of semantic descriptions and both holistic service and management information. When using semantics the interaction between systems named interactive entities is to reduce the reliance on technological dependencies for services support and increasing the interoperability between heterogeneous service and network management systems. A federated autonomic architecture must supports such interactions offering a full service lifecycle control by using federated autonomic mechanisms where relations and interactions for unified management operations are based on the use of formal mechanisms between different domains. This interaction relies on supporting end-user interface components ensuring high level management systems information exchange.

5.1   Federated Management Enforcement Process

The purpose of federation in autonomic is to manage *complexity* and *heterogeneity*. In a federated autonomic network, time-consuming manual tasks are mostly or completely automated; and decisions delegated, this dramatically reduces manually-induced configuration errors, and hence lowers operational expenditures when decision needs to be implemented. By representing high level business requirements in a formal manner, information and data can be integrated, and the power of machine-based learning and reasoning can be more fully exploited. Autonomic control loops and its formalisms [29][30], such as FOCALE [25] and AutoI [21][23] translate data from a device-specific form to a device- and technology-neutral form to facilitate its integration with other types of information. The key difference in the autonomic control loop, compared with a non-autonomic control loop, is the use of *semantic* information to guide the decision-making process. The key enabling federation is the process of semantic integration that can create associations and relationships among entities specified in different processes and formal representations acting as foundations into the system.

The following paragraphs are concentrated to describe the key logic domains which the federated autonomic architecture concentrates and which have been identified as main interest research topics around federation. The interactions between the logic areas are represented by cursives. As shown in figure 3. The design of a federated autonomic architecture enables transitions from high-level goals (as codified by service rules for example) to low-level as network policies. In a federated autonomic architecture, information is used to relate knowledge, rather than only map data, at different abstractions and domain levels co-relating independent events each other in a federated way. We envisage federation of networks, network management systems and service management applications at three levels of abstraction.

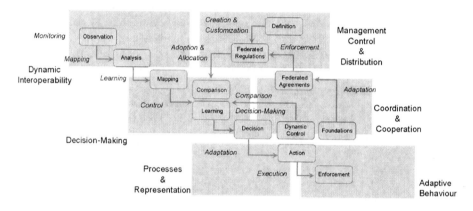

Fig. 3. Federated Management Enforcement Process

At the lowest level *adaptive behaviour* and *processes & representation*, at this level is the heterogeneous infrastructure where networks and devices coordinate self-management operations. Management systems should support self-management by local resources in a given domain ensuring that this self-managed behaviour is coordinated across management boundaries. In the top level, *Management Control & Distribution* and *Dynamic Interoperability* termed federation domain, the vision of federated autonomic management for end-to-end communications services orchestrate federated service management where management systems should semantically interoperate to support evolving value chains and the end-to-end delivery of services.

At the middle level, *coordination & cooperation* and *decision-making* where multiple management domains (service, application and networking) should interact and interoperate for configuring network resources in a consistent way, following with federated business goals, but in a manner that is consistent with configuration activity in neighbouring network domains participating in a value network. The federated autonomic architecture proposed can be seen itself as a federated system where exist regulations defining the performance or the behaviour in the heterogeneous infrastructure, such regulations are transformed using adaptations processes and formal representation to express and deploy the conditions comprised and described by the man-

agement distribution. Such regulations must be deployed with no further consideration from other systems or sub-systems; this feature is used when conflicts and negotiations need to be performed in the federation space.

In the federated architecture proposed the management control deal with federated
agreements necessaries to satisfy in one hand the enterprise requirements and in the
other hand the management system requirement as result of events coming from the
heterogeneous infrastructure. The events are expressed in a form of coordination and
cooperation functions, which have origins in mapping events between the diverse
enterprise processes and the heterogeneous infrastructure. The federation also acts as
mediator between autonomic and/or non-autonomic management sub-systems when
controlling each one independently their behaviour and a higher level mechanism is
necessary to act as a one in pursuing a common goal, in an heterogeneous service
deployment and support for example.

6   End-to-End Federated Service Management Scenarios

In this scenarios description section conversely, we also provide research results
about what information enterprise application management systems can provide to
federate management systems to allow the latter to more robustly and efficiently allocate network services. Brief scenario descriptions illustrate the possible challenges are
necessaries to tackle around the term federation and particularly on federated systems
and federated management applications.

6.1   Federation of Wireless Networks Scenario

generates more demand on management systems to be implemented satisfying diversity, capacity and service demand. Given the fact that in urban areas (shopping centres, apartment buildings, offices) generates more demand in deploying wireless
802.11-based mesh networks this expansion will be a patchwork of mesh networks;
challenges arise relating to how services can be efficiently delivered over these overlapping infrastructures. Challenges in wireless mesh networks relate to both resource
management within the network infrastructure itself and the way in which management systems of individual network domains can federate dynamically to support end-
to-end delivery of services to end-users. Furthermore, there are challenges relating to
securing the delivery of services across (possible multiple) wireless mesh infrastructure domains.

This research scenario opens work mainly for focusing on the specifics of resource
management within multi-provider mesh networks by using federation principles
(federated management). The exact nature of the mesh networks to be used in multi-
provider networks is not yet clear. However, from a management system perspective,
the scope of this scenario rely in the fact on how the use of semantic models capturing
knowledge relating to security functionality and the use of policy rules to control end-
to-end configuration of this functionality can provide a basis for the support flexible
trust distributed management across wireless meshes, (federated deployment).

6.2   Federation of Network and Enterprise Management Systems

Typical large enterprise systems contain thousands of physically distributed software components that communicate across different networks to satisfy client requests. Management and configuration is increasingly complex at both the network and enterprise application levels. The complex nature of user requests can result in numerous traffic flows within the networks that can not be correlated with each other, thus which it ideally should be treated in a federated fashion by the networks. Challenges in this scenario relies on how monitoring at the network level can provide knowledge that will enable enterprise application management systems to reconfigure software components to better adapt applications to prevailing network conditions. This reconfiguration may, for example, involve redeployment of application components in different locations by using different infrastructures (federation) in order to alleviate congestion detected within particular parts of the network. Conversely, the information enterprise application management systems can provide to network management systems allowing a more efficient and cost-effectively manage traffic flows relating to complex transactions (federated management).

This scenario carry on work for developing federated monitoring techniques that can be applied to record and analyse information and trends in both network management systems and enterprise application management systems, in a manner such that a coherent view of the communication needs and profile of different transaction types can be built. Enterprise application management systems must be specified to provide relevant application descriptions and behaviours (e.g., traffic profiles and QoS levels) to network management system allowing shared knowledge to be optimally used (federation) in network traffic management processes.

6.3   Federation of Customer Value Networks Scenario

Network service usage is increasingly billed as flat-rate Internet access. Within the "Web 2.0" development, online value is expanding from searching and e-consumerism applications, to participative applications including blogs, wikis, online social networks, RSS feeds, Instant Messaging, P2P applications, online gaming and increasingly pervasive VoIP applications. Particularly as users have been empowered to mix and match applications to create customised functionality (e.g. mash-ups). Similarly, at a business or organisational level, the knowledge sector of modern economies is increasingly focussed on value networks rather than on value chains. Value networks offer benefits that are more intangible (e.g. the value of professional contact networks). Value networks share with Web 2.0 application users a concern with value of interacting effectively with rest of the network community (federation).

In highly active value networks, the cost of interrupted interactions may be perceived as high; however, there is little visibility of the root responsible source of interaction breakdowns between the various communication services providers, application service hosts, or value network members themselves. Due to this lack of visibility, value networks have very limited avenues for taking remedial or preventative actions, such as recommending different network or service providers to their members. Value networks of customers can only properly be served by federated service providers, henceforth termed Service Provider Federation (SPF).

6.4   Federation of Home Area Networks Services and Applications

An emerging trend in communications networks is the growing complexity and heterogeneity of the "outer edge" domain – the point of attachment of Home Area Networks (HANs) and other restricted area networks to various access networks. Challenges on this scenario must address management of outer edge devices, such as femto base stations, home gateways, set-top boxes and of networks thereof. Today this task is provided on an piecemeal basis, with different devices having a wide range of management functionality, which is often proprietary or, at best, conforms one of a range of competing standards. Furthermore, management operations must be performed by end-users. Achieving this requires increased degrees of integration between telecommunications network management systems and devices. In particular, it is important to develop methods (management functions) through which network management systems can assume responsibility for appropriate configuration of HAN devices. A significant challenge in this regard is that as the diversity and capabilities of HAN devices increases.

To address these scenario requirements the use of distributed event processing and correlation techniques that can process relevant data in a timely and decentralised manner and relay it as appropriate to management federated making functions are necessaries to investigate (federation). This scenario discloses on aspects about federation and integrated management of outer edge network environments; delegation of management authority to network management systems and decentralised assurance of service delivery in a home area are important too.

7   Summary and Outlook

In the future Internet new designs ideas of Federated Management in Future Internet Architectures must consider high demands of information interoperability to satisfy service composition requirements being controlled by diverse, heterogeneous systems make more complex perform system management operations. The federated autonomic reference model approach introduced in this paper as a design practice for Future Internet architectures emerges as an alternative to address this complex problem in the Future Internet of networks and services.

We have studied how federation brings support for realisation on the investigated solution(s) for information interoperability and cross-domain information sharing controlling communication systems in the Future Internet. Additional issues such as service representation and networks information can facilitate service composition and management processes. Remaining research challenges regarding information model extensibility and information dissemination exist and would be conducted to conclude implementations, experiments composing services in some of the scenarios described in this paper.

7.1   Research Outputs as Rationale for Federation

- Techniques and agreements for composition/decomposition of federated control frameworks for physical and virtual resources and systems
- Techniques and mechanisms for controlling workflow for all systems across federated domains, ensuring bootstrapping, initialisation, dynamic reconfiguration, adaptation and contextualisation, optimisation, organisation, and closing down of service and network components.
- Mechanisms for dynamic deployment on-the-fly of new management functionality without running interruption of any systems across multiple and federated domains.
- Algorithms and processes to allow federation in enterprise application systems to visualize software components, functionality and performance.
- Techniques for analysis, filtering, detection and comprehension of monitoring data in federated enterprise and networks.
- Algorithms and processes to allow federated application management systems reconfigure or redeploy software components realizing autonomic application functionality.
- Guidelines and exemplars for the exchange of relevant knowledge between network and enterprise application management systems.

This paper makes references to design foundations for the development of federated autonomic management in architectures in the Future Internet. Scenarios has been shortlisted to identify challenges and provide research results about what information enterprise application management systems can provide to federate management systems by using an interoperability of information as final objective.

Acknowledgements. The work introduced in this paper is a contribution to SFI FAME-SRC (Federated, Autonomic Management of End-to-End Communications Services - Scientific Research Cluster). Activities are partially funded by Science Foundation Ireland (SFI) via grant 08/SRC/I1403 FAME-SRC (Federated, Autonomic Management of End-to-End Communications Services - Scientific Research Cluster) and by the UniverSELF EU project [31], grant agreement n° 257513, partially funded by the Commission of the European Union.

References

1. NSF-funded initiative to rebuild the Internet (Online: Oct. 2010),
   http://www.geni.net/
2. Clean Slate Program, Stanford University (Online: Nov. 2010),
   http://cleanslate.stanford.edu/
3. Architecture Design Project for New Generation Network (Online: Oct. 2010),
   http://akari-project.nict.go.jp/eng/index2.htm

4. Galis, A., et al.: Management and Service-aware Networking Architectures (MANA) for Future Internet Position Paper: System Functions, Capabilities and Requirements (Invited paper). In: IEEE 2009 Fourth International Conference on Communications and Networking in China (ChinaCom09) 26-28 August, Xi'an, China (2009)
5. Clark, D., et al.: NewArch: Future Generation Internet Architecture., NewArch Final Technical Report, http://www.isi.edu/newarch/
6. van der Meer, S., Davy, A., Davy, S., Carroll, R., Jennings, B., Strassner, J.: Autonomic Networking: Prototype Implementation of the Policy Continuum. In: Proc. of 1st IEEE International Workshop on Broadband Convergence Networks (BcN 2006), in conjunction with NOMS 2006, Canada, April 7, 2006, pp. 163–172. IEEE Press, Los Alamitos (2006)
7. van der Meer, S., Fleck II, J.J., Huddleston, M., Raymer, D., Strassner, J., Donnelly, W.: Technology Neutral Principles and Concepts for Autonomic Networking. In: Advanced Autonomic Networking and Communication, Birkhäuser, Basel (2008)
8. Raymer, D., van der Meer, S., Strassner, J.: From Autonomic Computing to Autonomic Networking: an Architectural Perspective. In: Proc. of 5th IEEE Workshop on Engineering of Autonomic and Autonomous Systems (EASe 2008), Co-located with ECBS, Belfast, United Kingdom, March 31–April 4 (2008)
9. Jennings, B., van der Meer, S., Balasubramaniam, S., Botvich, D., Foghlú, M.Ó., Donnelly, W., Strassner, J.: Towards Autonomic Management of Communications Networks. IEEE Comms Magazine, IEEE 45(10), 112–121 (2007)
10. Blumenthal, M., Clark, D.: Rethinking the design of the Internet: the end to end arguments vs. the brave new world. ACM Transactions on Internet Technology 1(1) (2001)
11. Subharthi, P., Jianli, P., Raj, J.: Architectures for the Future Networks and the Next Generation Internet: A Survey. Computer Communications (July 2010), 63 pp., http://www1.cse.wustl.edu/~jain/papers/ftp/i3survey.pdf
12. Curran, K., Mulvenna, M., Galis, A., Nugent, C.: Challenges and Research Directions in Autonomic Communications. International Journal of Internet Protocol Technology (IJIPT) 2(1) (2006)
13. Rubio-Loyola, J., Astorga, A., Serrat, J., Chai, W.K., Mamatas, L., Galis, A., Clayman, S., Cheniour, A., Lefevre, L., Fischer, A., Paler, A., Al-Hazmi, Y., de Meer, H.: Platforms and Software Systems for an Autonomic Internet. In: IEEE Globecom 2010, Miami, USA, 6-10 December (2010)
14. Jennings, B., Brennan, R., van der Meer, S., Lewis, D., et al.: Challenges for Federated, Autonomic Network Management in the Future Internet. In: ManFI workshop, June, NY, USA (2009)
15. Strassner, J.C., Foghlú, M.Ó., Donnelly, W., Serrat, J., Agoulmine, N.: Review of knowledge engineering requirements for semantic reasoning in autonomic networks. In: Ma, Y., Choi, D., Ata, S. (eds.) APNOMS 2008. LNCS, vol. 5297, pp. 146–155. Springer, Heidelberg (2008)
16. Strassner, J., Foghlú, M.Ó., Donnelly, W., Agoulmine, N.: Beyond the Knowledge Plane: An Inference Plane to Support the Next Generation Internet. In: IEEE GIIS 2007, 2-6 July (2007)
17. Galis, A., Denazis, S., Brou, C., Klein, C.: Programmable Networks for IP Service Deployment. Artech House Books, London (2004)
18. Serrano, M., Strassner, J., Foghlú, M.Ó.: A Formal Approach for the Inference Plane Supporting Integrated Management Tasks in the Future Internet. In: 1st IFIP/IEEE ManFI Intl Workshop, In conjunction 11th IEEE IM2009, Long Island, NY, USA, June 2009, IEEE Computer Society Press, Los Alamitos (2009)
19. Brennan, R., Feeney, K., Keeney, J., O'Sullivan, D., Fleck II, J., Foley, S., van der Meer, S.: Multi-Domain IT Architectures for Next Generation Communications Service Providers. IEEE Communications Magazine 48, 110–117 (2010)

20. Serrano, J.M., van der Meer, S., Holum, V., Murphy, J., Strassner, J.: Federation, A Matter of Autonomic Management in the Future internet. In: IEEE/IFIP Network Operations & Management Symposium, NOMS 2010, Osaka, Japan, 19-23 April (2010)
21. Bassi., A., Denazis., S., Galis., A., Fahy., C., Serrano., M., Serrat, J.: Autonomic Internet: A Perspective for Future Internet Services Based on Autonomic Principles. In: 2007 IEEE Management Week – ManWeek 2007 2nd IEEE MACE 2007 Workshop, San José, CA, USA, 29 Oct.–2 Nov (2007)
22. Rochwerger, B., et al.: An Architecture for Federated Cloud Computing. In: Cloud Computing: Principles and Paradigms, Wiley, ISBN: 0470887990 (April 2011)
23. Galis, A., et al.: Management Architecture and Systems for Future Internet Networks. In: Towards the Future Internet – A European Research Perspective, p. 350. IOS Press, Amsterdam (2009)
24. Feldmann, A.: Internet clean-slate design: what and why? ACM SIGCOM Computer Communication Review 37(3) (2007)
25. Strassner, J., Agoulmine, N., Lehtihet, E.: FOCALE – A Novel Autonomic Networking Architecture. ITSSA Journal 3(1), 64–79 (2007)
26. Foley, S.N., Zhou, H.: Authorisation Subterfuge by Delegation in Decentralised Networks. In: Proc. of the 13th International Security Protocols Workshop, Cambridge, UK (April 2005)
27. Jennings, B., et al.: Towards Autonomic Management of Communications Networks. IEEE Comms Magazine 45(10), 112–121 (2007)
28. Strassner, J.: Autonomic Networks and Systems: Theory and Practice. In: NOMS 2008 Tutorial, Brasil (April 2008)
29. Strassner, J., Kephart, J.: Autonomic Networks and Systems: Theory and Practice. In: NOMS 2006 Tutorial (Apr 2006)
30. Serrano, J.M.: Management and Context Integration Based on Ontologies for Pervasive Service Operations in Autonomic Communication Systems., PhD Thesis, UPC (2008)
31. UniverSELF Project (January 2011), http://www.univerself-project.eu/

An Architectural Blueprint for a Real-World Internet

Alex Gluhak[1], Manfred Hauswirth[2], Srdjan Krco[3], Nenad Stojanovic[4], Martin Bauer[5], Rasmus Nielsen[6], Stephan Haller[5], Neeli Prasad[6], Vinny Reynolds[2], and Oscar Corcho[8]

[1] University of Surrey, UK
[2] National University of Galway, Ireland
[3] Ericsson, Serbia
[4] FZI, Germany
[5] NEC, Germany
[6] Aalborg University, Denmark
[7] SAP, Switzerland
[8] Universidad Politécnica de Madrid, Spain

Abstract. Numerous projects in the area of Real-World Internet (RWI), Internet of Things (IoT), and Internet Connected Objects have proposed architectures for the systems they develop. All of these systems are faced with very similar problems in their architecture and design and interoperability among these systems is limited. To address these issues and to speed up development and deployment while at the same time reduce development and maintenance costs, reference architectures are an appropriate tool. As reference architectures require agreement among all stakeholders, they are usually developed in an incremental process. This paper presents the current status of our work on a reference architecture for the RWI as an architectural blueprint.

Keywords: Real-World Internet, Internet of Things, Internet Connected Objects, Architecture

1 Introduction

Devices and technologies ubiquitously deployed at the edges of the networks will provide an infrastructure that enables augmentation of the physical world and interaction with it, without the need for direct human intervention, thus creating the essential foundations for the Real-World Internet (RWI).

Leveraging the collective effort of several projects over the last number of years [SENSEI, ASPIRE, IOT-A, PECES, CONET, SPITFIRE, SemsorGrid4Env], this chapter presents the current status of the work aimed at definition of an RWI reference architecture. The core contribution of this paper is the distillation of an initial model for RWI based on an analysis of these state of art architectures and an understanding of the challenges. This is achieved by:

- An identification of a core set of functions and underlying information models, operations and interactions that these architecture have in common.
- A discussion on how these architectures realize the above identified functions and models and what features they provide.

J. Domingue et al. (Eds.): Future Internet Assembly, LNCS 6656, pp. 67–80, 2011.

2   The Real World Internet

Since the introduction of the terminology over a decade ago, the "Internet of Things (IoT)" has undergone an evolution of the underlying concepts as more and more relevant technologies are maturing. The initial vision was of a world in which all physical objects are tagged by Radio Frequency Identification (RFID) transponders in order to be uniquely identified by information systems. However, the concept has grown into multiple dimensions, encompassing sensor networks able to provide real world intelligence or the goal-oriented autonomous collaboration of distributed objects via local wireless networks or global interconnections such as the Internet.

Kevin Ashton, former Director of the Auto-ID Center, once famously formulated: "Adding radiofrequency identification and other sensors to everyday objects will create an Internet of Things, and lay the foundations of a new age of machine perception".

We believe that machine perception of the real world is still at the heart of the Internet of Things, no matter what new technologies have meanwhile become available to enable it. As such, one of the key roles of the Internet of Things is to bridge the physical world and its representation in the digital world of information systems, enabling what we refer to in part of the Future Internet Assembly (FIA) community as the so called Real World Internet (RWI).

The RWI is the part of a Future Internet that builds upon the resources provided by the devices [HAL] of the Internet of Things, offering real world information and interaction capabilities to machines, software artifacts and humans connected to it.

The RWI assumes that the information flow to and from IoT devices is taking place via local wired and wireless communication links between devices in their proximity and/or through global interconnections in the form of the current Internet and mobile networks or future fixed and mobile network infrastructures.

One important property of the RWI which distinguishes it from the current Internet is its heterogeneity, both regarding the types of devices as well as communication protocols used. IPv6 and in particular 6LoWPAN play an important role, but other proprietary wireless protocols will see continued use as well. To deal with this heterogeneity, services – in the form of standard Web Services and DPWS[1], but more likely using RESTful approaches and application protocols like CoAP – provide a useful abstraction. As services play a pivotal role in the Future Internet Architecture, the use of services for integrating the RWI also fits well into the overall architectural picture. One has to keep in mind though that RWI services have some different properties from common, enterprise-level services: They are of lower granularity, e.g., just providing simple sensor readings and, more importantly, they are inherently unreliable; such RWI services may suddenly fail and the data they deliver has to be associated with some quality of information parameters before further processing.

---

[1]   Device Profile for Web Services

3   Reference Architecture

In this section we present an initial model on which several of the current RWI archi-tecture approaches are based. While not as comprehensive as a reference architecture, it already identifies the major underlying system assumptions and architectural arti-facts of the current RWI approaches. The model has been developed through a careful analysis of the existing RWI architectures according to the following dimensions:

1. Underlying system assumptions,
2. functional coverage of the services provided by the architectures,
3. underlying information models in the architectures, and
4. operations and interactions supported in these architectures.

3.1 Underlying RWI Architecture Assumptions

Common to all RWI architectures is the underlying view of the world, which is di-vided into a real and a digital world as depicted in Fig. 1. The real world consists of the physical environment that is instrumented with machine readable identification tags, sensors, actuators and processing elements organized in domain specific islands in order to monitor and interact with the physical entities that we are interested in. The digital world consists of:

a) Resources which are representations of the instruments – Resource level,
b) Entities of Interest (EoI) which are representations of people, places and things – Entity level, and
c) Resource Users which represent the physical people or application software that intends to interact with Resources and EoI.

Providing the services and corresponding underlying information models to bridge the physical and the digital world by allowing users/applications to interact with the Re-sources and EoI is the main contribution of the RWI reference architecture towards a RWI. Typically, RWI architectures provide two abstraction levels for such interac-tions: resource level and entity level.

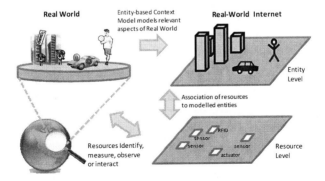

Fig. 1. World-view of RWI systems

On the resource level, resource users directly interact with resources. Such interactions are suitable for certain types of RWI applications where the provided information or interaction does not need any context (e.g., an understanding of how information is related to a real-world entity).

On the entity level, some RWI architectures offer the option to applications to use an inherent context model which is centered around the EoI. For these EoIs, relevant aspects like the activity of a person or the current location of a car are modeled as context attributes. Applications can base their requests on EoI and context attributes. The underlying requirement is that the resources providing information are associated with the respective entities and attributes, so that the services offered by the RWI architectures can find the required resources for the entity-level requests. Therefore, architectural components exist that enable contextualized information retrieval and interaction, as well as dynamic service composition.

Besides the above assumptions, various architectures take also socio-economic aspects into consideration, as they consider various actors in one or more business roles within the context of the RWI eco-system created around their architecture, forming the so-called RWI communities. The main roles in these communities are:

1. Resource Providers who own the resources,
2. Framework Providers who own the architectural framework components, and
3. Resource Users who are the main users of the resources or architectural services.

3.1   Functional Coverage of RWI Architectures

This section explores the different functional features provided by the service functions of the existing architectures to support the interactions between resources and resource users and the corresponding business roles inside the RWI ecosystem.

Resource discovery is one of the basic services RWI architectures provide for resource-level access. It allows resource users to lookup and find resources that are made available by resource providers in an RWI community. Resource users specify characteristics of a resource, e.g., the identifier or type they are interested in, and receive (references to) one or more resources that match the requested criteria.

Context information query is a more advanced functionality provided by some RWI architectures for entity level access. It allows resource users to directly access context information in the RWI concerning EoIs or find resources from which such information can be obtained. Unlike resource discovery, context information queries involve semantic resolution of declarative queries and require resources and entities to be adequately modeled and described.

Actuation and control loop support is another advanced functionality providing access to RWI resources at entity level. It allows resource users to declaratively specify simple or complex actuation requests or expected outcomes of actuations on an EoI. The respective functions ensure that resource users are provided with an adequate set of resources able to achieve the specified objectives or that appropriate actions are executed according to the specified outcomes.

Dynamic resource creation is an advanced functionally of some architectures and mainly relates to virtual resources such as processing resources. It enables the dynamic instantiation of resources (e.g., processing services) on resource hosts in order to satisfy context information requests and actuation requests.

Session management functionality is provided to support longer lasting interactions between resources and resource users, in particular if these interactions span multiple resources. Longer lasting interactions may require adaptation of the interactions to system dynamics, such as change of availability of resources, e.g., the replacement of one or more resource endpoints during the lifetime of the interaction, shielding this complexity from the resource user.

Access control functionality is essential to ensure that only authorized resource users are able to access the resources. It typically involves authentication of resource users at request time and subsequent authorization of resource usage. Another aspect of resource access is access arbitration, if concurrent access occurs by multiple authorized users. This requires mechanisms to resolve contention if multiple conflicting requests are made including pre-emption and prioritization.

Auditing and billing functionality are necessary to provide accounting and accountability in an RWI architecture. Based on the accounting model, resource users can be charged for the access to resources or provided information and actuation services. Accountability and traceability can be achieved by recording transactions and interactions taking place at the respective system entities.

3.2   Smart Object Model

At its core, the proposed architectural model defines a set of entities and their relationships, the Smart Object Model. The entities form the basic abstractions on which the various system functions previously described operate. The object model reflects a clear separation of concerns at the various system levels and their real-world interrelationships according to the assumptions described in Section 2.1.

A central entity in the Smart Object Model is the concept of a resource. Conceptually, resources provide unifying abstractions for real-world information and interaction capabilities comparable to web resources in the current web architecture. In the same way as a web user interacts with a web resource, e.g., retrieve a web page, the user can interact with the real-world resources, e.g., retrieve sensor data from a sensor. However, while the concept of the web resource refers to a virtual resource identified by a Universal Resource Identifier (URI), a resource in the RWI context is an abstraction for a specific set of physical and virtual resources.

The resources in the Smart Object Model abstract capabilities offered by real-world entities such as sensing, actuation, processing of context and sensor data or actuation loops, and management information concerning sensor/actuator nodes, gateway devices or entire collections of those. Thus a resource has a manifestation in the physical world, which could be a sensor, an actuator, a processing component or a combination of these. In the latter case we refer to it as a composite resource. A resource is unique within a system (domain) and is described by an associated resource description, whose format is uniform for all resources across systems and domains. This uniform resource description format enables and simplifies the reuse of existing resources in different contexts and systems and is a major contribution of the proposed architectural model.

The Smart Object Model distinguishes between the (physical) instances of system resources and the software components implementing the interaction endpoints from the user perspective (Resource End Point – REP). Furthermore, the model distinguishes between the devices hosting the resources (Resource Host) and the network devices hosting the respective interaction end points (REP Host). This separation enables the various system functions described in the previous section to deal with real-world dynamics in an efficient and adequate manner and facilitates different deployment models of a system. Fig. 2 shows the Smart Object Model in terms of entities and their inter-relationships.

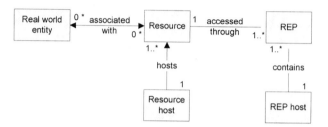

Fig. 2. Key entities and their relationships in the RWI system model

A REP is a software component that represents an interaction end-point for a physical resource. It implements one or more Resource Access Interfaces (RAIs) to the resource. The same resource may be accessible via multiple REPs, through the same RAI or different ones. In comparison to the current web architecture, REPs can be considered equivalent to web resources, which are uniquely identified by a URI.

The device hosting a resource is referred to as the Resource Host. Sensor nodes are typical examples for resource hosts, but there can be arbitrary devices acting in this role, for example, mobile phones or access points that embed resources. A REP Host is a device that executes the software process representing the REP.

As mentioned before, the resources and REPs are conceptually separated from their hosts to facilitate different deployment options. In some cases a REP host and a resource host can be co-located on the same physical device, e.g., in the case of a mobile phone. Similarly, there may be cases where the REP is not hosted on the resource host itself, for example, a computer in the network or an embedded server may act as the REP host for a resource, which is physically hosted on a sensor node connected to it. This distinction is important when mobility, disconnections and other system dynamics come into play, as it provides a conceptual model to effectively keep the system state consistent for the correct operation of an overall system. Moreover, this separation of concerns provides a means of protecting low-capability resources, e.g., low-power sensor nodes, from attacks by hosting their REPs on more powerful hardware.

Unlike other models, the Smart Object Model considers also real-world entities in its model and manages the dynamic associations between the real world entities and the sensors/actuators that can provide information about them/act upon them. Examples of the real-world entities – also known as Entities of Interest or EoIs – are persons, places, or objects of the real world that are considered relevant to provide a service to users or applications. A resource in the Smart Object Model thus provides (context) information or interaction capabilities concerning associated real-world entities.

3.3 Interaction Styles

The classes of system functions described in Section 2.1 may be realized through different interaction styles which can be classified along the following dimensions:

- Synchronous or asynchronous: Does the operation block the thread of control and wait for a result (blocking) or is it executed in parallel (non-blocking)?
- Session context: If an interaction depends on previous interactions, then the system must store and maintain the state of a "conversation".
- One-shot or continuous: The interaction may either return a single result immediately or run continuously and return results as they come along.
- Number of participants: Interactions among resources and resource users can be 1:1, 1:n or m:n.

Well-known styles can easily be mapped to these dimensions, for example, synchronous-continuous would be "polling", whereas asynchronous-continuous would be "event-driven".

Each style can be implemented in various ways, depending on the specific system and its requirements. A continuous interaction can e.g. be implemented through a pub/sub service; a complex event processing system via polling in regular intervals or by a simple asynchronous callback mechanism, etc. The different choices determine not only the resource consumption and communication stress on the underlying infrastructure but also the flexibility, extensibility, dependability, determinism, etc. of the implemented system. However, the interfaces to these choices at the implementation architecture level should be uniform as this allows the exchange of one communication infrastructure by another without requiring major recoding efforts of an application and also enables an n-system development and deployment.

Also, the interaction patterns manifest themselves in communication flows of different characteristics. In order to effectively support these flows, different types of communication services may be required from the underlying communication service layer. Table 1 shows a simple way to assess and compare interaction styles of different architectures by arranging the possible combinations in a two-dimensional grid.

Table 1. Classification of interactions

Architecture name		Synchronicity		Session Context	
		sync	async	yes	no
Duration	One-shot				
	Continuous				
Participants	1:1				
	1:n				
	m:n				

4   Analysis of Existing Architectures

In this section we briefly review five of the most relevant RWI architecture approaches with respect to the functional coverage provided in the context of the above defined reference architecture. These approaches have been recently developed in the ASPIRE, FZI Living Lab AAL (Ambient Assisted Living), PECES, SemsorGrid4Env and SENSEI European research projects. Following this, a number of other relevant architectures are identified and a table at the section's end summarizes the functional coverage of the five main architectures.

4.1   ASPIRE

The ASPIRE architecture [ASPIRE] is based on EPGglobal [EPC] with a number of objective-specific additions. In a Radio Frequency Identification (RFID) based scenario, the tags act as hosts for the resources in form of Electronic Product Codes (EPCs), IDs or other information as well as for value-added information in form of e.g. sensor data. The resource hosts are abstracted through the RFID readers due to the passive communication of the tags. The Object Naming Service (ONS) corresponds to the Entity Directory that returns the URLs of relevant resources for the EPC in question – this is the White Pages service. The EPC Information Service (EPCIS) implements the Resource Directory by storing more rich information of the resource. The information stored covers WHAT, WHERE, WHEN and WHY for an EPC and can be used as a Yellow Pages service. The Application Layer Event (ALE) functionality implements the functionality of a Semantic Query Resolver (SQR). The ALE operates through an Event Cycle specification (ECspec) where resources are defined. ASPIRE introduces a Business Event Generator (BEG) which implements additional logic for interactions using semantics of the specific RFID application. Query planning is done through the definition of an ECspec and can be mapped into the SQR. In addition, three request modes are standardized corresponding to interactions. "Subscribe" issues a standing request for asynchronous reporting of an ECspec and is defined as a continuous request with no one-shot scenario. "Poll" issues a standing request for synchronous reporting of an ECspec and maps to the synchronous interaction, but again only for continuous requests. "Immediate" maps to the synchronous one-shot interaction and requires no ECspec as it focuses on one-shot customized reporting.

4.2   FZI Living Lab AAL

The FZI Living Lab AAL architecture [LLAAL] represents a combination of the service-oriented provision of AAL services and event-driven communication between them, in order to enable a proactive reaction on some emergent situations in the living environment of elderly people. The system is based on the OSGi service middleware and consists of two main sub systems: the service platform openAAL and the ETALIS Complex event processing system (icep.fzi.de). It provides generic platform services like context management for collecting and abstracting data about the environment, workflow based specifications of system behaviour and semantically-enabled service discovery. Framework and platform services are loosely coupled by operating and communicating on shared vocabulary (most important ontologies: AAL domain, Sensor-ontology). The architecture can be mapped on the RWI Reference Architecture as follows. RWI sensors and RWI actuators are analogous to the AAL sensors and actuators. AAL AP (assisted person) corresponds to the RWI Resource User and RWI Entities of Interest (Entities of Interest) are analogous to the contextual information provided by AAL contextual manager. ETALIS (CEP engine) and Pub-sub service correspond to the RWI CEP Resource and RWI pub-sub service, respectively. Both one-shot and continuous interactions are supported between components, whereas the primary way of interaction is the asynchronous-continuous, i.e. event-driven one.

4.3   PECES

The PECES architecture [PECES] provides a comprehensive software layer to enable the seamless cooperation of embedded devices across various smart spaces on a global scale in a context-dependent, secure and trustworthy manner. PECES facilitates the easy formation of communities and collaboration across smart spaces, thus supporting nomadic users and remote collaboration among objects in different smart spaces in a seamless and automatic way. The PECES middleware architecture enables dynamic group-based communication between PECES applications (Resources) by utilizing contextual information based on a flexible context ontology. Although Resources are not directly analogous to PECES middleware instances, gateways to these devices are more resource-rich and can host middleware instances, and can be queried provided that an application-level querying interface is implemented. Entities of Interest are analogous to the contextual information underlying PECES. These entities are encapsulated as any other contextual information, a model abstraction which can include spatial elements (GIS information), personal elements (personal profiles) and devices and their profiles. The PECES Registry component implements a Yellow Pages directory service, i.e., services are described through attributes, modeled as contextual information, and a range of services (resources). Any service (resource) matching that description may be returned by the registry. Although no "session context" is required, a pre-requirement exists that interacting PECES applications, whether they are entities or resources, must be running the PECES middleware before any interaction may occur. Both one-shot and continuous interactions are supported between components and PECES provides the grouping and addressing functionality and associated security mechanisms that are required to enable dynamic loosely-coupled systems. The number of participants can be m:n, as PECES primarily targets group-based communication scenarios.

4.4   SemsorGrid4Env

The SemSorGrid4Env architecture [SSG4Env] provides support for the discovery and use of sensor-based, streaming and static data sources in manners that were not necessarily foreseen when the sensor networks were deployed or the data sources made available. The architecture may be applied to almost any type of real world entity, although it has been used mainly with real world entities related to natural phenomena (e.g, temperature, humidity, wave length). The types of resources considered are: sensor networks, off-the-shelf mote-based networks or ad-hoc sensors; streaming data sources, normally containing historical information from sensors; and even relational databases, which may contain any type of information from the digital world (hence resource hosts are multiple). These resources are made available through a number of data-focused services (acting as resource endpoints), which are based on the WS-DAI specification for data access and integration and which are supported by the SemSor-Grid4Env reference implementation. These services include those focused on data registration and discovery (where a spatio-temporal extension of SPARQL – stSPARQL -, is used to discover data sources from the SemSorGrid4Env registry), data access and query (where ontology-based and non-ontology-based query languages are provided to access data: SPARQL-Stream and SNEEql – a declarative continuous query language over acquisition sensor networks, continuous streaming data, and traditional stored data), and data integration (where the ontology-based SPARQL-Stream language is used to integrate data from heterogeneous and multi-modal data sources). Other capabilities offered by the architecture are related to supporting synchronous and asynchronous access modes, with subscription/pull and push-based capabilities, and actuating over sensor networks, by in-network query processing mechanisms that take declarative queries and transform them into code that changes the behavior of sensor networks. Context information queries are supported by using ontologies about roles, agents, services and resources.

4.5   SENSEI

The SENSEI architecture [SENSEI] aims at integrating geo-graphically dispersed and internet interconnected heterogeneous WSAN (Wireless Sensor and Actuator Networks) systems into a homogeneous fabric for real world information and interaction. It includes various useful services for both providers and users of real world resources to form a global market space for real world information and interaction. SENSEI takes a resource oriented approach which is strongly inspired by service oriented principles and semantic web technologies. In the SENSEI architecture each real world resource is described by a uniform resource description, providing basic and semantically expressed advanced operations of a resource, describing its capabilities and REP information. These uniform descriptions provide the basis for a variety of different supporting services that operate upon. On top of this unifying framework SENSEI builds a context framework, with a 3 layer information model. One of the key support services is a rendezvous mechanism that allows resource users to discover and query resources that fulfill their interaction requirements. At lower level this is realized by a federated resource directory across different administrative domains. On top of it, the architecture provides a semantic query support, allowing resource users to declaratively express context information or actuation tasks. Using a semantic query resolver and the support of an entity directory (in which bindings of real world resources and entities are maintained) suitable sensor, actuator and processing services can be identified and dynamically combined in order to provide request context information or realize more complex actuation loops. In order to increase flexibility at run-time, dynamic resource creation functionality allows for the instantiation of processing resources that may be required but not yet deployed in the system. The SENSEI architecture supports both one-time and longer lasting interactions between resource users and resource providers, that can be streaming or event based and provides mechanism through the execution manager to maintain a desired quality of information and actuation despite system dynamics. A comprehensive security framework provides functions for the realization of a variety of different trust relationships. This is centered on a security token service for resource users and AAA (Authentication, Authorization and Accounting) service to enforce access at the access controlled entities covering resources and framework functions. Furthermore AAA services perform accounting and auditing for authorized use of real world resources.

4.6   Other Architectures

A number of projects focus on aspects beyond the architectural blueprint presented in this chapter, the most prominent being SPITFIRE [SPITFIRE] and IoT-A [IoT-A]. As these projects have just started and have not produced architectures yet, they can only be included in the future work on an RWI reference architecture.

SPITFIRE aims at extending the Web into the embedded world to form a Web of Things (WoT), where Web representations of real-world entities offer services to access and modify their physical state and to mash up these real-world services with traditional services and data available in the Web. SPITFIRE extends the architectural model of this chapter by its focus on services, supporting heterogeneous and resource-constrained devices, its extensive use of existing Web standards such as RESTful interfaces and Linked Open Data, along with semantic descriptions throughout the whole architecture.

The IoT-A project extends the concepts developed in SENSEI further to provide a unified architecture for an Internet of Things. It aims at the creation of a common architectural framework making a diversity of real world information sources such as wireless sensor networks and heterogeneous identification technologies accessible on a Future Internet. While addressing various challenges [ZGL+], it will provide key building blocks on which a future IoT architecture will be based, such as a global resolution infrastructure that allows IoT resources to be dynamically resolved to entities of the real world to which they can relate.

4.7   Summary of Project Realizations

Table 2a. Functional coverage of current RWI architecture approaches

	SENSEI	ASPIRE	PECES	SemSorGrid4Env	LLAAL
Resource discovery	Using a standalone or federated (peered) resource directory as a rendezvous point that stores resource descriptions	Using ONS and EPCIS for id-based or information centric resource discovery	Distributed registry, information centric (yellow pages)	Using an RDF-based registry of data sources, with spatio-temporal query (stSPARQL) capabilities	Using an RDF-based registry
Context information query	SPARQL based query interface, semantic query resolution involving ED (Entity Directory) and RD (Resource Directory)	Query of the EPCIS which stores WHAT, WHERE, WHEN and WHY for all resources	Via registry or queries to resources via roles	Role, agent, service and resource ontologies as information models and corresponding stSPARQL queries	Contextual Manager provides an ontology-based information storage that captures sensor information and user input
Actuation and control loops	Actuation task model, support for execution of control loops through execution manager	Actuation based on applications and business events in the related components.	Implicit in application code	In-network query processing capabilities (SNEE) with mote-based sensor networks	Procedural Manager manages and executes easy to define and installation independent workflows which react to situations of interest
Dynamic resource creation	Dynamic instantiation of processing resources using predefined templates	Incorporated in ALE and BEG based on level of application interaction	Dynamic roles and dynamic smart spaces	Data services are generated dynamically according to WS-DAI (Web Services Data Access and Integration) indirect access mode	Composer analyses services which are available in a certain installation and selects and combines those services to achieve the (abstract) service goals

Table 2b. Functional coverage of current RWI architecture approaches

	SENSEI	ASPIRE	PECES	SemSorGrid4Env	LLAAL
Session management	Execution manager responsible for main-tenance of long last-ing requests	ECspecs, filtering and collection as well as read cycles.	Implicit via middleware	Limited management, through WS-DAI indirect access mode	N/A
Access control	Security token service for resource users and AAA service to en-force access at the access controlled entity, resource access proxy service for cross-domain access	Role-based access control for individual middleware compo-nents	Expressive (based on ontologies), enforce-able	N/A	N/A
Auditing and billing	auditing for AAA service to perform accounting and au-thorized use	N/A	N/A	N/A	N/A
Underlying Resource model	SENSEI resource description and ad-vanced resource descriptions with semantics	EPC and value-added sensing	PECES role ontology	According to W3C Semantic Sensor Network Ontology	LL AAL Sensor-level ontology. It supports integration of sensors and AAL services
Underlying context model	3 layer information model (raw data, observation & meas-urement, ontology based context model)	EPCIS standard	PECES context ontology	Observation&Measuremen t, role, agent, service and resource ontologies	Context Ontology: low- and top-level. It supports context reasoning from a low-level sensor-based model to a high-level service-oriented model

5   Concluding Remarks

The chapter presents a blueprint for design of systems capable of capturing information from and about the physical world and making it available for usage in the digital world. Based on the inputs and analysis of several research projects in this domain, it provides an outline of the main architectural components, interactions between the components and a way to describe the information and capabilities of the components in a standardized manner. Although not the final RWI reference architecture, the blueprint already captures the main features of such systems well as can be seen from the analysis of architectures designed in five different projects.

The work on the IoT reference architecture will continue to be driven by the RWI group of the FIA in collaboration with the FP7 IOT-i coordinated action project (http://www.iot-i.eu) and the IERC, the European Research Cluster on the Internet of Things (http://www.internet-of-things-research.eu/). The results will be contributed to the FIA Architecture track. It is expected that the final architecture will be ready by the end of 2011.

References

[ASPIRE]   Advanced Sensors and lightweight Programmable middleware for Innovative RFID Enterprise applications, FP7, http://www.fp7-aspire.eu/

[CONET]   Cooperating Objects NoE, FP7, http://www.cooperating-objects.eu/

[EPC]   EPCGlobal: The EPCglobal Architecture Framework 1.3 (March 2009), available from: http://www.epcglobalinc.org/

[HAL]   S. Haller: The Things in the Internet of Things. Poster at the Internet of Things Conference, Tokyo (IoT, 2010) (2010), available at http://www.iot-a.eu/public/news/resources/TheThingsintheInternetof Things_SH.pdf [Accessed Jan. 24, 2011]

[IoT-A]   EU FP7 Internet of Things Architecture project, http://www.iot-a.eu/public

[LLAAL]   FZI Living Lab AAL, http://aal.fzi.de/

[PECES]   PErvasive Computing in Embedded Systems, FP7, http://www.ict-peces.eu/

[SemsorGrid4Env] Semantic Sensor Grids for Rapid Application Development for Environmental Management, FP7, http://www.semsorgrid4env.eu/

[SENSEI]   Integrating the Physical with the Digital World of the Network of the Future, FP7, http://www.ict-sensei.org

[SPITFIRE]   Semantic-Service Provisioning for the Internet of Things using Future Internet Research by Experimentation, FP7, http://www.spitfire-project.eu/

[ZGL+]   Zorzi, M., Gluhak, A., Lange, S., Bassi, A.: From Today's INTRAnet of Things to a Future INTERnet of Things: A Wireless- and Mobility-Related View. IEEE Wireless Communications 17(6) (2010)

Towards a RESTful Architecture for Managing a Global Distributed Interlinked Data-Content-Information Space

Maria Chiara Pettenati, Lucia Ciofi, Franco Pirri, and Dino Giuli

Electronics and Telecommunications Department, University of Florence, Via Santa Marta, 3
50139 Florence, Italy
{mariachiara.pettenati, lucia.ciofi, franco.pirri,
dino.giuli}@unifi.it

Abstract. The current debate around the future of the Internet has brought to front the concept of "Content-Centric" architecture, lying between the Web of Documents and the generalized Web of Data, in which explicit data are embedded in structured documents enabling the consistent support for the direct manipulation of information fragments. In this paper we present the InterDataNet (IDN) infrastructure technology designed to allow the RESTful management of interlinked information resources structured around documents. IDN deals with globally identified, addressable and reusable information fragments; it adopts an URI-based addressing scheme; it provides a simple, uniform Web-based interface to distributed heterogeneous information management; it endows information fragments with collaboration-oriented properties, namely: privacy, licensing, security, provenance, consistency, versioning and availability; it glues together reusable information fragments into meaningful structured and integrated documents without the need of a pre-defined schema.

Keywords: Web of Data; future Web; Linked Data; RESTful; read-write Web; collaboration.

1 Introduction

There are many evolutionary approaches of the Internet architecture which are at the heart of the discussions both in the scientific and industrial contexts: Web of Data/Linked Data, Semantic Web, REST architecture, Internet of Services, SOA and Web Services and Internet of Things approaches. Each of these approaches focus on specific aspects and objectives which underlie the high level requirements of being a driver towards "a better Internet" or "a better Web".

Three powerful concepts present themselves as main drivers of the Future Internet [1][2]. They are: a user-centric perspective, a service-centric perspective and a content-centric perspective. The user-centric perspective emphasizes the end-user experience as the driving force for all technological innovation; the service-centric perspective is currently influenced in enterprise IT environment and in the Web2.0 mashup culture, showing the importance of flexibly reusing service components to build efficient applications. The Content-Centric perspective leverages on the importance of creating, pub-

J. Domingue et al. (Eds.): Future Internet Assembly, LNCS 6656, pp. 81–90, 2011.

lishing and interlinking content on the Web and providing content-specific infrastructural services for (rich media) content production, publication, interlinking and consumption. Even if it is very difficult to provide a strict separation of approaches because either they are often positioned or have evolved touching blurred areas between Content, Services and User perspectives, a rough schema in Table 1 can provide highlights the main, original, driving forces of such approaches.

Table 1. Rough classification of main driving forces in current Future Network evolutionary approaches

	Content-centric	Service-centric	Users-centric
Approaches	Web of Data/	Internet of	Web 2.0,
	Linked Data	Services	Web 3.0,
		WS-*	Semantic Web
	REST	SOA	Internet of Things

The three views can be interpreted as emphasizing different aspect rather than expressing opposing statements. Hence, merging and homogenizing towards an encompassing perspective may help towards the right decision choice for the Future Internet. Such an encompassing perspective has been discussed in terms of high-level general architecture in [1] and has been named "Content-Centric Internet". At the heart of this architecture is the notion of Content, defined as "any type and volume of raw information that can be combined, mixed or aggregated to generate new content and media" which is embedded in the Content Object, "the smallest addressable unit managed by the architecture, regardless of its physical location". In such an high-level platform, Content and Information are separate concepts [3] and Services are built as a result of a set of functions applied to the content, to pieces of information or services. As a consequence of merging the three views (user, content, service-oriented) the Future Internet Architecture herewith described essentially proposes a Virtual Resources abstraction required for the Content-Centric approach. Another view of "Content-centric Internet architecture" is elaborated in [2] by Danny Ayers, based on the assumption that "what is missing is the ability to join information pieces together and work more on the level of knowledge representation". Ayers' proposal is therefore a "Transitional Web" lying between the Web of Documents and the generalized Web of Data in which explicit data are embedded in documents enabling the consistent support for the direct manipulation of information as data without the limitation of current data manipulation approaches. To this end, Ayers identifies the need to find and develop technologies allowing the management of "micro-content" i.e. sub-document-sized chunks (information/document fragments), in which content being managed and delivered is associated with descriptive metadata.

Abstracting from the different use of terms related to the concepts "data", "content" and "information" which can be found in literature with different meanings [4], the grounding consistency that can be highlighted is related to the need of providing an evolutionary direction to the network architecture hinging on the concept of a small, Web-wide addressable data/content/information unit which should be organized according a specific model and handled by the network architecture so as to provide basic Services at an "infrastructural level" which in turn will ground the development of Applications fulfilling the user-centric needs and perspectives. Among the different paths to the Web of Data the one most explored is adding explicit data to content. Directly treating content as data has instead had little analysis.

In this paper we discuss evolution of InterDataNet (IDN) an high-level Resource Oriented Architecture proposed to enable the Future Internet approaches (see [5] [6] and references therein).

InterDataNet is composed of two main elements: the IDN-Information Model and the IDN-Service Architecture. Their joint use is meant to allow:

1. the management of addressable and reusable information fragment
2. their organization into structured documents around which different actors collaborate
3. the infrastructural support to collaboration on documents and their composing information fragments
4. the Web-wide scalability of the approach.

The purpose of this paper is to show that InterDataNet can provide a high-level model of the Content-Centric Virtualized Network grounding the Future Internet Architecture. For such a purpose InterDataNet can provide a Content-Centric abstraction level (the IDN-Information Model) and a handling mechanism (the IDN Service Architecture), to support enhanced content/information-centric services for Applications, as highlighted in Figure 1.

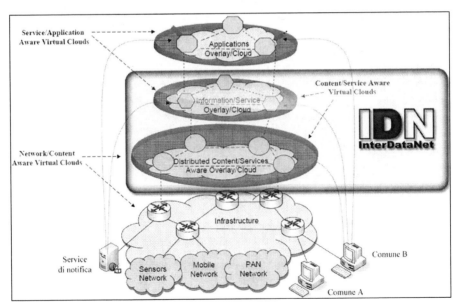

Fig. 1. InterDataNet architecture situated with respect to the Future Internet architecture envisaged in [7].

Referring to Table 1, current Content-centric network approaches and architectures, though aiming at dealing with distributed granular content over the Web, suffer from a main limitation: the more we get away from the data and move into the direction of information, the fewer available solutions are there capable of covering the following requirements:

— the management of addressable and reusable information fragment
— their organization into structured documents around which different actors collaborate
— the infrastructural (i.e. non application-dependent) support to collaboration on above documents and their composing information fragments
— the uniform REST interaction with the resources
— the Web-wide scalability of the approach.

This consolidates the need to look for and provide solutions fitting the visionary path towards a content-information/abstraction levels as illustrated in Figure 1, to which we provide our contribution, with the technological solution described in the following section.

2   The InterDataNet Content-Centric Approach

InterDataNet main characteristics are the following:

1. IDN deals with globally identified, addressable and reusable information fragments (as in Web of Data)
2. IDN adopts an URI-based addressing scheme (as in Linked Data)
3. IDN provides simple a uniform Web-based interface to distributed heterogeneous data management (REST approach)
4. IDN provides - at an infrastructural level - collaboration-oriented basic services, namely: privacy, licensing, security, provenance, consistency, versioning and availability
5. IDN glues together reusable information fragments into meaningful structured and integrated documents without the need of a pre-defined schema.

This will alleviate application-levels of sharing arbitrary pieces of information in ad-hoc manner while providing compliancy with current network architectures and approaches such as Linked Data, RESTful Web Services, Internet of Service, Internet of Things.

2.1   The InterDataNet Information Model and Service Architecture

IDN framework is described through the ensemble of concepts, models and technologies pertaining to the following two views (Fig. 2):

IDN-IM (InterDataNet Information Model). It is the shared information model representing a generic document model which is independent from specific contexts and technologies. It defines the requirements, desirable properties, principles and structure of the document to be managed by IDN.

IDN-SA (InterDataNet Service Architecture). It is the architectural layered model handling IDN-IM documents (it manages the IDN-IM concrete instances allowing the users to "act" on pieces of information and documents). The IDN-SA implements the reference functionalities defining subsystems, protocols and interfaces for IDN document collaborative management. The IDN-SA exposes an IDN-API (Application Programming Interface) on top of which IDN-compliant Applications can be developed.

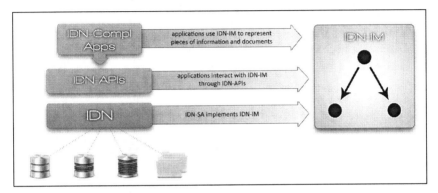

Fig. 2. InterDataNet framework

The InterDataNet Information Model. The Information Model is the graph-based data model (see Figure 3) to describe interlinked data representing a generic document model in IDN and is the starting point from which has been derived the design of IDN architecture.

Generic information modeled in IDN-IM is formalized as an aggregation of data units. Each data unit is assigned at least with a global identifier and contains generic data and metadata; at a formal level, such data unit is a node in a Directed Acyclic Graph (DAG). The abstract data structure is named IDN-Node. An IDN-Node is the "content-item" handled by the "content-centric" IDN-Service Architecture. The degree of atomicity of the IDN Nodes is related to the most elementary information fragment whose management is needed in a given application. The information fragment to be handled in IDN-IM compliant documents, has to be inserted into a container represented by the IDN-Node structure, i.e. a rich hierarchically structured content. The IDN Node is addressable by an HTTP URI and its contents (name, type, value and metadata) are under the responsibility of a given entity. The node aggregated into an IDN-IM document is used (read, wrote, updated, replicated, etc.) under the condition applied/specified on it by its responsible. An IDN-document structures data units is composed of nodes related to each other through directed "links". Three main link types are defined in the Information Model:

— *aggregation links*, to express relations among nodes inside an IDN-document;
— *reference links*, to express relations between distinct IDN-documents;
— *back links*, to enable the mechanism of notification of IDN-nodes updates to parent-nodes (back propagation).

The different types of links envisaged in IDN-IM are conceived to enable collaboration extending the traditional concept of *hyperlink*. Indeed the concept of link expressed by the "href" attribute in HTML tags inherently incorporate different "meanings" of the link: either inclusion in a given document, such as it is the case of the tag <img> image, or reference to external document, such as it is the case of the <a> anchor tag. Explicitly separating these differences allow to make the meaning explicit and consequently enable the different actions, e.g. assign specific authorizations on the resources involved in the link. Actually, if two resources are connected through an aggregation link then the owner of each resources has the capability to write and read both of them. Instead the reference link implies only the capability to read the resource connected. In Figure 3 aggregation, reference and back links are illustrated.

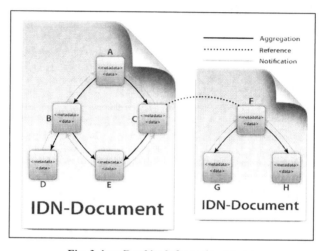

Fig. 3. InterDataNet Information-Model

InterDataNet Service Architecture. IDN Service Architecture (IDN-SA) is made up of four layers: Storage Interface, Replica Management, Information History and Virtual Resource. Each layer interacts only with its upper and lower level and relies on the services offered by IDN naming system. IDN-Nodes are the information that the layers exchange in their communications. In each layer a different type of IDN-Node is used: SI-Node, RM-Node, IH-Node and VR-Node. Each layer receives as input a specific type of IDN-Node and applies a transformation on the relevant metadata to obtain its own IDN-Node type. The transformation (adding, modifying, updating and deleting metadata) resembles the encapsulation process used in the TCP/IP protocol stack. The different types of IDN-Nodes have different classes of HTTP-URI as identifiers.

Storage Interface (SI) provides the services related to the physical storage of information and an interface towards legacy systems. It offers a REST interface enabling CRUD operations over SI-Nodes. SI-Nodes identifiers are URLs.

Replica Management (RM) provides a delocalized view of the resources to the upper layer. It offers a REST interface enabling CRUD operations over RM-Nodes. RM-Nodes identifiers are HTTP-URI named PRI (Persistent Resource Identifier). RM presents a single RM-Node to the IH layer hiding the existence of multiple replicas in the SI layer.

Information History (IH) manages temporal changes of information. It offers a REST interface enabling CRUD operations over IH-Nodes. IH Nodes identifiers are HTTP-URI named VRI (Versioned Resource Identifier). The IH layer presents the specific temporal version of the Node to the VR layer.

Virtual Resource (VR) manages the document structure. It offers to IDN compliant applications a REST interface enabling CRUD operations on VR-Nodes. VR-Nodes identifiers are HTTP-URIs named LRIs (Logical Resource Identifiers).

On top of the four layers of the Service Architecture, the IDN-compliant Application layer uses the documents' abstraction defined in the Container-Content principle. Interfacing to the VR layer, the application is entitled to specify the temporal instance of the document handled.

The communications between IDN-SA layers follows the REST [8] paradigm through the exchange of common HTTP messages containing a generic IDN-Node in the message body and IDN-Node identifier in the message header.

IDN architecture envisages a three layers naming system (figure 5): in the upper layer are used Logical Resource Identifier (LRI) to allow IDN-application to identify IDN-nodes. LRI are specified in a human-friendly way and minted by the IDN-Applications. Each IDN-Node can be referred to thanks to a global unique canonical

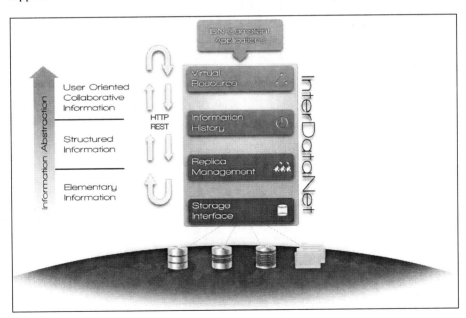

Fig. 4. InterDataNet Service Architecture

name and one or more "aliases. In the second layer are used Persistent Resource Identifiers (PRI) in order to obtain a way to unambiguously, univocally and persistently identify the resources within IDN-middleware independent of their physical locations; in the lower layer are used Uniform Resource Locators (URL) to identify resource replicas as well as to access them. Each resource can be replicated many times and therefore many URLs will correspond to one PRI. The distributed pattern adopted in the IDN naming system resembles the traditional DNS system in which the load corresponding on a single hostname is distributed on several IP addresses. Analogously we distributed a PRI (HTTP-URI) on a set of URLs (HTTP-URIs), using a DNS-like distributed approach.

The implementations of IDN-SA are a set of different software modules, one module for each layer. Each module, implemented using an HTTP server, will offers a REST interface. The interaction between IDN-compliant applications and IDN-SA follows the HTTP protocol as defined in REST architectural style too. CRUD operations on application-side will therefore be enabled to the manipulation of data on a global scale within the Web.

REST interface has been adopted in IDN-SA implementation as the actions allowed on IDN-IM can be translated in CRUD style operations over IDN-Nodes with the assumption that an IDN-document can be thought as an IDN-Node resources collection. The resources involved in REST interaction are representations of IDN-Nodes. As introduced earlier in this Section, there are several types of Nodes all of which are coded in an "IDN/XML format" (data format defined with XML language). Every resource in such format must be well formed with respect to XML syntax, and valid with respect to a specific schema (coded in XML Schema) defined according to this purpose. The schema for "IDN/XML format" uses both an ad-hoc built vocabulary to describe terms which are peculiar to the adopted representation, as well as standard vocabularies from which it borrows set of terms. Each IDN-Node resource is identified by an HTTP-URI. Through its HTTP-URI it is possible to interact with the resource in CRUD style, using the four primitives POST, GET, PUT, DELETE.

It is worth highlighting that the IDN-Service Architecture is designed to allow inherent scalability also in the deployment of its functions. According to the envisaged steps the architecture can offer and deploy specific functionalities of the architecture

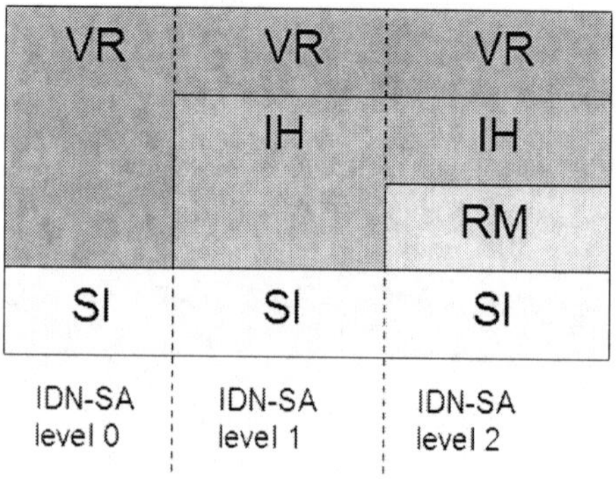

VR	VR	VR
	IH	IH
		RM
SI	SI	SI
IDN-SA level 0	IDN-SA level 1	IDN-SA level 2

Fig. 5. InterDataNet Service Architecture scalability features

without the need to achieve the complete development of the architecture before its adoption. In this way, specific IDN-Applications developed on top of IDN-SA level 0/1/2 would thus leverage on a subset of IDN characteristics, and functions while keeping the possibility to upgrade to the full IDN on a need or case basis and once further releases will be available, as it is illustrated in Figure 5.

3   Conclusion

The kind of Content-centric interoperability we aim to provide in IDN is related to the exchange and controlled use (discovery, retrieve, access, edit, publish, etc.) of the distributed information fragments (contents) handled through IDN-IM documents. The presented approach is not an alternative to current Web of Data and Linked Data approaches rather it aims at viewing the same data handled by the current Web of Data from a different perspective, where a simplified information model, representing only information resources, is adopted and where the attention is focused on collabo-ration around documents and documents fragments either adopting the same global naming convention or suggesting new methods of handling data, relying on standard Web techniques.

InterDataNet could be considered to enable a step ahead from the Web of Docu-ment and possibly grounding the Web of Data, where an automated mapping of IDN-IM serialization into RDF world is made possible using the Named Graph approach [9]. Details on this issue are beyond the scope of the present paper.

The authors are aware that the IDN vision must be confronted with the evaluation of the proposed approach. Providing demonstrable contribution to such a high level goal is not an easy task, as it is demonstrated by the state of the art work defending this concept idea which, as far as we know, either do not provide concrete details about the possible implementation solutions to address it [1][3] or circumvent the problem adopting a mixed approach, while pinpointing the main constraint of the single solutions. However, several elements can be put forward to sustain our pro-posal, with respect to three main elements: a) the adoption of layered architecture approach to break down the complexity of the system problem [10]; b) using HTTP URIs to address information fragments to manage resources "in" as well as "on" the Web [11]; c) the adoption of a RESTful Web Services, also known as ROA – Re-source Oriented Architecture to leverage on REST simplicity (use of well-known standards i.e. HTTP, XML, URI, MIME), pervasive infrastructure and scalability. The current state of InterDataNet implementation and deployment, is evolving along two directions: a) the infrastructure; the Proof of Concept of the implementation of the full IDN-Service Architecture is ongoing. Current available releases of the Archi-tecture implement all the layers except for the Replica Management layer, while the implementation of the three-layers naming system is being finalized; b) applica-tion/working examples: an IDN-compliant applications has been developed on top of a simplified instance of the IDN-SA, implementing only the IDN-VR and IDN-SI layers. This application is related to the online delivery of Official Certificates of Residence. The implemented Web application allows Public Officers to assess current citizens' official residence address requesting certificates to the entitled body, i.e. the Municipality.

InterDataNet technological solution decreases the complexity of the problems at the Application level because it offers infrastructural enablers to Web-based interoperation without requiring major preliminary agreements between interoperating parties thus providing a contribution in the direction of taking full advantage of the Web of Data potential.

Acknowledgments. We would like to acknowledge the precious work of Davide Chini, Riccardo Billero, Mirco Soderi, Umberto Monile, Stefano Turchi, Matteo Spampani, Alessio Schiavelli and Luca Capannesi for the technical support in the implementation IDN-Service Architecture prototypes.

References

1. Zahariadis, T., Daras, P., Bouwen, J., Niebert, N., Griffin, D., Alvarez, F., Camarillo, G.: Towards a Content-Centric Internet. In: Tselentis, G., Galis, A., Gavras, A., Krco, S., Lotz, V., Simperl, E., Stiller, B., Zahariadis, T. (eds.) Towards the Future Internet - Emerging Trends from European Research, pp. 227–236. IOS Press, Amsterdam (2010)
2. Ayers, D.: From Here to There. IEEE Internet Comput 11(1), 85–89 (2007)
3. European Commission Information Society and Media. Future Networks The way ahead! European Communities: Belgium (2009)
4. Melnik, S., Decker, S.: A Layered Approach to Information Modeling and Interoperability on the Web. In: Proceedings ECDL'00 Workshop on the Semantic Web, Lisbon (September 2000)
5. Pettenati, M.C., Innocenti, S., Chini, D., Parlanti, D., Pirri, F. (2008) Interdatanet: A Data Web Foundation For The Semantic Web Vision. Iadis International Journal On Www/Internet 6(2) (December 2008)
6. Pirri, F., Pettenati, M.C., Innocenti, S., Chini, D., Ciofi, L.: InterDataNet: a Scalable Middleware Infrastructure for Smart Data Integration, in D. In: Giusto, D., et al. (eds.) The Internet of Things: 20th Tyrrhenian Workshop on Digital Communications, Springer, Heidelberg (2009), doi: 10.1007/978-1-4419-1674-7_12
7. Zahariadis, T., Daras, P., Bouwen, J., Niebert, N., Griffin, D., Alvarez, F., Camarillo, G.: Towards a Content-Centric Internet Plenary Keynote Address. Presented at Future Internet Assembly (FIA) Valencia, SP, 15-16 April (2010)
8. Richardson, L., Ruby, S.: RESTful Web Services; O'Reilly Media, Inc.: Sebastopol, CA, USA (2007)
9. Carroll, J.J., Bizer, C., Hayes, P., Stickler, P.: Named graphs, provenance and trust. In: Proceedings of the 14th international conference on World Wide Web - WWW '05. Presented at the 14th international conference, Chiba, Japan, p. 613. Chiba, Japan (2005), doi:10.1145/1060745.1060835
10. Zweben, S.H., Edwards, S.H., Weide, B.W., Hollingsworth, J.E.: The Effects of Layering and Encapsulation on Software Development Cost and Quality. IEEE Trans. Softw. Eng. 21(3), 200–208 (1995)
11. Hausenblas, M.: Web of Data. "Oh – it is data on the Web" posted on April 14, 2010; accessed September 8, 2010, http://webofdata.wordpress.com/2010/04/14/oh-it-is-data-on-the-web/

A Cognitive Future Internet Architecture

Marco Castrucci[1], Francesco Delli Priscoli[1], Antonio Pietrabissa[1], and
Vincenzo Suraci[2]

[1] University of Rome "La Sapienza", Computer and System Sciences Department
Via Ariosto 25, 00185 Rome, Italy
{castrucci, dellipriscoli, pietrabissa}@dis.uniroma1.it
[2] Università degli studi e-Campus
Via Isimbardi 10, 22060 Novedrate (CO), Italy
vincenzo.suraci@uniecampus.it

Abstract. This Chapter proposes a novel Cognitive Framework as reference architecture for the Future Internet (FI), which is based on so-called *Cognitive Managers*. The objective of the proposed architecture is twofold. On one hand, it aims at achieving a full interoperation among the different entities constituting the ICT environment, by means of the introduction of *Semantic Virtualization Enablers*, in charge of virtualizing the heterogeneous entities interfacing the FI framework. On the other hand, it aims at achieving an inter-network and inter-layer cross-optimization by means of a set of so-called *Cognitive Enablers*, which are in charge of taking consistent and coordinated decisions according to a fully cognitive approach, availing of information coming from both the transport and the service/content layers of all networks. Preliminary test studies, realized in a home environment, confirm the potentialities of the proposed solution.

Keywords: Future Internet architecture, Cognitive networks, Virtualization, Interoperation.

1   Introduction

Already in 2005, there was the feeling that the architecture and protocols of the Internet needed to be rethought to avoid Internet collapse [1]. However, the research on Future Internet became a priority only in the last five years, when the exponential growth of small and/or mobile devices and sensors, of services and of security requirements began to show that current Internet is becoming itself a bottleneck. Two main approach have been suggested and investigated: the radical approach [2], aimed at completely re-design the Internet architecture, and the evolutionary approach [3], trying to smoothly add new functionalities to the current Internet towards.

Right now, the technology evolution managed to cover the lacks of current Internet architecture, but, probably, the growth in Internet-aware devices and the always more demanding requirements of new services and applications will require radical architecture enhancements very soon. This statement is proved by the number of financed projects both in the USA and in Europe.

J. Domingue et al. (Eds.): Future Internet Assembly, LNCS 6656, pp. 91–102, 2011.

In the USA, there are significant initiatives. NeTS [4] (Networking Technology and Systems) was a program of the National Science Foundation (NSF) on networking research with the objectives of developing the technology advances required to build next generation networks and improve the understanding of large, complex and heterogeneous networks. The follow-up of NeTS, NetSE [5] proposes a clean-state approach to properly meet new requirements in security, privacy and economic sustainability. GENI [6] (Global Environment for Network Innovations) is a virtual laboratory for at scale experimentation of network science, based on a 40 Gbps real infrastructure. Stanford Clean Slate [7] proposes a disruptive approach by creating service platforms available to the research and user communities.

In Europe, Future Internet research has been included as one of the topics in FP6 and FP7. European initiatives appear less prone to a completely clean-state approach with respect of USA ones, and tries to develop platforms which support services and applications by utilizing the current Internet infrastructure. For instance, G-Lab [8] (Design and experiment the network of the future, Germany), is the German national platform for Future Internet studies, includes both research studies of Future Internet technologies and the design and setup of experimental facilities. GRIF [9] (Research Group for the Future Internet, France) and Internet del Futuro [10] (Spain) promotes cooperation based on several application areas (e.g., health) and technology platforms. FIRE [11] is an EU initiative aimed at the creation of an European Experimental Facility, which is constructed by progressively connecting existing and upcoming testbeds for Future Internet technologies.

The contribution of this Chapter is the proposal of a Future Internet architecture which seamlessly cope with the evolutionary approach but is also open to innovative technologies and services. The main idea is to collect and elaborate all the information coming from the whole environment (i.e., users, contents, services, network resources, computing resources, device characteristics) via virtualization and data mining functionalities; the metadata produced in this way are then input of intelligent cognitive modules which provide the applications/services with the required functionalities in order to maximize the user Quality of Experience with the available resources.

The Chapter is organized as follows: Section 2 is devoted to the description of the concepts underlying the proposed architecture; Section 3 describes the Future Internet platform in detail; experimental results showing the potential of the platform are described in Section 4; finally, Section 5 draws the conclusions.

2   Architecture Concept

A more specific definition of the entities involved in the Future Internet, as well as of the Future Internet target, can be as follows:

- Actors represent the entities whose requirement fulfillment is the goal of the Future Internet; for instance, Actors include users, developers, network providers, service providers, content providers, etc.;

- <u>Resources</u> represent the entities that can be exploited for fulfilling the Actors' requirements; example of Resources include services, contents, terminals, devices, middleware functionalities, storage, computational, connectivity and networking capabilities, etc.;
- <u>Applications</u> are utilized by the Actors to fulfill their requirements and needs exploiting the available resources.

In the authors' vision, the Future Internet target is to allow Applications to transparently, efficiently and flexibly exploit the available Resources, thus allowing the Actors, by using such Applications, to fulfill their requirements and needs. In order to achieve this target, the Future Internet should overcome the following main limitations.

(i) A first limitation is inherent in the traditional layering architecture which forces to keep algorithms and procedures, laying at different layers, independent one another. In addition, even in the framework of a given layer, algorithms and procedures dealing with different tasks are often designed independently one another. These issues greatly simplify the overall design of the telecommunication networks and greatly reduce processing capabilities, since the overall problem of controlling the telecommunication network is decoupled in a certain number of much simpler sub-problems. Nevertheless, a major limitation of this approach derives from the fact that algorithms and procedures are poorly coordinated one another, impairing the efficiency of the overall telecommunication network control. The issues above claim for a stronger coordination between algorithms and procedures dealing with different tasks.

(ii) A second limitation derives from the fact that, at present, most of the algorithms and procedures embedded in the telecommunication networks are open-loop, i.e. they are based on off-line "reasonable" estimation of network variables (e.g. offered traffic), rather than on real-time measurements of such variables. This limitation is becoming harder and harder, since the telecommunication network behaviours, due to the large variety of supported services and the rapid evolution of the service characteristics, are becoming more and more unpredictable. This claims for an evolution towards closed-loop algorithms and procedures which are able to properly exploit appropriate real-time network measurements. In this respect, the current technology developments, which assure cheap and powerful sensing capabilities, favours this kind of evolution.

(iii) The third limitation derives from the large variety of existing heterogeneous Resources which have been developed according to different heterogeneous technologies and hence embedding technology-dependent algorithms and procedures, as well as from the large variety of heterogeneous Actors who are playing in the ICT arena. In this respect, the requirement of integrating and virtualizing these Resources and Actors so that they can be dealt with in an homogeneous and virtual way by the Applications, claims for the design of a technology-independent, virtualized framework; this framework, on the one hand, is expected to embed algorithms and procedures which, leaving out of consideration the specificity of the various networks, can be based on abstract advanced methodologies and, on the other hand, is expected to be provided with proper virtualizing interfaces which allow all Applications to benefit from the functionalities offered by the framework itself.

The concept behind the proposed Future Internet architecture, which aims at overcoming the three above-mentioned limitations, is sketched in Fig. 1. As shown in the figure, the proposed architecture is based on a so-called "Cognitive Future Internet Framework" (in the following, for the sake of brevity, simply referred to as "Cognitive Framework") adopting a modular design based on middleware "enablers". The enablers can be grouped into two categories: the *Semantic Virtualization Enablers* and the *Cognitive Enablers*. The Cognitive Enablers represent the core of the Cognitive Framework and are in charge of providing the Future Internet control and management functionalities. They interact with Actors, Resources and Applications through *Semantic Virtualization Enablers*.

The Semantic Virtualization Enablers are in charge of virtualizing the heterogeneous Actors, Resources and Applications by describing them by means of properly selected, dynamic, homogeneous, context-aware and semantic aggregated metadata.

The Cognitive Enablers consist of a set of modular, technology-independent, interoperating enablers which, on the basis of the aggregated metadata provided by the Semantic Virtualization Enablers, take consistent control and management decisions concerning the best way to exploit and configure the available Resources in order to efficiently and flexibly satisfy Application requirements and, consequently, the Actors' needs. For instance, the Cognitive Enablers can reserve network resources, compose atomic services to provide a specific application, maximize the energy efficiency, guarantee a reliable connection, satisfy the user perceived quality of experience and so on.

The control and management decisions taken by the Cognitive Enablers are handled by the Semantic Virtualization Enablers, in order to be actuated involving the proper Resources, Applications and Actors.

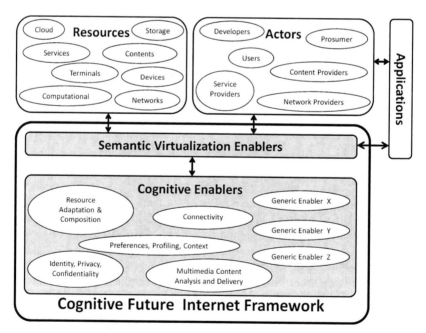

Fig. 1. Proposed Cognitive Future Internet Framework conceptual architecture

Note that, thanks to the aggregated semantic metadata provided by the Semantic Virtualization Enablers, the control and management functionalities included in the Cognitive Enablers have a technology-neutral, multi-layer, multi-network vision of the surrounding Actors, Resources and Applications. Therefore, the information enriched (fully cognitive) nature of the aggregated metadata, which serve as Cognitive Enabler input, coupled with a proper design of Cognitive Enabler algorithms (e.g., multi-objective advanced control and optimization algorithms), lead to cross-layer and cross-network optimization.

The Cognitive Framework can exploit one or more of the Cognitive Enablers in a dynamic fashion: so, depending on the present context, the Cognitive Framework activates and properly configures the needed Enablers.

Furthermore, in each specific environment, the Cognitive Framework functionalities have to be properly distributed in the various network entities (e.g. Mobile Terminals, Base Stations, Backhaul network entities, Core network entities). The selection and the mapping of the Cognitive Framework functionalities in the network entities is a critical task which has to be performed case by case by adopting a transparent approach with respect to the already existing protocols, in order to favour a smooth migration.

It should be evident that the proposed approach allows to overcome the three above-mentioned limitations:

(i) Concentrating control and management functionalities in a single Cognitive Framework makes much easier to take consistent and coordinated decisions. In particular, the concentration of control functionalities in a single framework allows the adoption of algorithms and procedures coordinated one another and even *jointly* addressing in a one-shot way, problems traditionally dealt with in separate and uncoordinated fashion.

(ii) The fact that control decisions can be based on properly selected, aggregated metadata describing, in real time, Resources, Actors and Applications allows closed-loop control, i.e. networks become *cognitive*. In particular, the Cognitive Enablers can, potentially, perform control elaborations availing of information coming from all the layers of the protocol stack and from all networks. Oversimplifying, according to the proposed *fully cognitive* approach, potentially, all layers benefit from information coming from all layers of all networks, thus allowing to perform a full cross-layer, cross-network optimization.

(iii) Control decisions, relevant to the best exploitation of the available Resources, can be made in a technology independent and virtual fashion, i.e. the specific technologies and the physical location behind Resources, Actors and Applications can be left out of consideration. In particular, the decoupling of the Cognitive Framework from the underlying technology transport layers on the one hand, and from the specific service/content layers on the other hand, allows to take control decisions at an abstract layer, thus favouring the adoption of advanced control methodologies (e.g. constrained optimization, adaptive control, robust control, game theory...) which can be closed-loop thanks to the previous issue. In addition, interoperation procedures among heterogeneous Resources, Actors and Applications become easier and more natural.

3   Cognitive Future Internet Framework Architecture

The Cognitive Framework introduced in the previous section consists of a conceptual framework that can be deployed as a distributed functional framework. It can be realized through the implementation of appropriate Cognitive Middleware-based Agents (in the following referred to as Cognitive Managers) which will be transparently embedded in appropriate network entities (e.g. Mobile Terminals, Base Stations, Backhaul Network entities, Core Network entities). There not exist a unique mapping between the proposed conceptual framework over an existing telecommunication network. However we proposed a proof-of-concept concrete scenario in section 4, where the conceptual framework has been deployed in a real home area network test case. Indeed the software nature of the Cognitive Manager allows a transparent integration in the network nodes. It can be deployed installing a new firmware or a driver update in each network element. Once the Cognitive Manager is executed, that network node is enhanced with the Future Internet functionalities and become part of the Future Internet assets.

Fig. 2 outlines the high-level architecture of a generic Cognitive Manager, showing its interfacing with Resources, Actors and Applications.

Fig. 2 highlights that a Cognitive Manager will encompass five high-level modular functionalities, namely the Sensing, Metadata Handling, Elaboration, Actuation and API (Application Protocol Interface) functionalities. The Sensing, Actuation and API functionalities are embedded in the equipment which interfaces the Cognitive Manager with the Resources (*Resource Interface*), with the Actors (*Actor Interface*) and with the Applications (*Application Interface*); these interfaces must be tailored to the peculiarities of the interfaced Resources, Actors and Applications.

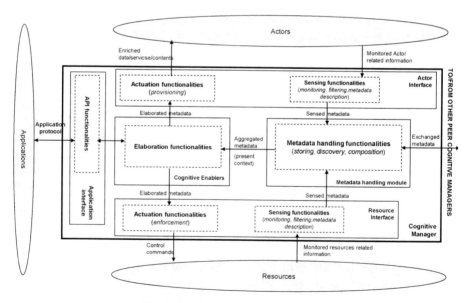

Fig. 2. Cognitive Manager architecture

The Metadata Handling functionalities are embedded in the so-called *Metadata Handling module*, whilst the *Elaboration functionalities* are distributed among a set of *Cognitive Enablers*. The Metadata Handling and the Elaboration functionalities (and in particular, the Cognitive Enablers which are the core of the proposed architecture) are independent of the peculiarities of the surrounding Resources, Actors and Applications.

With reference to Fig. 2, the Sensing, Metadata Handling, Actuation and API functionalities are embedded in the Semantic Virtualization Enablers, while the Elaboration functionalities are embedded in the Cognitive Enablers. The roles of the above-mentioned functionalities are the following.

Sensing functionalities are in charge of (i) the *monitoring* and preliminary *filtering* of both Actor related information coming from service/content layer (Sensing functionalities embedded in the Actor Interface) and of Resource related information (Sensing functionalities embedded in the Resource Interface); this monitoring has to take place according to transparent techniques, (ii) the formal description of the above-mentioned heterogeneous parameters/data/services/contents in homogeneous metadata according to proper ontology based languages (such as OWL – Web Ontology Language).

Metadata Handling functionalities are in charge of the storing, discovery and composition of the metadata coming from the sensing functionalities and/or from metadata exchanged among peer Cognitive Managers, in order to dynamically derive the aggregated metadata which can serve as inputs for the Cognitive Enablers; these aggregated metadata form the so-called *Present Context*; it is worth stressing that such Present Context has an highly dynamic nature.

Elaboration functionalities are embedded in a set of Cognitive Enablers which, following the specific application protocols and having as key inputs the aggregated metadata forming the Present Context, produce *elaborated metadata* aiming at (i) controlling the Resources, (ii) providing enriched data/services/contents to the Actors. In addition, these enablers control the sensing, metadata handling, actuation and API functionalities (these control actions, for clarity reasons, are not represented in Fig. 2).

Actuation functionalities are in charge of (i) actuation of the Cognitive Enabler control decisions over the Resources (*Enforcement functionalities* embedded in the Resource Interface; see Fig. 2); the decision enforcement has to take place according to transparent techniques, (ii) provisioning to the appropriate Actors the enriched data/contents/services produced by the Cognitive Enablers (*Provisioning functionalities* embedded in the Actor Interface; see Fig. 2).

Finally, *API functionalities* are in charge of interfacing the protocols of the Applications managed by the Actors with the Cognitive Enablers.

A so-called *Supervisor and Security Module* (not shown for clarity reason in Fig. 2) is embedded in each Cognitive Manager supervising the whole Cognitive Manager and, at the same time, assuring the overall security of the Cognitive Manager itself (e.g., including end-to-end encryption, Authentication, Authorization and Accounting (AAA) at user and device level, Service Security, Intrusion Detection, etc.). Another key role of this module is to dynamically decide, consistently with the application protocols, the Cognitive Manager functionalities which have to be activated to handle the applications, as well as their proper configuration and activation/deactivation timing.

The proposed approach and architecture have the following key efficiency and flexibility advantages which are hereinafter outlined in a qualitative way:

Advantages Related to Efficiency

(1)   The Present Context, which is the key input to the Cognitive Enablers, includes multi-Actor, multi-Resource information, thus potentially allowing to perform the Elaboration functionalities availing of a very "rich" feedback information.

(2)   The proposed architecture (in particular, the technology independence of the Elaboration functionalities, as well as the valuable input provided by the Present Context) allows to take all decisions in a cognitive, abstract, coordinated and co-operative fashion within a set of strictly cooperative Cognitive Enablers. The concentration of the control functionalities in such Cognitive Enablers allows the adoption of multi-object algorithms and procedures which *jointly* address problems traditionally dealt with in a separate and uncoordinated fashion at different layers. So, the proposed architecture allows to pass from the traditional layering approach (where each layer of each network takes uncoordinated decisions) to a scenario in which, potentially, all layers of all networks benefit from information coming from all layers of all networks, thus, potentially, allowing a full cross-layer, cross-network optimization.

(3)   The rich feedback information mentioned in the issue (1), together with the technology independence mentioned in the issue (2), allow the adoption of innovative and abstract *closed-loop* methodologies (e.g. constrained optimization, data mining, adaptive control, robust control, game theory, operation research, etc.) for the algorithms and rules embedded in the Cognitive Enablers, which are expected to remarkably improve efficiency.

Advantages Related to Flexibility

(4)   Thanks to the fact that the Cognitive Managers have the same architecture and work according to the same approach regardless of the interfaced heterogeneous Applications/Resources/Actors, interoperation procedures become easier and more natural.

(5)   The transparency and the middleware (firmware based) nature of the proposed Cognitive Manger architecture makes relatively easy its embedding in any fixed/mobile network entity (e.g. Mobile Terminals, Base Station, Backhaul network entities, Core network entities): the most appropriate network entities for hosting the Cognitive Managers have to be selected environment by environment. Moreover, the Cognitive Managers functionalities (and, in particular, the Cognitive Enabler software) can be added/upgraded/deleted through remote (wired and/or wireless) control.

(6)   The modularity of the Cognitive Manager functionalities allows their ranging from very simple SW/HW/computing implementations, even specialized on a single-layer/single-network specific monitoring/elaboration/actuation task, to

complex multi-layer/multi-network/multi-task implementations. In particular, for each Cognitive Manger, the relevant Actuation/Sensing functionalities, the aggregated information which form the Present Context, as well as the relevant Elaboration functionalities have to be carefully selected environment-by-environment, trading-off the advantages achieved in terms of efficiency with the entailed additional SW/HW/computation complexity.

(7) Thanks to the flexibility degrees offered by issues (4)-(6), the Cognitive Managers could have the same architecture regardless of the interfaced Actors, Resources and Applications. So, provided that an appropriate tailoring to the considered environment is performed, the proposed architecture can actually scale from environments characterized by few network entities provided with high processing capabilities, to ones with plenty of network entities provided with low processing (e.g. Internet of Things).

(8) The above-mentioned flexibility issues favours a *smooth migration* towards the proposed approach. As a matter of fact, it is expected that Cognitive Manager functionalities will be gradually inserted starting from the most critical network nodes, and that control functionalities will be gradually delegated to the Cognitive Modules.

Summarizing the above-mentioned advantages, we propose to achieve Future Internet revolution through a smooth evolution. In this evolution, Cognitive Managers provided with properly selected functionalities are gradually embedded in properly selected network entities, aiming at gradually replacing the existing open-loop control (mostly based on traditional uncoordinated single-layer/single-network approachs), with a cognitive closed-loop control trying to achieve cross-optimization among heterogeneous Actors, Applications and Resources. Of course, careful, environment-by-environment selection of the Cognitive Manager functionalities and of the network entities in which such functionalities have to be embedded, is essential in order to allow scalability and to achieve efficiency advantages which are worthwhile with respect to the increased SW/HW/computing complexity.

The following section shows an example of application of the above-mentioned concepts. Much more comprehensive developments are being financed in various frameworks (EU and national projects), which are expected to tailor the presented approach to different environments aiming at assessing, in a quantitative way, the actual achieved advantages in terms of flexibilty (scalability) and efficiency; nevertheless, in the authors' vision such advantages are already evident, in a qualitative way, in the concepts and discussions presented in this section.

4   Experimental Results

The proposed framework has been tested in a home scenario for a preliminary proof-of-concept, but the same results can be obtained even in wider scenarios involving also Access Networks and/or Wide Area Networks. We consider a hybrid home network, where connectivity among devices is provided using heterogeneous wireless (e.g., WiFi, UWB) and wired (e.g., Ethernet, PLC) communication technologies. For

testing purposes, only a simplified version of the Cognitive Manager has been implemented in each node of the network, which includes the following functionalities:

- the *Service and Content adapter*: a QoS adapter module has been implemented,
  able to acquire information about the characteristics of the flow that has to be
  transmitted over the network, in terms of Traffic Specifications (TSpec) and QoS
  requirements, and to map them into pre-defined flow identifiers;
- the *Command and measurement adapter*: a Monitoring Engine has been implemented in order to acquire information about the topology of the network and the
  status of the links, while an Actuator module has been used to enforce elaboration
  decision over the transport network, in particular in order to modify the forwarding
  table used by the node to decide the network interface to be used for the transmission of the packets;
- the *Metadata handling storing functionality*: all the heterogeneous information
  collected by the Service/content adapter and by the Command and measurement
  adapter are translated using a common semantic and stored in proper database,
  ready to be used by elaboration functionalities;
- a *Cognitive connectivity enabler*: it has been implemented to perform technology
  independent resource management algorithms (e.g., layer 2 path selection), in order
  to guarantee that flow's QoS requirements are satisfied during the transmission of
  its packets over the network. In particular, a Connection Admission Control algorithm, a Path selection algorithm and a Load Balancing algorithm has been considered in our tests.

The framework has been implemented as a Linux Kernel Module and it has been
installed in test-bed machines and in a legacy router[1] for performance evaluation. Fig. 3
shows three nodes connected together by means of a IEEE 802.11n link at 300 Mbit/s,
and two IEEE 802.3u links at 100 Mbit/s.

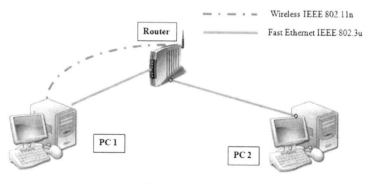

Fig. 3. Test scenario

---

[1] We have modified the firmware of a Netgear router (Gigabit Open Source Router with
Wireless-N and USB port; 453 MHz Broadcom Processor with 8 MB Flash memory and 64
MB RAM; a WAN port and four LAN up to 1 Gigabit/s) and "cross-compiled" the code, to
run the framework on the Router.

To test the technology handover performances a FTP download session (file size 175 MB) has been conducted on the Ethernet link. After approximately 10s, one extremity of the Ethernet cable has been physically disconnected from its socket and the flow has been automatically redirected onto the wireless link thanks a context-aware decision taken by the Cognitive connectivity enabler. Switching on the Wi-Fi link causes more TCP retransmissions and an increased transfer time. This is natural, since Ethernet and Wi-Fi have different throughputs. Without the cognitive framework, it is evident that the FTP session would not be terminated at all. As shown in Fig. 4, the experimented handover time is around 240 ms, during which no packet is received. The delay is influenced by the processing time that the framework module spends in triggering and enforcing the solutions evaluated by the path selection routines implemented in the cognitive connection enabler.

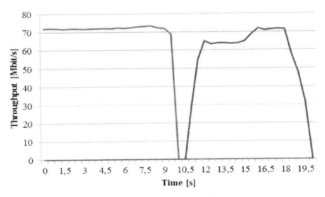

Fig. 4. Technology handover

5   Conclusions

This paper proposes a novel reference architecture for the Future Internet, with the aim to provide a solution to overcome current Internet limitations. The proposed architecture is based on Cognitive Modules which can be transparently embedded in selected network entities. These Cognitive Modules have a modular organization which is claimed to be flexible and scalable, thus allowing a smooth migration towards the Future Internet and, at the same time, allowing to implement only the needed functionalities in a give scenario. Interoperation among heterogeneous entities is achieved by means of their virtualization, obtained thanks to the introduction of Semantic Virtualization Enablers. At the same time, the Cognitive Enablers, which are the core of the Cognitive Managers, can *potentially* benefit from information coming from all layers of all networks and can take consistent and coordinated context-aware decisions impacting on all layers. Clearly, which Cognitive Enabler have to be activated, which input information has to be provided to the Cognitive Enabler, the algorithms the Cognitive Enabler will be based on, have all to be carefully selected case by case; nevertheless, the proposed architecture has an inherent formidable point of strength in the concentration of all management and control tasks in a single technology/service/content independent layer, opening the way, in a natural fashion, to inter-network, inter-layer cross-optimizations.

References

1. Talbot, D.: The Internet is broken, Technology Review, December 2005-January 2006 (2006), http://www.technologyreview.com/article/16356/
2. FISS: Introduction to content centric networking, Bremen, Germany, 22 June (2009), http://www.comnets.uni-bremen.de/typo3site/uploads/media/vjCCN-FISS09.pdf
3. Miller., R.: Vint Cerf on the Future of the Internet. The Internet Today, The Singularity University (2009), http://www.datacenterknowledge.com/archives/2009/10/12/vint-cerf-on-the-future-of-the-internet/
4. National Science Foundation: Networking Technology and Systems, NeTS (2008), http://www.nsf.gov/pubs/2008/nsf08524/nsf08524.htm
5. National Science Foundation: Network Science and Engineering, NetSE (2010), http://www.nsf.gov/funding/pgm_summ.jsp?pims_id=503325
6. BNN Technologies: GENI: Exploring networks of the future, http://www.geni.net/ (2010)
7. Stanford Clean Slate: http://cleanslate.stanford.edu/ (2011)
8. G-Lab: http://www.german-lab.de/home/ (2008)
9. Jutand, F.: National Future Internet Initiatives - GRIF (France), http://www.francenumerique2012.fr/ (2010)
10. AETIC: Internet del Futuro, http://www.idi.aetic.es/esInternet/ (2008)
11. ICT FP7 Research: Future Internet Research & Experimentation (FIRE), http://cordis.europa.eu/fp7/ict/fire/ (2010)

Title Model Ontology for Future Internet Networks

Joao Henrique de Souza Pereira[1], Flavio de Oliveira Silva[1],
Edmo Lopes Filho[2], Sergio Takeo Kofuji[1], and Pedro Frosi Rosa[3]

[1] University of Sao Paulo, Brazil
joaohs@usp.br, flavio@pad.lsi.usp.br, kofuji@pad.lsi.usp.br
[2] Algar Telecom, Brazil
edmo@algartelecom.com.br
[3] Federal University of Uberlandia, Brazil
pedro@facom.ufu.br

Abstract. The currently Internet foundation is characterized on the interconnection of end-hosts exchanging information through its network interfaces usually identified by IP addresses. Notwithstanding its benefits, the TCP/IP architecture had not a bold evolution in contrast with the augmenting and real trends in networks, becoming service-aware. An Internet of active social, mobile and voracious content producers and consumers. Considering the limitations of the current Internet architecture, the envisaged scenarios and work efforts for Future Internet, this paper presents a contribution for the interaction between entities through the formalization of the Entity Title Model.

Keywords: Entity, Future Internet, Ontology, Title Model

Introduction

The Internet of today has difficulties to support the increasing demand for resources and one of the reasons is related to the restricted evolution of the TCP/IP architecture since the 80s. More specifically, the evolution of the layers 3 and 4, as discussed in [23]. The commercial usage of Internet and IP networks was a considerable obstacle to the improvements in the intermediate layers in this architecture.

The challenges to Future Internet Networks are the primary motivation to this paper and the cooperation in the evolution of computer networks, specifically in the TCP/IP intermediate layers, is another one. The purpose is to present the Entity Title Model formalization, using the OWL (Web Ontology Language), to collaborate with one integrated reference model for the Future Internet, including others projects efforts.

This paper is organized as follows: Section 1 presents works in the area of Future Internet and ontology in computer systems. Section 2 describes the concepts of the Entity Title Model and the ontology at network layers. Finally, section 3 presents some concluding remarks and suggestions for future works.

J. Domingue et al. (Eds.): Future Internet Assembly, LNCS 6656, pp. 103–114, 2011.

1   Future Internet Works

A Future Internet full of services requirements demands networks where the necessary resources to service delivery are orchestrated and optimized efficiently. In this research area there are extensive number of works and projects for the Future Internet and some of these are being discussed in collaboration groups like FIA, FIND, FIRE, GENI and others [10, 11, 14, 31, 32].

At this moment, several research groups are working towards a Future Internet reference architecture and the Title Model ontology is a contribution to this area. Projects, among others, like 4WARD, ANA, PSIRP and SENSEI proposes new network architectures which contains collaborative relations to the model proposed by this paper [1] [3] [8] [30] [33]. The 4WARD Netinf concept is related to the Domain Title Service (DTS) proposed in [26] and its horizontal addressing can leverage Netinf concept. The DTS can deal with the information and with the context of the consumers taking into account their communication needs at each context, supporting their change over time.

The Entity Title Model concepts can be used at the communications layer to the real world architecture envisaged by SENSEI [33] project, besides that, the concept of addressing by use of a Title is suitable for real world Internet and its sensor networks. The title concept can be used at the publish and subscribe view proposed by PSIRP [30] and used in conjunction with its proposed patterns providing new important inputs to the content-centric view of Future Internet.

1.1   Some Other Future Internet and Ontology Works

Studies and proposals for development of the intermediate layers of the TCP/IP architecture are being discussed since the 80s, but there is still no clear and definite perspective about which standard will be used in the evolution of this architecture.

In the area of the evolution of intermediate layers of the TCP/IP there are proposals as LISP (Locator Identifier Separation Protocol), which seek alternatives to contribute to the evolution of computer networks. In the proposed implementation of LISP there is low impact on existing infrastructure of the Internet since it can use the structure of IP and TCP, with the separation of Internet addresses into Endpoint Identifiers (EID) and Routing Locators (RLOC) [9].

In the area of next generation Internet there is also the works of Landmark developed by Tsuchiya, that proposed hierarchical routing in large networks and Krioukov work on compact routing for the Internet. Pasquini proposes changes in the use of Landmark with RoFL (Routing on Flat Labels), and flat routing in binary identity space. He also proposes the use of domain identifiers for a next-generation Internet architecture [21] [22].

Previous studies in RoFL were presented by Caesar who also made proposals in IBR (Identity-based routing) and VRR (Virtual Ring Routing) [7]. In the area of mobility on a next-generation Internet Wong proposes solutions that include support for multi-homing [36]. In this area, there are also proposals by Ford, who specifies the UIP/UIA (Unmanaged Internet Protocol) and UIA (Unmanaged Internet Architecture) [12].

Related to ontology, there are extensive studies in philosophy, whose concept of this term is assigned to Aristotle, who defines it as the study of "being as being". However, the name ontology was first used only in the seventeenth century by Johannes Clauberg [2]. In the area of technology its initial use was performed by Mealy in 1967 [20] and expanded especially in areas of artificial intelligence, database, information systems, software engineering and semantic web. In the technology area one of the most commonly used definitions is from Tom Gruber, who defines it as "the explicit representation of a conceptualization" [15].

In technology, the use of ontology is also associated with formalizations that allow technological systems to exchange concepts. For these formalizations there are extensive literature which defines different languages and tools. As examples of languages used there are DAML, OIL, KIF, XSLT, KM, Predicate Calculus of First Order, Propositional Logic, Ontolingua, Loom, and Semantic Web languages (RDF, RDFS, DAML+OIL, OWL SPARQL, GRDDL, RDFa, SHOE AND SKOS), among others [13].

For communication between network elements, ontology is usually used in the application layer, without extending to the middle and lower layers of computer networks. In this research area, this paper aims to contribute to advancing the use of ontology to the intermediate layers as a collaborative proposal for the Future Internet.

2   Ontology at Network Layers

Ontologies can use layer model or distinct architectures, however, in general, they remain restricted to the application layer. For example, the architecture of the Web Ontology Language defined by W3C, presented in Fig. 1 extracted from [17], is confined in the application layer of the TCP/IP architecture.

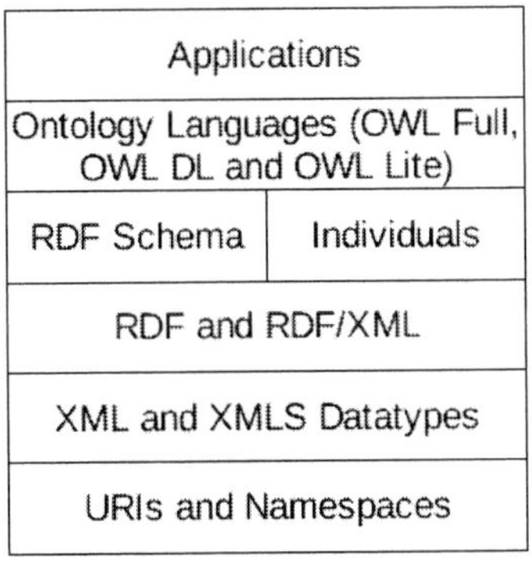

Applications
Ontology Languages (OWL Full, OWL DL and OWL Lite)
RDF Schema / Individuals
RDF and RDF/XML
XML and XMLS Datatypes
URIs and Namespaces

Fig. 1. Architecture of Web Ontology Language [17].

In the use of TCP/IP, there are limitations concerning the application layer informing its needs to the transport layer. This occurs because in the TCP/IP architecture there are rules defined in the specification of the transport and network layers protocols to establish communication among the network elements. For example, the applications can select the protocol UDP or TCP, according to delivery guarantee, but they cannot tell the transport layer its needs of encryption or mobility.

It is possible to change the paradigm of client-server communication and the structure of the intermediate layers of the TCP/IP, so that the communication networks have expansion possibilities to support the needs of the upper layer. For so, one solution is to use an intermediate layer conceptually capable of communicating semantically with the top layer and translating these needs in the communication with and between the lower layers. A possibility proposed by the Entity Title Model.

2.1   Entity Title Model Concepts and Semantics

The use of ontology for model formalization needs clear definitions of the used concepts to build properly the ontology of the approached model. Thus, for the Entity Title Model its main terms concepts are:

Entity: Element whose communication needs can be semantically understood and supported by the service layer and subsequent lower Link and Physical layers. Examples of real world entities in the title model are: application, content, host, user, cloud computing and sensor networks. The notion of entity in the Title Model differs from the notion of resources in some relevant literature, as the entity here is a communication element and not one resource in a network. In this concept, the entities in the Title Model are not obligated to provide resources and can consume them. For example, one user, that demands resources, is one communication entity in the Title Model. Also, applications that do not offer resources, but demand some ones, are entities.

However, for an ontology there is correlation of the terms "Entity" and "Thing", as described in [13], where "Entity" or "Thing" in an ontology refers to its first class, which is the superclass of all other classes.

For the taxonomy of the ontology, the classification of an entity in the Entity Title Model can expand the categories as application, content, cloud, sensor, host, user. Also can be created other kinds of classification, such as hardware, software and network, among others. Some one of them (not all) can be used as resources in others relevant literature.

As the root superclass of one ontology is "Entity" or "Thing" the Entity Title Model ontology designates a conceptually different "Entity" of this model, which in turn is an communication element that have its communication needs understood and supported by computer networks. For example, in this taxonomy the class "layer" is a subclass of "Thing" and neither this class nor its subclasses are entities to the Entity Title Model, although the class "Layer" is an entity to the concept of ontology, in general.

Title: It is the only designation to ensure an unambiguous identification. An unique identity. The Entity Title Model proposes that the use of titles of applications, specified in the ISO-9545/X.207 recommendation, be extended to the other communication entities of the computer networks. According to this recommendation, the ASO-title (Application Service Object-title), which are used to identify with no ambiguity the ASO in an OSI environment, consists of AP-title (Application Process title) which, by nature, addresses the applications horizontally [16].

This work broadens the use of the title from the applications with the unification of addresses by using the AP-title and also suggests that the intermediate layers support the needs of the entities in a better way, with the purpose of improving the addressing of internet architecture by horizontal addressing and facilitate communication among the entities and with the other layers [24]. Not to use a separate classification for "user title", "host title" and "application title", which would reduce the flexibility of its use in other addressing needs (eg, grid title, cluster title and sensor network title), this model defines de use of the single designation "entity title" or simply "title", whose goal is to identify an entity, regardless of which one it is.

Entity Title: It is the sole designation to ensure the unambiguous identification of a communication element whose needs may be semantically understood and supported by the service layer and subsequent lower link and physical layers. Examples of entity title are: Digital Signature, DNA, e-mail address and hash.

Layer: It designates the concept to explain the general ideas of abstraction of the complexities of a problem under its responsibility. A layer deals internally with the details under its responsibility and has an interface with the adjacent neighboring layers. The Entity Title Model layers are: Physical, Link, Service and Entity.

Entity Title Model: It is the 4-layer model that defines the entity layer as the upper layer, whose communication needs are semantically understood and supported by the service layer (intermediate layer) that has the physical and link layers as subsequent lower ones.

Link: It is the connection between two or more entities.

Physical: It is a tangible material in a computer network, such as: cables, connectors, general optical distributor, antenna, base station and air interface.

Service: It is the realization of the semantics of the need of a communication element, based on "service concept" presented by Vissers, where users communicate with each other through a "Service Provider", whose interface is accomplished by a "Service Access Point" (SAP) [34]. In the Entity Title Model the "entity" is the "user" of the Vissers service model and the "Service Layer" is the "Service Provider". In the Entity Title Model the SAP is formalized with the use of ontology, which in this work was built in OWL.

Needs: They are functionality or desirable technological requirements, essential or indispensable.

Entity Needs: They are functionality or desirable technological requirements, essential or indispensable for the communication elements whose needs can be semantically understood and supported by the service layer and subsequent lower link and physical layers. Examples of needs of the entities are: Low latency, low jitter, bandwidth, addressing, delivery guarantee, management, mobility, QoS and security.

The changing needs of the entities may vary depending on the context of the entities in communication, and also the context of communication itself. The contexts can be influenced by space, time, specific characteristics of entities, among other forms of influence. Discussion on the changing needs are presented in [24], where associations between elements of communication may vary according to their desired needs and their variation in time.

Regardless of the time, the nature of communication can also influence the desired values for the facets. For example, to transfer data from a file, or content of email / instant message, it is necessary to have delivery guarantee in communication. On the other hand, for an audio or video communication in real time, it will not necessarily be important the delivery guarantee, as other needs will be most desirable, such as low jitter and low latency.

Horizontal Addressing: Possibility of having neighborhoods regardless of physical or logical location of entities in computer networks, without the need of reserved bandwidth, networks segmentation, specific physical connections or virtual private network.

Entity Layer: This is the layer that has the responsibility on the part of the problem corresponding to the elements of communication, whose needs can be understood and semantically supported by the service layer and subsequent lower link and physical layers.

Service Layer: This is the layer that has the responsibility to understand the needs of the entity layer and translate them into functionality in computer networks.

Link Layer: This is the layer that has the responsibility to establish the link between two or more entities and ensure that data exchange occurs at the link level and takes place according to the understanding made by the service layer.

Physical Layer: This is the layer that has the responsibility of the complexities of real-world tangible materials. For example, this layer has responsibility for: The levels of electrical, optical and electromagnetic signals, shape of connectors and attenuation.

Domain Title Service (DTS): It is a domain able to understand and record instances of entities and their properties and needs, facilitating communication services among them. This domain has world-wide coverage and hierarchical scalability formed by elements of local communication, masters and slaves, similar to DNS (Domain Name System). The DTS does the orchestration of the entities communication, as showed in Fig. 2.

2.2   Cross Layer Ontology for Future Internet Networks

For intermediate semantic layer, this work did the creation of an ontology for the Entity Title Model, considering others works and projects efforts for Future Internet, as 4WARD, Content-Centric, User-Centric, Service-Centric and AutoI

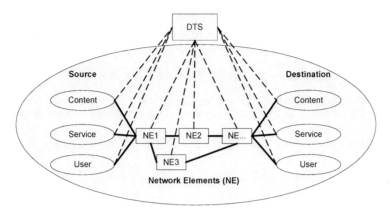

Fig. 2. Entities Communication Orchestrated by the DTS.

[4] [28]. This ontology also supports the proposal of Horizontal Addressing by Entity Title, presented in [26], as well as the semantic approaching cross layers for the Future Internet.

The Horizontal Addressing by Entity Title has limitations related with the communications needs formalization and standardization, and also has limitations with the collaboration with others Future Internet projects efforts. The reason is because the solution for horizontal addressing and communication needs was represented and supported using the Lesniewski Logic [18] [29]. The benefits for the use of the propositional logic for network formalization is the implementation facility in software and hardware. However, in a collaborative effort to others Future Internet works, the Entity Title Model has better contributions by the use of a more expressive and standardized representation language.

Also, this Model is more complete than the solution for just the horizontal addressing, as it formalize the concepts to the intermediate layers interwork and support to approaches like the Content, Service and User Centric. In addition, it permits semantic communication cross layers to contribute with, for example, the autonomic management, as the AutoI works. These are also limitations from the previous Horizontal Addressing by Entity Title works with value added by the Entity Title Model.

Others actual researches show the use of ontologies at different network layers like: OVM (Ontology for Vulnerability Management) to support security needs [35]; NetQoSOnt (Network QoS Ontology) to meet the needs of service quality [27]; OOTN (Ontology for Optical Transport Networks) for use in the lower layers [6]; Ontology for management and governance of services [5]. However, these studies does not use the ontology to the formalization of concepts for replacement of the intermediate layers of the TCP/IP (including its major protocols such as IP, UDP and TCP).

In the Entity Title Model, entities, regardless of their categories, are supported by a layer of services. It is very important to highlight that the name "service" in the "service layer", does not intend to conflict with the traditional meaning of "service concept" as, in general, the layers also expose services to other layers. In its concept, the service layer is able to understand and meet the entities needs. Fig. 3 shows the Entity Title Model layers compared with the TCP/IP and the extension of the semantic power, cross layers, enabled by the Entity Title Model.

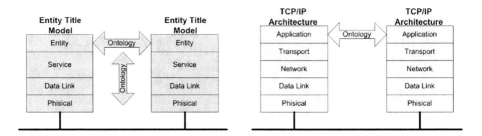

Fig. 3. Semantic Extension Cross Layers in the Title Model and the Semantic in TCP/IP.

The relationship between Entity, Services and Data Link layers are made by the use of concepts directly represented in OWL. For the communication between the layers running in a Distributed Operating System, without the traditional sockets used in TCP/IP, is used the Raw Socket to enable the communication [19].

The following OWL sample code shows one use case example for distributed programming, where the application entity with title Master-USP-1 sends its needs to the Service Layer. These needs include: Communication with Slave-USP-A; Payload Size Control equal to 84 Bytes; and; Delivery Guarantee request. In this context, this need is informed, to the Service Layer, by the direct use of the Raw Socket to communicate with the Distributed Operating System, without the use of IP, TCP, UDP and SCTP.

```
<owl:Thing rdf:about="&TitleModel;Distributed_Programming_LAM_MPI">
      <rdf:type rdf:resource="&TitleModel;Entity"/>
      <rdf:type rdf:resource="&owl;NamedIndividual"/>
      <Application_Title>Master_USP_1</Application_Title>
      <Slave_Title>Slave_USP_A</Slave_Title>
      <Payload_Size_Control>84 Bytes</Payload_Size_Control>
      <DeliveryGuarantee rdf:datatype="&xsd;boolean">Yes</
          DeliveryGuarantee>
      <rdfs:comment>Example of the Entity Title Model to support
          distributed programming needs.
      </rdfs:comment>
      <Has_Need rdf:resource="&TitleModel;
          Distributed_Programming_LAM_MPI"/>
</owl:Thing>
```

By this semantic information, the Service and Data Link layers can support the distributed programming communication using different approaches, as the addressing proposal presented in [25]. However, the use of the Entity Title Model is independent from the addressing way used. For example, the works related to Generic Path, Information Channels, RoFL and LISP can use it, but some of them, as RoFL and LISP, should change their structure to semantically support the entities needs and identification, unified in title, and not only the addressing of hosts or applications. Others works as, for example, 4WARD, AutoI OSKMV planes (Orchestration, Service Enablers, Knowledge, Management and Virtualisation planes) and the Content-Centric can use this model collaboratively.

The context name in the Content-Centric project is expanded by the title concept in the Entity Title Model, as in this model it is possible to address contents and also others entities. This can benefits the Content-Centric works to address the content by name (or title) as, in some situations, one user may need the Content directly from Services or from other Users (thoughts). In this perspective, the Entity Title Model and its ontology can contribute to converge some Future Internet projects, as the Content, Service and User Centric works, monitored and managed by the OSKMV planes using semantics cross layers, and not only in the application layer as happen in the TCP/IP architecture.

In this example for the contribution with the Content, Service and User Centric works, in the Title Model it is possible the unification of the different entities address in the Future Internet. This means that application, content, host and user can have its needs supported and can be located by its title.

By this possibilities, this work aims to contribute with the discussions for a collaborative reference model in the Future Internet, that includes different categories of communication entities, and its needs. One basic sample of the taxonomy for this "Entity" concept is showed in the Fig. 4, extracted from the Title Model ontology built in Protégé.

In this taxonomy "title" is one facet of the concept Entity and one individual of Entity has "title".

For the service layer to support semantically the entities needs this work uses the Web Ontology Language, so that the Entity layer can communicate semantically with the Service layer, which translates this communication in functionality

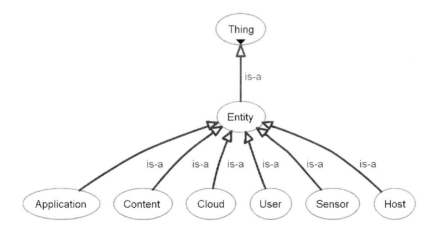

Fig. 4. Entity Taxonomy in the Title Model.

through the Physical and Link layers. OWL was used because of its significant use in current and future trend, since its adoption and recommendation by the W3C [17].

By the Entity Title Model, some current needs of the applications are to be met in a more natural and less complex way. For example, once the title addresses the entities horizontally, the mobility on the Internet becomes natural, since there is no longer the hierarchy of segments of the network and sub network that occurs in the IP address with the use of masks. By this, the coupling between the neighborhoods are reduced, so, an entity and its neighbors can be naturally distributed anywhere in the world.

Besides reducing the complexity of the multiple addresses used in the current architecture, the use of the Entity Title Model solves the problem of the number of possible addresses, as it makes an unlimited number of addresses, since in this proposal each entity has an unique identification, through its title, without defining the amount of possible characters, or bits.

3   Conclusion

Studies in ontology in the technology area are used, in most part, in the application layer of TCP/IP architecture, with few studies in the lower and middle layers of this architecture. In this scenario, this work contributes to the use of ontology in the middle layers of the Internet, with the proposal of semantic formalization, in computer networks, for the Entity Title Model.

Therefore, it is possible the approaching between the upper and lower layers. As a result there is improvement in the exchange of meanings between the layers through the use of Entity and Service layers. This is a possible contribution to the Future Internet efforts and projects like AutoI, Content-Centric, User-Centric, Service-Centric, 4WARD and others. Also, is a possibility for the collaborative discussions about the reference model related to these, and others, Future Internet efforts.

As future work there will be continued the development of this ontology and its collaborative perspective with others Future Internet efforts and projects. It is suggested to extend discussions and studies concerning the unique identification of the entities and the formalization of security mechanisms for the Entity Title Model. Also, the interoperability, scalability and stability test cases for this model.

It is also suggested the continuity of studies and discussions on the use of semantic representation languages in place of protocols in the lower and middle layers of computer networks, thereby defining the communication architecture whose study go over the definitions in the area of protocols architecture.

Acknowledgment. Part of the results of this work received contributions from the MEHAR Project researches. The authors would like to thank the MEHAR Project members for all discussions and collaboration.

References

[1] 4WARD: 4WARD. European Union IST 7th Framework Programme, http://www.4ward-project.eu (2011)

[2] Abbagnano, N.: Dicionário de Filosofia, Trad. Alfredo Bosi. Martins Fontes (2000)

[3] Project, A.A.: - Autonomic Network Architectures. European Union IST 6th Framework Programme, http://www.ana-project.org (2011)

[4] AUTOI: Autonomic Internet Project. European Union IST 7th Framework Programme (2011)

[5] Baiôco, G., Costa, A., Calvi, C., Garcia, A.: IT Service Management and Governance - Modeling an ITSM Configuration Process: a Foundational Ontology Approach. Symposium on Integrated Network Management - SINM/IFIP/IEEE (2009)

[6] Barcelos, P.: et al.: OOTN - An Ontology Proposal for Optical Transport Networks. International Conference on Ultra Modern Telecommunications, IEEE Xplore, Print ISBN: 978-1-4244-3942-3 (2009)

[7] Caesar, M.: Identity-based routing. Technical Report No. UCB/EECS-2007-114 (2007)

[8] D'Ambrosio, M., Marchisio, M., Vercellone, V., Ahlgren, B., Dannewitz, C.: Second NetInf architecture description, http://www.4ward-project.eu (2010)

[9] Farinacci, D., Fuller, V., Meyer, D., Lewis, D.: LISP Project - Locator/ID Separation Protocol. Network Working Group, draft-ietf-lisp-06 (2010)

[10] FIND: Future Internet Design Program. National Science Foundation, http://www.nets-find.net (2011)

[11] FIRE: FIRE White Paper. Future Internet Research and Experimentation (2009)

[12] Ford, B.: UIA: A Global Connectivity Architecture for Mobile Personal Devices. Ph.D. Thesis. Massachusetts Institute of Technology, Dep. of Electrical Eng. and Comp. Sci (2008)

[13] Gasevic, D., Djuric, D., Devedzic, V.: Model Driven Engineering and Ontology Development, 2nd edn. Springer, Heidelberg (2009)

[14] GENI: Global Environment for Network Innovation Program. National Science Foundation, http://www.geni.net (2011)

[15] Gruber, T.: Toward Principles for the Design of Ontologies Used for Knowledge Sharing. International Journal of Human and Computer Studies, 43(5–6): 907–928 (1995)

[16] ITU-T: Information Technology - Open Systems Interconnection - Application Layer Structure. Recommendation X.207 - ISO/IEC 9545:1993 (1993)

[17] Lacy, L.: OWL: Representing Information Using the Web Ontology Language. Trafford (2005)

[18] Lesniewski, S.: Comptes rendus des séances de la Société des Sciences et des Lettres de Varsovie. pp. 111–132 (1930)

[19] Malva, G.R., Dias, E.C., Oliveira, B.C., Pereira, J.H.S., Kofuji, S.T., Rosa, P.F.: Implementação do Protocolo FINLAN. In: 8th International Information and Telecommunication Technologies Symposium (2009)

[20] Mealy, G.: Another look at data. In: Proceedings of the Fall Joint Computer Conference. AFIPS November 14-16, Volume 31, pp. 525–534. Thompson Books, Washington and Academic Press, London (1967)

[21] Pasquini, R., Paula, L., Verdi, F., Magalhães, M.: Domain Identifiers in a Next Generation Internet Architecture. In: IEEE Wireless Communications and Networking Conference - WCNC (2009)

[22] Pasquini, R., Verdi, F., Magalhães, M.: Towards a Landmark-based Flat Routing. In: 27th Brazilian Symposium on Computer Networks and Distributed Systems - SBRC (2009)

[23] Pereira, J.H.S., Kofuji, S.T., Rosa, P.F.: Distributed Systems Ontology. In: IEEE/IFIP New Technologies, Mobility and Security Conference (2009)

[24] Pereira, J.H.S., Kofuji, S.T., Rosa, P.F.: Horizontal Address Ontology in Internet Architecture. In: IEEE/IFIP New Technologies, Mobility and Security Conference (2009)

[25] Pereira, J., Sato, L., Rosa, P., Kofuji, S.: Network Headers Optimization for Distributed Programming. In: 9th International Information and Telecommunication Technologies Symposium (2010)

[26] Pereira, J.H.S., Kofuji, S.T., Rosa, P.F.: Horizontal Addressing by Title in a Next Generation Internet. In: IEEE International Conference on Networking and Services p. 7 (2010)

[27] Prudêncio, A., Willrich, R., Diaz, M., Tazi, S.: NetQoSOnt: Uma Ontologia para a Especificação Semântica de QoS em Redes de Computadores. In: 14o Workshop de Gerência e Operação de Redes e Serviços - WGRS-SBRC (2009)

[28] Rubio-Loyola, J., Serrat, J., Astorga, A., Chai, W.K., Galis, A., Clayman, S., Mamatas, L., Abid, M., Koumoutsos, G.: et al.: Autonomic Internet Framework Deliverable D6.3. Final Results of the AutonomicI Approach. AutoI Project (2010)

[29] Souza, J.: Lógica para Ciência da Computação. Campus (2008)

[30] Trossen, D., Nikander, P., Visala, K., Burbridge, T., Botham, P., Reason, C., Sarela, M., Lagutin, D., Koptchev, V.: Publish Subscribe Internet Routing Paradigm - PSIRP. Final Updated Architecture, Deliverable D2.5 (2010)

[31] Tselentis, G., et al.: Towards the Future Internet - A European Research Perspective. IOS Press, Amsterdam (2009)

[32] Tselentis, G., et al.: Towards the Future Internet - Emerging Trends from European Research. IOS Press, Amsterdam (2010)

[33] Tsiatsis, V., Gluhak, A., Bauge, T., Montagut, F., Bernat, J., Bauer, M., Villa-longa, C., Barnaghi, P., Krco, S.: The SENSEI architecture-Enabling the Real World Internet. In: Towards the Future Internet, pp. 247–256. IOS Press, Amsterdam (2010)

[34] Vissers, C., Logrippo, L.: The Importance of the Service Concept in the Design of Data Communications Protocols. In: Proceedings of the IFIP WG6 1, 3 (1986)

[35] Wang, J., Guo, M., Camargo, J.: An Ontological Approach to Computer System Security. Information Security Journal: A Global Perspective (2010)

[36] Wong, W.: et al.: An Architecture for Mobility Support in a Next Generation Internet. In: The 22nd IEEE International Conference on Advanced Information, Networking and Applications - AINA (2008)

Part II:

Future Internet Foundations: Socio-economic Issues

Introduction

Information and Communication Technologies (ICT) provide in recent years solutions to the sustainability challenge by, *e.g.*, measuring impacts and benefits of economic activity via integrated environmental monitoring and modeling, by managing consequences, and by enabling novel low-impact economic activities, such as virtual industries or digital assets. In turn, ICT enables novel systems – in terms of technologies and applications – encouraging and generating socio-economic values. Additionally, these models address in many cases free-market forces, which may be likely ruled by governmental and regulatory acts. Thus, the inter-dependencies between global markets have never been greater and the global connectivity principle, underpinning the technology of the Internet, is particularly responsible for this accelerating trend.

Particularly, controlling and monetizing the evolution of the Internet and its vast application range is seen as a critical goal for most economic regions. Therefore, socio-economic aspects determine a highly important set of influencing factors, which are required to be understood for an in-depth and in-detail investigation of the economic viability and the social acceptability of modern technology and applications. While pure economic research as well as pure social research has been undertaken for decades, the combination of the two and its application to the new Internet – the one, which is rooted in the commercialization of the native research Internet of the early 90's – becomes an important element in investigating, estimating, and understanding the risks, challenges, and usability aspects of this network of networks.

As collected by the FISE (Future Internet Socio-economics) working group within the FIA on its wiki, the following general aspects of socio-economics, particularly in networking, are considered to be important: (1) The study of the relationship between any sort of economic activity (here networking in the areas of Internet-based and telecommunications-based communications for a variety of lower-level network/telecommunication as well as application-based services) and the social life of user (here, mainly addressing private customers of such services and providers offering such services); (2) Markets of Internet Service Providers (ISP) and Telecommunication Providers; (3) ISPs peering agreements and/or transit contracts; (4) Customer usage behavior and selections of content; (5) The investigation of emerging technologies and disruptive technologies, which effect the user/customer-to-provider relation; (6) The investigation of (European) regulation for e-services markets and security regulations; (7) The investigation of the physical environment of e-services in terms of availability, world-wide vs. highly focused (cities), and dependability for commercial services; and (8) The determination (if possible) of the growth of the Gross Domestic Product (GDP), providers' revenue maximization, and customers' benefits. While this collection cannot be considered complete, it clearly outlines that a combination of social and economic viewpoints on pure Internet-based networking is essential.

Thus, the full understanding and modeling of these socio-economic impacts on Internet communications particularly and the Internet architecture generally challenges networking research and development today. Economic effects of technical mechanisms in a given setup and topology needs to be investigated and benefits obtained by optimizing or even changing existing protocols may lead to more cost-effective approaches. Furthermore, the users' perspectives need to be taken into close consideration, since detailed and specific security demands, electronic identities, or Quality-of-Experience (QoE) will outline societal requirements to be met by technological support means, while being at the same time in contrast to simplicity and ease-of-operations of a variety of Internet-based services.

In this emerging area of research the specific view on the networking and transmission domain of the Internet had been taken as one starting point of socio-economic research for this FIA book. Thus, the content of these chapters on socio-economics of the Future Internet contains three views, where the decision of inclusion was based on two rounds of abstract reviews and on subsequent reviews of complete chapter proposals. The submission of in total six chapter proposals, addressing the socio-economics domain, has shown that the interest of such cross-disciplinary work and its relevance increases slowly. While the first socio-economic chapter addresses aspects (1), (2), (4), and (8) as above, the second one works on (5) and (8). Finally, the last chapter tackles aspects (2), (3), and (8).

The first chapter by I. Papafili et al. is entitled "Assessment of Economic Management of Overlay Traffic – Methodology and Results". Due to the fact that overlay applications as of today still generate large volumes of data, Internet Service Providers (ISP) need to address the problem of expensive interconnection-charges. Thus, a reduction of inter-AS traffic (Autonomous Systems), which crosses domain boundaries of competitors, was tackled by an incentive-based approach, since traditional traffic management approaches do not deal with overlay traffic effectively. To ensure a mutually beneficial situation for all stakeholders in a Future Internet scenario, the "TripleWin" investigations determine the key goal of Economic Traffic Management (ETM) mechanisms developed. Thus, this chapter outlines the methodology developed, applied, and evaluated under a variety of constraints, which results in a detailed discussion of various ETM mechanisms.

The second chapter by E. Eardly et al. was submitted with the title "Deployment and Adoption of Future Internet Protocols". Based on the assumption that many well-designed protocols designed for the Future Internet will fail – as it happened for the traditional Internet –, however, badly-designed ones are successful. Thus, the problem of protocols' deployability is addressed. In order to do so, a framework had been developed, which includes the investigation possibilities for deployment effects, adoption characteristics, and their respective mechanisms. In a case-based study, the Multipath TCP (Transmission Control Protocol) and the Congestion Exposure approach are evaluated applying the framework developed. In turn, this chapter concludes that a careful consideration of certain parameters can increase the likelihood that a newly developed protocol, as it happens currently for the Future Internet, can get adopted.

Finally, the third chapter by C. Kalgoris et al. is on "An Approach to Investigating Socio-economic Tussles Arising from Building the Future Internet". Based on the assumption that the Internet has evolved into a world-wide social and economic platform with a variety of stakeholders involved, the key motivations of each of them and their behavior has changed over the recent past dramatically. In turn, conflicts have emerged, which are determined by opposing and contradicting interests. While this general problem had been characterized as "tussle" in the literature, it was decided to investigate, classify, and develop an analysis framework for such tussles in the socio-economic domain of Internet stakeholders. In turn, the chapter outlines a new methodology, with which tussles are analyzed. Although a survey reveals that many tussles are known, neither of them are modeled in full nor even solved. Therefore, existing Future Networks projects in the FP7 program are identified for inputting existing tussles in order to provide for a structured analysis of social and economic aspects in a coherent and integrated manner on real-world examples.

Burkhard Stiller

Assessment of Economic Management of Overlay Traffic: Methodology and Results

Ioanna Papafili[1], George D. Stamoulis[1], Rafal Stankiewicz[2], Simon Oechsner[3], Konstantin Pussep[4], Robert Wojcik[2], Jerzy Domzal[2], Dirk Staehle[3], Frank Lehrieder[3], and Burkhard Stiller[5]

[1] Athens University of Economics and Business, Athens, Greece
[2] AGH University of Science and Technology, Krakow, Poland
[3] Julius-Maximilian Universität Würzburg, Würzburg, Germany
[4] Technische Universität Darmstadt, Germany
[5] University of Zürich, Zürich, Switzerland

Abstract. Overlay applications generate huge amounts of traffic in the Internet, which determines a problem for Internet Service Providers, since it results in high charges for inter-domain traffic. Traditional traffic management techniques cannot deal successfully with overlay traffic. An incentive-based approach that employs economic concepts and mechanisms is required in order to deal with the overlay traffic in a way that is mutually beneficial for all stakeholders of the Future Internet. This "TripleWin" situation is the target of Economic Traffic Management (ETM). A wide variety of techniques are employed by ETM for optimizing overlay traffic management considering performance requirements of overlay and underlay networks together with cost implications for ISPs. However, the assessment of ETM requires an innovative methodology. In this article this methodology is described and major results are presented as obtained accordingly from the evaluation of different ETM mechanisms.

Keywords: Economic Traffic Management; socio-economics; TripleWin; performance; cost; incentives; Internet Service Providers; overlays.

1   Introduction

Applications such as peer-to-peer (P2P) file-sharing and video-streaming generate huge volumes of traffic in the Internet due to their high popularity and large size of the files exchanged. This typically underlay-agnostic overlay traffic results in high inter-domain traffic, which implies significant charges for the Internet Service Providers (ISP). Individual optimization in the overlay (decisions made either at random or taking partly into account underlay information) and in the underlay (as in traditional Traffic Engineering) may lead to a sub-optimal situation, *e.g.*, involving traffic oscillations and degraded Quality-of-Experience (QoE) for the end users [1]. Therefore, an incentive-based approach is required that employs economic concepts and mechanisms to deal with the overlay traffic in a way that is incentive compatible for all parties involved, and, thus, leads the system to a situation that is mutually

J. Domingue et al. (Eds.): Future Internet Assembly, LNCS 6656, pp. 121–131, 2011.

beneficial for all end users, overlay providers and ISPs. The so-called "TripleWin" situation is the main target of Economic Traffic Management (ETM) [2] proposed by the SmoothIT project [3]. ETM aims at dealing with the performance requirements of traffic at both overlay and underlay levels, and at reducing ISP inter-connection costs.

SmoothIT has proposed a wide variety of ETM mechanisms aiming at this incentive-based optimization. The entire set of these mechanisms and the synergies identified among them are included in the framework called ETM System (ETMS). ETMS and its particular design choices and implementations offer several alternatives for addressing a selected set of the ALTO (Application-layer Traffic Optimization) requirements [4], formulated in the corresponding IETF working group. Thus, besides providing effective solutions for Internet at present ETM is deemed as applicable to the Future Internet, both conceptually and concerning specific ideas and mechanisms.

All approaches proposed within ETMS are classified in three main categories:

- *Locality Promotion* enables peers of an ISP domain to receive ratings of their overlay neighbors by an entity called SmoothIT Information Service (SIS). The rating is based on ISP-related factors, such as underlay proximity, and link congestion. An example is locality promotion based on BGP routing data.
- *Insertion of Additional Locality-Promoting Peers/Resources* involves (a) the insertion of *ISP-owned Peers (IoPs)* in the overlay or (b) the enhancement of the access rate of *Highly Active Peers (HAPs)* aiming at both the promotion of locality and faster content distribution. Both approaches, due to the offering of extra capacity resources, exploit the native self-organizing incentive-based mechanisms of overlays to increase the level of traffic locality within ISPs.
- *Inter-Domain Collaboration*, where collaboration with other domains, either source or destination ones, results in making better local decisions to achieve the aforementioned objectives, due to the extra information made available.

SmoothIT has investigated all of these categories and evaluated them with respect to their performance, reliability, and scalability properties. Here, the focus is laid on the first two categories; note that mechanisms and results discussed here constitute a subset of SmoothIT's investigations. The assessment of ETM requires an innovative methodology (cf. Section 2). The rest of the article is organized as follows: Section 3 deals with the assessment of locality promotion, Section 4 with the insertion of locality-promoting peers/resource, while in Section 5 presents concluding remarks.

2   Methodology of Assessment

The detailed studies undertaken to assess ETMS deployment evaluate how all three stakeholders (end users, service providers, and ISP) would benefit thereby and under which circumstances "TripleWin" arises. To attain this, an innovative methodology of assessment for the ETM mechanisms has been developed. The main constituents of this methodology are as follows:

- The methodology does not consider the optimization of a "total cost" metric for each case; separate objective metrics are used for each player to reflect their diverse requirements and related incentives to employ an ETM mechanism.
- Several evaluation scenarios have been defined to cover a possibility of different ETM deployment degree, popularity of ETM among end users, various swarm sizes, peer distribution among network domains, network topologies etc.
- A game-theoretic analysis is applied to study interactions of ISPs, regarding decisions whether or not to employ an ETM mechanism and the associated ISPs' interactions by taking into account the end user benefit as well.

From the ISP point of view, the ultimate confirmation of "win" is a monetary benefit from ETM deployment. It is, however, not possible to quantify this benefit, since many factors contribute to the overall balance, including deployment and operational costs for an ETM mechanism, savings resulting from inter-domain traffic reduction and the structure of interconnection tariffs, business models, marketing factors. Thus, the assessment methodology focused on another quantifiable metric, namely the inter-domain traffic reduction, since ISPs benefit mostly thereby. In experiments, the inter domain traffic reduction was measured directly (upstream and downstream traffic is evaluated separately) or assessed by a metric called Missed Local Opportunities (MLO). It is a measure of a degree of traffic localization. If a piece of content (*e.g.*, a chunk) is downloaded from a remote domain, although it is available locally, an event of MLO is counted. The less MLOs observed, the better localization is provided by ETM and, thus, an ISP "wins". This metric is used in an ongoing external trial with real users.

As a measure of "win" for end users QoE metrics are used. For file-sharing P2P applications the most important perceivable parameter is download time (or download speed). It can be strongly influenced by ETM mechanisms, both favorably and adversely. Users will use a given mechanism if this improves or, at least, preserves their download time. Ideally, this should be guaranteed on a per *individual* user basis. However, it is most often analyzed by comparing the average values of the metrics with and without an ETM mechanism. Such averages are taken over all peers in a swarm, or over subsets (*e.g.*, those that belong to the same AS). Main QoE metrics associated with video streaming applications taken into account in assessment of "win" are the probability of playback stalling, stalling time, and start-up delay. These can be influenced by ETM mechanisms as well.

Assessment of "win" for service providers is based on the content availability in the whole overlay (swarm). Another dimension of a service provider's "win" is a decreased traffic volume from its own content servers and reduced load of the servers, as well as an improved performance of the application, which should translate into increased popularity of the service. In reality, these issues often coincide with the objectives of the end users and possibly of the ISPs. Thus, this paper focuses hereafter on assessing win-win for the ISP and end users.

To obtain a reliable assessment of ETM mechanisms several evaluation scenarios have been defined:

- Various network topologies: triangle-shaped, star-shaped, "bike"-shaped, and reflecting a part of the real Internet topology, with a subset of ASes and inter-domain connections;

- Semi-homogeneous and heterogeneous distribution of peers belonging to a single swarm among ASes; i.e. both "small" and "large" ASes, based on measurements, are considered;
- Varied ETM deployment degree, and interaction between peers located in ASes with ETM deployed and peers located in ASes without ETM;
- Varied end user interest in and adoption of ETM: coexistence of users employing ETM and ones declining support, even within a single swarm;
- Different swarm sizes; and
- Various type of content: files and video of different sizes.

It was assessed whether each player achieves a *win, lose*, or *no-lose* situation. It was shown that in certain scenarios a player may benefit or lose depending on whether it implements an ETM mechanism or not, and it was argued that the outcome for a player may depend also on the decisions of other players. The assessment has been carried out by means of simulations. All simulations generate quantitative results for those scenarios considered, but mainly should lead to qualitative results regarding the efficiency of the ETM mechanisms. Nevertheless, certain approximate theoretical models have been defined and investigated numerically, all of which pertain to the simplest evaluation scenarios ([5], [6]). Thus, the attention is directed to simulations.

3   Locality Promotion

As a selected example for a locality promotion ETM mechanism, the BGP-based one uses the Border Gateway Protocol (BGP) routing information to rate potential overlay neighbors according to their BGP routing distance to the local, querying peer. To this end, metrics like the AS (Autonomous System) hop distance, the local preference value and, if implemented, the MED (Multi Exit Discriminator) value are used. This rating is supported at ETM-enabled clients with mechanisms of Biased Neighbor Selection (BNS) [7] and Biased Unchoking (BU) [8]. The respective mechanism is specified fully and implemented in the ETMS and its client releases. The results from the evaluation of this ETM mechanism have partly already been reported in [8-11].

Here, it shows results from a larger simulation study published in [11]. Specifically, a heterogeneous peer distribution is considered and it is based on live BitTorrent swarm measurements [12], leading to the evaluation of the effect of locality-awareness on each of the different user groups separately. This is a new methodology in contrast to related work, where average results or a cumulative density function for all peers is shown, and mainly homogeneous distributions are used.

In SmoothIT's evaluations, a star topology consisting of 20 stub-Autonomous Systems (AS) is applied. Peers and the ETMS servers, providing rating information, are located in these stub-ASes, which are interconnected via a hub-AS containing the initial seed. The access links of peers, which share a file of size 150 MB, have a typical ADSL connection capacity of 16 MBit/s downlink and 1 MBit/s uplink. The average number of concurrently online peers is between 120 and 200 depending on the scenario, which is a typical swarm size according to the measurements. Peers are distributed hyperbolically from AS 1 to AS 20 according to the distribution in [11], with AS 1 holding the largest share of the swarm, and AS 20 the smallest.

The SmoothIT client implementations of the locality-aware mechanisms BNS and BU as described in [8] are applied, comparing the performance of regular BitTorrent [13] with BGP-based locality promotion using both BNS and BU (BGPLoc). For more details about the scenario and the used simulator refer to [11].

In order to assess the performance from an ISP's perspective, the amount of inter-domain traffic is considered. In particular, all traffic entering or leaving an AS is considered as inter-AS traffic of the ISP the AS belongs to. This traffic was measured in intervals of one minute during the whole simulation and then averaged over one simulation run. Download times of peers for each AS are presented, mainly to judge the overlay performance from the user's point of view. Here, download times of all peers are averaged in one AS in one simulation run. For each parameter setting 20 simulation runs happened, and the average value over all runs for all observed variables are depicted. The confidence intervals for a confidence level of 95% are calculated and shown for all scenarios. Fig. 1 and 2 present those results observed.

Fig. 1 outlines that the locality-aware ETM reduces the inter-AS traffic for all ASes, most significantly for the large ASes. Here, it can be stated that a clear win for all providers exist, since they save costs. The amount of these costs depends on the actual agreement between the ISPs and the number of peers of a swarm the ISP holds.

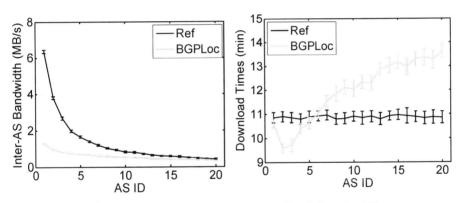

Fig. 1. Mean Inter-AS Bandwidth   **Fig. 2.** Download Times

On the other hand, the situation is not as simple when considering end users, cf. Fig. 2. Typically no-lose situations are the result in related work. However, this is true only on average. Due to a shift in upload capacity, which results from applying locality-awareness to a heterogeneously distributed swarm, there are large groups of peers that take longer to download the file in comparison to regular BitTorrent. From this perspective, it is not given that users would accept such a mechanism, since they cannot be sure not to lose from it. Furthermore, in another set of experiments investigating the dynamics of BGP-locality adoption, it can be seen that, if large ASes start employing the mechanism one-by-one, smaller ASes will be forced, too. However, to remedy this situation, another mechanism has been developed [11]: smaller ASes are grouped into a meta-AS, with peers in this meta-AS considered as local or at least as closer than the rest of the swarm by all peers in the group. Taking this additional mechanism into account, it can be concluded that the locality promotion mechanism in the ETMS may lead to a win-no lose situation, *i.e.*, a reduction in traffic and at least no reduction in the performance for the user, even if considering a more realistic scenario than typically used in related studies.

4   Insertion of Additional Locality-Promoting Peers/Resources

The insertion of additional locality-promoting peers/resources implies provision of resources by the ISP in terms of bandwidth and/or storage to increase the level of traffic locality within an ISP, and thus reduce traffic redundancy on its inter-domain links, and to improve performance experienced by the users of peer-to-peer applications. Two approaches, the insertion of ISP-owned Peers (IoPs), and the promotion of Highly Active Peers (HAPs), specifics of methodology employed for their assessment and major qualitative results obtained are described below.

4.1   Insertion of ISP-Owned Peers

The ISP-owned Peer (IoP) [14] is a centralized entity equipped with abundant resources, managed and controlled by the ISP. The IoP runs the overlay protocol [13] with minor modifications. It participates actively in the overlay, storing content, and aiming at subsequently seeding this content. Within the seeding phase, the IoP acts as a transparent and non-intercepting cache. As such it can employ existing cache storage management and content replacement policies already used by the ISP. Besides this, the IoP comprises a set of different mechanisms [9] to improve its effectiveness. Extensive performance evaluation has been conducted for all of them [15]. The main focus here is on selected results for the Unchoking Policy and the Swarm Selection. For the latter, a simulation setup with more than one swarm was utilized; this is rarely done in the literature and highlights SmoothIT's assessment methodology.

IoP without and with Unchoking Policy. In the underlay, a simple two AS topology is considered, where the two ASes connect with each other through a hub AS that does not have end users. The tracker and an original seeder are connected to the hub AS, while the IoP is always inserted in AS 1. Values for access bandwidth are similar to ones reported above; only the IoP offers 40 Mbit/s up and down. In the overlay, a single BitTorrent swarm is considered, with a 150 MB file size and about 120 peers simultaneously online in steady state. Peers arrive according to a Poisson distribution and they serve as seeds for a random time duration that follows the exponential distribution. For further details about the scenario and the simulator refer to [9]. Results are presented as average values over 10 simulation runs along with their corresponding 95% confidence intervals.

Three cases are evaluated: (1) no IoP insertion (no IoP), (2) IoP, and (3) IoPUP (with Unchoking Policy). Fig. 3 shows high savings on incoming inter-domain traffic of AS1 due to the IoP, but an increase of the outgoing traffic due to the data exchange also with remote peers; however, IoPUP achieves also outgoing traffic reduction which could imply monetary savings under more charging schemes compared to the previous case. In Fig. 4, peers' QoE is significantly improved when the IoP is inserted; however, when an IoPUP is inserted, performance is slightly degraded but still improved compared to the no IoP case. Finally, it should be noted that it has been verified (by running couples simulations) that users benefit with respect to performance not just on the average but 95% of them on an individual basis too.

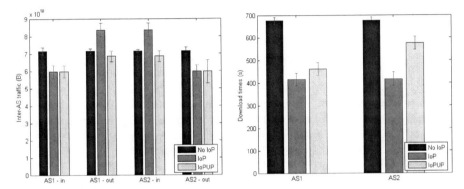

Fig. 3. Mean Inter-AS Bandwidth   **Fig. 4.** Download Times

Swarm Selection. In the underlay considered the same setup was applied; only a higher capacity for the IoP, i.e. 50 Mbit/s, is assumed. The overlay assumes two swarms, A and B, that are specified by the three set-up parameters file size, mean inter-arrival time, and mean seeding time. Peers from both swarms exist either in AS 1, or in AS2, or in both. The default file size is 150 MB, the leechers' mean inter-arrival time is meanIAT = 100 s, and the mean seeding time meanST = 600 s. To study the impact of the three factors, those parameters are tuned only for swarm A as reported in Table 1. For swarm B always default values are employed. For brevity reasons, the case of IoP with policy is not presented here; only results for incoming inter-domain traffic are shown. Further details on this scenario and the simulator are available in [15], [16].

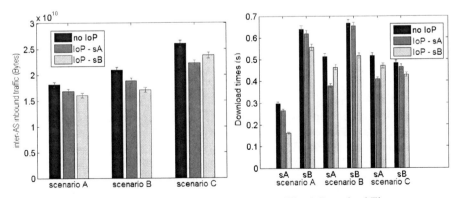

Fig. 5. Mean Inter-AS Bandwidth   **Fig. 6.** Download Times

Table 1. Evaluation Scenarios for the Swarm Selection

Scenario	A	B	C
Modified parameters	File Size: 50 MB	meanIAT: 300.0 s	meanST: 200.0 s

Fig. 5 and Fig. 6 present results for the inter-domain traffic and for the peers' download times, respectively. It can be observed that the impact on inbound inter-AS traffic of AS1 is higher when the IoP has joined a swarm with either i) larger file size, or ii) lower mean inter-arrival time, or iii) lower mean seeding time; namely a swarm with higher capacity demand. Performance is always improved since no policy is employed; in each case, higher improvement is observed for the peers of the swarm that the IoP joins. It can be concluded that the IoP provides a win-win situation, if it only uploads locally. Due to the additional upload capacity in the swarm, users benefit from this ETM mechanism, while the inter-AS traffic of the ISP employing the IoP is reduced. However, the efficiency of the mechanism depends on the characteristics of the swarm. Since the IoP can only provide limited resources, the decision about which swarms to join is an important aspect of this ETM mechanism.

A similar issue regarding the Bandwidth Allocation to swarms the IoP has joined has been observed. The effectiveness of this module depends on the number of local leechers and their bandwidth demand, thus, making it important to take this additional metric into account.

4.2   Promotion of Highly Active Peers

The Highly Active Peer (HAP) ETM mechanism [17] aims at promoting a more co-operative behavior among overlay peers. If a peer cooperates with the operator by being locality-aware (e.g., by following SIS-recommendations), the operator may increase the upload and download bandwidth of this peer. Additionally, the operator may consider other factors when selecting the peers to promote, such as their contribution to the overlay. In order to provide extra upload and download bandwidth to the peers, the ISP should employ NGN capabilities in its network.

Here, a dynamic HAP operation is considered, which implies instantaneous measurements and an instantaneous reaction of the ISP regarding peer promotion. A static case operates at a longer time range of several hours or even days and was evaluated in a peer-assisted video-on-demand scenario as described in [12].

The main assumption of the HAP ETM mechanism is that by promoting locality-aware highly active peers can achieve a win-win situation. The main evaluation goal is to show by simulations that the HAP ETM mechanism allows for decreasing download times of peers that want to become HAPs (due to the extra download bandwidth offered to them). Thus, it is shown that this ETM mechanism provides the right incentive for peers to behave cooperatively, according to ISP goals.

To evaluate the mechanism a triangle topology (three ASes connected with one another) is employed. In each AS, there are 100 peers. AS2 contains 6 initial seeders. The HAP mechanism is deployed only in AS1. The other ASes do not use ETM at all. Further details about the scenario and the used simulator are available in [15].

Fig. 7 presents the calculated mean download times of a certain content for the peers in AS1 with respect to the number of HAPs present in the network. The first case, *i.e.*, *No SIS* is the standard overlay scenario with no ETM mechanisms. '0 HAPs' means that Highly Active Peers were not present; however, the locality promotion mechanism provided by the SIS was used by all peers in AS1. As it can be seen in Fig. 7, the largest difference from the peer point of view is observed if the locality concept is implemented, but do not promote any peer to HAP. Just by introducing the basic locality promotion mechanism, the mean download time decreases significantly (see difference between 'No SIS' and '0 HAPs'); this is due to the fact that peers in AS1 share their resources among themselves, in contrast to the rest of peers, which share their upload capacity with the complete swarm. Afterwards, the download time can be further reduced by increasing the number of active HAPs. This phenomenon can be justified, since each HAP injects more bandwidth to the network, and, therefore, more bandwidth is overall available for peers. This is especially visible when comparing the '100 Peers/AS' with '200 Peers/AS' scenarios (left and right bars, respectively). In the former case, the mean download time is reduced more than in the latter. This is due to the fact that the injection of, say, 20 HAPs increases the available bandwidth by 20% if the number of peers in the AS is 100, but only by 10% when the number of peers in the AS is 200. Therefore, the relative increase in bandwidth is more significant when less peers are present in the AS, hence the difference in download time. The results in Fig. 7 clearly show that end-users benefit from the introduction of HAP ETM mechanism. Not only can they become an HAP and have gain additional download bandwidth, but also, with their extra upload bandwidth HAPs lead to the significant reduction of the average download time too.

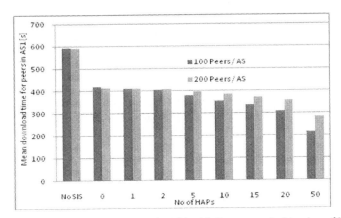

Fig. 7. Mean Download Times for Peers in AS1 with Respect to the Number of HAPs

When an HAP is implemented along with locality-awareness mechanisms, the operator benefits from reduced inter-domain traffic [17]. It allows for reducing costs for ISPs and confirms the advantages of the HAP ETM mechanism and the fact that it achieves win-win. The possibility of becoming an HAP also attracts more peers to use the provided locality mechanisms, therefore, further contributing to the ISP's win.

5   Concluding Remarks

This article presented an innovative methodology of assessment developed within the SmoothIT project especially for the evaluation of a variety of ETM mechanisms and, more specifically, ISP and end users interrelations in the context of such mechanisms. The application of this methodology has been outlined and related evaluation results in three representative mechanisms proposed by SmoothIT have been discussed: (a) the BGPLoc, (b) the insertion of the IoPs, and (c) the promotion of HAPs. Furthermore, this methodology has been employed to assess ETM mechanisms of another category identified, namely the inter-domain collaboration. Moreover, it should be noted that except for static and basic BGP information, an ETM approach can also employ other underlay parameters measured in networks, such as flow throughput, flow delays, and usage of inter- or intra-domain links. Implementation-wise for an operational prototype, the Admin component of the SmoothIT Information Service (SIS) has been designed as a Web-based tool for the ISP to administrate, monitor, and fine-tune the operation of the entire ETMS.

Finally, the extension and application of the methodology for other traffic types (not only P2P) generated according to trends in the Future Internet is an interesting and promising direction for future research.

Acknowledgments. This work has been accomplished in the framework of the EU ICT Project SmoothIT (FP7-2007-ICT-216259). The authors would like to thank all SmoothIT partners for useful discussions on the subject of the paper.

References

1. Liu, Y., Zhang, H., Gong, W., Towsley, D.: On the Interaction Between Overlay and Underlay Routing. In: IEEE INFOCOM 2005, Barcelona, Spain (April 2005)
2. Oechsner, S., Soursos, S., Papafili, I., Hoßfeld, T., Stamoulis, G.D., Stiller, B., Callejo, M.A., Staehle, D.: A framework of economic traffic management employing self-organization overlay mechanisms. In: Hummel, K.A., Sterbenz, J.P.G. (eds.) IWSOS 2008. LNCS, vol. 5343, pp. 84–96. Springer, Heidelberg (2008)
3. SmoothIT Project (December 2010), http://www.smoothit.org
4. Application-Layer Traffic Optimization (ALTO) Requirements: http://tools.ietf.org/html/draft-ietf-alto-reqs-06 (work in progress) (October 2010)
5. Papafili, I., Stamoulis, G.D.: A Markov Model for the Evaluation of Cache Insertion on Peer-to-Peer Performance. In: EuroNF NGI Conference, Paris (June 2010)
6. Lehrieder, F., Dán, G., Hoßfeld, T., Oechsner, S., Singeorzan, V.: The Impact of Caching on BitTorrent-like Peer-to-peer Systems. In: IEEE International Conference on Peer-to-Peer Computing P2P 2010, Delft, The Netherlands (August 2010)
7. Bindal, R., Cao, P., Chan, W., Medval, J., Suwala, G., Bates, T., Zhang, A.: Improving Traffic Locality in BitTorrent via Biased Neighbor Selection. In: 26th IEEE International Conference on Distributed Computing Systems, Montreal, Canada (June 2006)

8. Oechsner, S., Lehrieder, F., Hoßfeld, T., Metzger, F., Pussep, K., Staehle, D.: Pushing the Performance of Biased Neighbor Selection through Biased Unchoking. In: 9th International Conference on Peer-to-Peer Computing (P2P'09), Seattle, USA (September 2009)

9. The SmoothIT Project: Deliverable 2.3 — ETM Models and Components and Theoretical Foundations (Final) (October 2009)

10. Racz, P., Oechsner, S., Lehrieder, F.: BGP-based Locality Promotion for P2P Applications. In: 19th IEEE International Conference on Computer Communications and Networks (ICCCN 2010), Zürich, Switzerland (August 2010)

11. Lehrieder, F., Oechsner, S., Hoßfeld, T., Staehle, D., Despotovic, Z., Kellerer, W., Michel, M.: Mitigating Unfairness in Locality-Aware Peer-to-Peer Networks. International Journal of Network Management (IJNM), Special Issue on Economic Traffic Management (2011)

12. Hoßfeld, T., Lehrieder, F., Hock, D., Oechsner, S., Despotovic, Z., Kellerer, W., Michel, M.: Characterization of BitTorrent Swarms and their Distribution in the Internet, to appear in the Computer Networks (2011)

13. Cohen, B.: Incentives Built Robustness in BitTorrent. In: Kaashoek, M.F., Stoica, I. (eds.) IPTPS 2003. LNCS, vol. 2735, Springer, Heidelberg (2003)

14. Papafili, I., Soursos, S., Stamoulis, G.D.: Improvement of BitTorrent Performance and Inter-domain Traffic by Inserting ISP-owned Peers. In: ICQT'09, Aachen, Germany (May 2009)

15. SmoothIT Project: Deliverable 2.4 — Performance, Reliability, and Scalability Investigations of ETM Mechanisms (August 2010)

16. Papafili, I., Stamoulis, G.D., Lehrieder, F., Kleine, B., Oechsner, S.: Cache Capacity Allocation to Overlay Swarms. In: Bettstetter, C., Gershenson, C. (eds.) IWSOS 2011. LNCS, vol. 6557, Springer, Heidelberg (2011)

17. Pussep, K., Kuleshov, S., Groß, C., Soursos, S.: An Incentive-based Approach to Traffic Management for Peer-to-Peer Overlays. In: 3rd Workshop on Economic Traffic Management (ETM 2010), Amsterdam, The Netherlands (September 2010)

Deployment and Adoption of Future Internet Protocols

Philip Eardley[1], Michalis Kanakakis[2], Alexandros Kostopoulos[2], Tapio Levä[3],
Ken Richardson[4], and Henna Warma[3]

[1] BT Innovate & Design, UK
philip.eardley@bt.com
[2] Athens University of Economics and Business, Greece
{kanakakis, alexkosto}@aueb.gr
[3] Aalto University, School of Electrical Engineering, Finland.
{tapio.leva, henna.warma@aalto.fi}
[4] Roke Manor Research, UK
ken.richardson@roke.co.uk

Abstract. Many, if not most, well-designed Future Internet protocols fail, and
some badly-designed protocols are very successful. This somewhat depressing
statement illustrates starkly the critical importance of a protocol's deployability.
We present a framework for considering deployment and adoption issues, and
apply it to two protocols, Multipath TCP and Congestion Exposure, which we
are developing in the Trilogy project. Careful consideration of such issues can
increase the chances that a future Internet protocol is widely adopted.

Keywords: Protocol Deployment, Adoption Framework, Multipath TCP, Congestion Exposure.

1   Introduction

New protocols and systems are often designed in near isolation from existing protocols and systems. The aim is to optimise the technical solution, in effect for a greenfield deployment. The approach can be very successful, a good example being GSM but there are many more examples of protocols that are well-designed technically but where deployment has failed or been very difficult, for example interdomain IP multicast and QoS protocols. On the other hand there are several examples of protocols that have been successfully deployed despite a weak technical design (by general consent), such as WEP (Wired Equivalent Privacy).

Several attempts have been made at studying the adoption of consumer products [1] and new Internet protocols, including [2], [3], [4], [5], [6] and [7], which we build on. The adoption of Internet protocols is tricky because the Internet is a complex system with diverse end-systems, routers and other network elements, not all of whose aspects are under the direct control of the respective end users or service providers.

In this Chapter we propose a new framework for a successful adoption process (Section 2), and apply it to two emerging protocols, Multipath TCP (Section 3) and Congestion Exposure (Section 4).

J. Domingue et al. (Eds.): Future Internet Assembly, LNCS 6656, pp. 133–144, 2011.

The framework is not a "black box" where candidate protocols are the inputs and the output is the protocol that is certain to be adopted. Rather, it is a structured way of thinking, useful at the design stage to improve the chances that the new protocol will be widely deployed and adopted.

2   A Framework for the Deployment and Adoption of Future Internet Protocols

We propose a new framework (Figure 1) for a successful adoption process, with several key features:

- It asks two key questions at each stage: what are the benefits and costs? And is it an incremental process?
- It distinguishes an initial scenario from one where adoption is widespread
- It distinguishes implementation, deployment and adoption

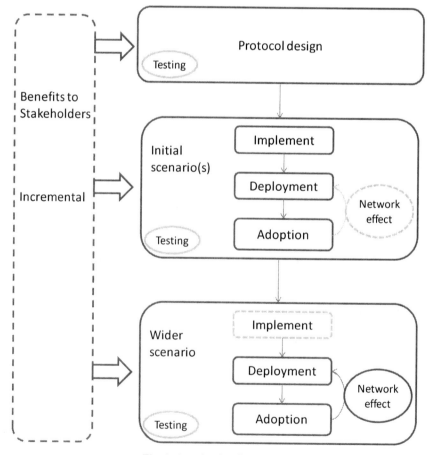

Fig. 1. An adoption framework

A version of the framework has been applied in two papers, [8] and [9]. The framework is intended to be generally applicable to Internet protocols.

The first key question is: *what are the benefits (and costs) of the protocol?* There must be a "positive net value (meet a real need)" [2]. Further, the benefits and costs should be considered for each specific party that needs to take action, as "the benefit of migration should be obvious to (at least) the party migrating" [3]. For example, browsers and the underlying http/html protocols give a significant benefit to both the end users (a nice user interface for easy access to the web) and to the content providers (their content is accessed more; new opportunities through forms etc). As another example, a NAT (Network Address Translator) allows an operator to support more users with a limited supply of addresses, and has some security benefit. As a counter-example, IPv6 deployment has a cost to the end host to support the dual stack, but the benefit is quite tenuous ('end-to-end transparency') and long-term. Deploying a new protocol may have knock-on costs, for example a new application protocol may require changes to the accounting and OSS (Operations Support Systems).

The second key question is: *can the changes required be adopted incrementally?* This is similar to the guideline "Contain coordination and Constrain contamination" [3], meaning that the scope of changes should be restricted in terms of (respectively) the number of parties involved and the changes required within one party. Backwards compatibility is also important. Successful examples include: https, which doesn't require deployment of an infrastructure to distribute public keys; and NATs, which can be deployed by an ISP without coordinating with anyone else. As a counter-example, IPv6 requires at least both ends and preferably the network to change.

Combining these two key factors leads to the idea of an incremental process, where the aim at each step is to bring a net benefit for the party(s) migrating. So commercial, not technical, considerations should determine what the right step size is – it adds sufficient functionality to meet a specific business opportunity. If each step is the same, this is equivalent to saying that there should be a benefit for earlier adopters. However, often the steps will be different, as typically a protocol gets deployed and adopted in a specific use case, later widening out if the protocol proves its utility. Then each step may involve different stakeholders, for example BitTorrent was initially adopted by application developers (and their end-users) to transfer large files, later widening out to some Content Providers, such as Tribler, to distribute TV online. Hence the framework distinguishes initial scenarios from a widespread one.

At each step of the framework careful consideration is needed of benefits and incrementalism. But such consideration should not wait until the initial scenario is about to start. Instead, during the design a mental experiment should be performed to think about a narrow use case and about the final step of widespread deployment and adoption. The more specific thinking may reveal new factors for the design to handle.

Finally, at each step the framework makes a distinction between the concepts of implementation, deployment and adoption:

- *Implementation* refers to the coding of the new protocol by developers
- *Deployment* refers to the protocol being put in the required network equipment and/or end-hosts
- *Adoption* is dependent upon deployment, with the additional step that the protocol is actually used.

For network equipment the distinction between implementation and deployment is particularly important because different stakeholders are involved – equipment vendors implement, whilst network operators deploy; their motivations are not the same.

No further implementation may be needed at the "wider scenario" stage, since the software has already been developed for the initial scenario and it is simply a matter of deploying and adopting it on a wider scale. Perhaps an enhanced version of the protocol can include lessons learnt from the initial use case. But for some protocols the wider scenario requires extra critical functionality – for example, security features, if the initial scenario is within a trusted domain. Also, at the wider scenario stage, "network externalities" are likely to be important: the benefit to an adopter is bigger the greater the numbers who have already adopted it [5].

Testing of the new protocol is included within each of these stages. For example, vendors will validate their implementation of the new protocol, operators will check that it works successfully if deployed on their network, and users will complain if they adopt it and it breaks something.

Note that it is not possible to prove that the framework is necessary or sufficient to guarantee the adoption of a protocol – the framework is not a "black box" with an input of a candidate protocol and an output of yes/no as to whether it will be adopted. The framework also ignores factors such as risks (deployment is harder if the associated risk is higher), regulatory requirements and the role of hype and "group think".

When there are competing proposals (which should be selected for deployment?) it is important to think through the issues in the framework, otherwise an apparently superior protocol may be selected that proves to be not readily deployable. It may be better instead to incorporate some of its ideas, perhaps in a second release.

The main message of this Chapter is that implementation, deployment and adoption need to be thought about carefully during the design of the protocol - for example, mental experiments performed for narrow and widespread scenarios.

3   Multipath TCP

Multipath TCP (MPTCP) enables a TCP connection to use multiple paths simultaneously. The current Internet's routing system only exposes a single path between a source-address pair, and TCP restricts communications to a single path per transport connection. But hosts are often connected by multiple paths, for example mobile devices have multiple interfaces.

MPTCP supports the use of multiple paths between source and destination. When multiple paths are used, MPTCP will react to failures by diverting the traffic through paths that are still working and have available capacity. Old and new paths can be used simultaneously. Since MPTCP is aware of the congestion in each path, the traffic distribution can adapt to the available rate of each path – the congestion controllers for the paths are coupled. This brings the benefits of resilience, higher throughput and handles more efficiently sudden increases in demand for bandwidth.

MPTCP is currently under development at the IETF [10]. The MPTCP design [11] provides multipath TCP capability when both endpoints understand the necessary extensions to support MPTCP. This allows endpoints to negotiate additional features between themselves, and initiate new connections between pairs of addresses on multi-homed endpoints.

The basic idea of building multipath capability into TCP has been re-invented multiple times [12] [13] [14] [15] [16] [17] [18]. However, none of these proposals made it into the mainstream. The detailed design of MPTCP strives to learn the lessons from these proposals, for instance:

- It builds on the breakthrough of [19] [20], who showed theoretically that the right coupled congestion controller balances congestion across the sub-flows in a stable manner. Stability is required for the benefits (of resilience and throughput) to be worthwhile.
- It seeks to be equitable with standard TCP, essentially meaning that at a bottleneck link MPTCP consumes the same bandwidth as TCP would do; again, this helps persuade the IETF that it is safe to deploy on the internet. Also an operator might otherwise be tempted to block MPTCP to prevent the degradation of the throughput of its "legacy" users.
- It is designed to be application-friendly: it just uses TCP's API so it looks the same to applications. This, plus the following two bullets, help MPTCP be incrementally deployable.
- It automatically falls back to TCP if the MPTCP signalling fails, hence an MPTCP user can still communicate with legacy TCP users and can still communicate if the signalling is corrupted by a middlebox.
- It is designed to be middlebox-friendly (be it a NAT, firewall, proxy or whatever), in order to increase the chances that MPTCP works when there are middleboxes en route:
  − MPTCP appears "on the wire" to be TCP
  − The signalling message that adds a new sub-flow includes an Address ID field, which allows the sender and receiver to identify the sub-flow even if there is a NAT
  − Either end-host can signal to add a new path (in case one end-host's signalling is blocked by a middlebox).
  − MPTCP's signalling is in TCP-Options, because signalling in the payload is more likely to get traumatised by some middleboxes.
  − There is a separate connection-level sequence number, in addition to the standard TCP sequence number on each sub-flow; if there was only a connection-level sequence number, on one sub-flow there would be gaps in the sequence space, which might upset some proxies and intrusion detection systems.
- NAT behaviour is unspecified and so, despite our care in designing MPTCP with NATs in mind, their behaviour may be quite surprising. So we are now working on a NAT survey to probe random paths across the Internet to test how operational NATs impact MPTCP's signalling messages [21].

There are also various proposals for including multipath capability in other transport protocols, such as SCTP [22], RTP [23] and HTTP [24].

For MPTCP, our current belief is that a data centre is the most promising initial scenario (Figure 2). Within a data centre, one issue today is how to choose what path to use between two servers amongst the many possibilities - MPTCP naturally spreads traffic over the available paths.

- Benefits: Simulations show there are significant gains in typical data centre topologies [25], perhaps increasing the throughput from 40% to 80% of the theoretical maximum. However, the protocol implementation should not impact hardware offloading of segmentation and check-summing. One reason that MPTCP uses TCP-Options for signalling (rather than the payload) is that it should simplify offloading by network cards that support MPTCP, due to the separate handling of MPTCP's signalling and data.
- Incremental: the story is good, as only one stakeholder is involved viz the data centre operator.

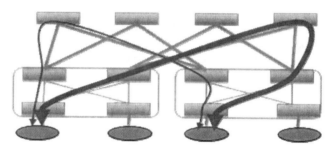

Fig. 2. Potential MPTCP deployment scenario, in a data centre. In this example, traffic between the two servers (at the bottom) travels over two paths through the switching fabric of the data centre (there are four possible paths).

Another potential initial scenario would be a mobile user using MPTCP over multiple interfaces. The scenario reveals a potential distinction between deployment (which involves the OS vendor updating their stack) and adoption (which means that MPTCP is actually being used and requires the consumer to have multiple links) – so in theory it would be possible for MPTCP to be fully deployed but zero adopted. (Note there's little distinction between implementation and deployment, since it is only in end-hosts and deployment is mainly decided by the OS (Operating System) vendor and not the end user.)

Therefore we believe that a more promising initial scenario is an end user that accesses content, via wireless LAN and 3G, from a provider that controls both end user devices and content servers [26] – for example, Nokia or Apple controls both the device and the content server, Nokia Ovi or Apple App Store.

- Benefits: MPTCP improves resilience - if one link fails on a multi-homed terminal, the connection still works over the other interface. But it is a prerequisite, and cost, that devices are multihomed.
- Incremental: Both the devices and servers are under the control of one stakeholder, so the end user 'unconsciously' adopts MPTCP. However, there may be NATs on the data path, and MPTCP's signalling messages must get through them.

The wider scenario of widespread deployment and adoption is again worth thinking about this even during the design of the protocol.

- Benefits: Several stakeholders may now be involved. For instance, it is necessary to think about the benefits and costs for OS vendors, end users, applications and ISPs (Internet Service Providers). Here also we see the importance of network effects. For instance, as soon as a major content provider, such as Google, deploys MPTCP – perhaps as part of a new application with better QoS - then there is a much stronger incentive for OSs to deploy it as well, as the network externality has suddenly increased.
- Incremental: Existing applications can use MPTCP as though it was TCP, ie the API is unaltered (although there will also be an enhanced API for MPTCP-aware applications). MPTCP is an extension for end-hosts – it doesn't require an upgrade to the routing system; if both ends of the connection have deployed MPTCP, then it "just works" (NATs permitting).

4   Congestion Exposure

The main intention of Congestion Exposure (Conex) is to make users and network nodes accountable for any congestion that is caused by the traffic they send or forward. This gives the right incentives to promote cooperative traffic management, so that (for example) the network's resources are efficiently allocated, senders are not unnecessarily rate restricted, and an operator has a better incentive to invest in new capacity.

Conex introduces a mechanism so that any node in the network has visibility of the whole-path congestion – and thus also rest-of-path congestion (since it can measure congestion-so-far, by counting congestion markings or inferring lost packets), Figure 3. A Conex-enabled sender adds signalling, in-band at the IP layer, about the congestion encountered by packets earlier in the same flow, typically 1 round trip time earlier.

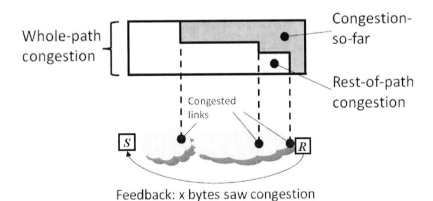

Fig. 3. Conex gives all nodes visibility of the whole path congestion, and thus also rest-of-path congestion.

(In today's internet this information is only visible at the transport layer, and hence not available inside the network without packet sniffing.)

Conex is currently under development at the IETF [27].

Today, without Conex, a receiver reports to the sender information about whether packets have been received or whether they have been lost (or received ECN-marked). The former causes a Conex-sender to flag packets it sends as "Conex-Not-Marked", and the latter to flag packets as "Conex-Re-Echo". By counting "Conex-Re-Echoes", any node has visibility of the whole-path congestion.

Conex also requires, by default, two types of functionality in the network. Firstly, an auditor to catch those trying to cheat by under-declaring the congestion that their traffic causes; the auditor checks (within a flow or aggregate) that the amount of traffic tagged with Conex-Re-Echo is at least equal to the amount of traffic that is lost (or ECN-marked). Secondly, a policer to enforce policy specifically related to the user being served. A user pays, as part of its contract, to be allowed to cause a certain amount of congestion. The policer checks the user is within its allowance by counting the Conex-Re-Echo signals. Similarly, a policer at a network's border gateway checks that a neighbouring ISP is within its contractual allowance.

Conex's default requirement for a policer and auditor, as well as a Conex-enabled sender, is problematic as it requires several stakeholders to coordinate their deployment [9]. Since this is likely to be difficult, we seek an initial scenario that is more incrementally deployable.

We believe that the most promising initial scenario for Conex is as part of a "premium service" by an operator who runs both an ISP and a CDN (Content Distribution Network); network operators are increasingly seeking to run their own CDN, to reduce their interconnect charges and to decrease the latency of data delivery. The CDN server sends "premium" packets (perhaps for IPTV) as Conex-Not-Marked or Conex-Re-Echo. Conex traffic is prioritised by the operator ("premium service"). To a first order of approximation, the only point of contention is the backhaul – where the operator already has a traffic management box, typically doing per end user (consumer) volume caps and maybe deep packet inspection, to provide all users with a "fair share". The operator upgrades its traffic management box so that it drops Conex traffic with a lower probability. However, the operator does not need to deploy a policer or auditor, since it is also running the CDN and therefore trusts it.

In the initial scenario the CDN offers a range of deals to Content Providers, with the more expensive deals allowing a Content Provider to send more Conex traffic and more Conex-Re-Echoes (ie to cause more congestion) – effectively the CDN offers different QoS classes. In turn, the content provider (presumably) charges consumers for premium content.

- Benefits: The CDN offers a premium service to its Content Providers. Also, the Conex (premium) traffic is not subject to per end user caps or rate limits by the ISP.
- Incremental: Only one party has to upgrade, ie the combined CDN-ISP. The Content providers and consumers don't know about Conex. Note that the receiver doesn't need to be Conex-enabled, and the network doesn't need to support ECN-marking.

One way this scenario could widen out is that the content provider is now informed about the Conex-Re-Echoes and upgraded to understand them. The benefit is that, at a time of congestion, the content provider can manage its premium service as it wants - effectively it can choose different QoS classes for different users.

Another way this scenario could develop is that the operator offers the service to all CDNs, again as a premium service. However, the ISP can no longer trust that the CDN is well-behaved – it might never set packets as Conex-Re-Echo in order to try to lower its bill. Therefore the ISP needs to upgrade two things. Firstly its traffic management box: it needs to do occasional auditing spot-checks, to make sure that after it drops a packet then it hears (a round trip time later) a Conex-Re-Echo packet from the CDN sender. Secondly, its gateway with the new CDN needs to count the amount of Conex traffic and the amount of Conex-Re-Echo, to make sure the CDN stays within its contractual allowance.

Eventually the scenario could widen out to end hosts (consumers) so that the ISP also offers them the premium service. Most likely the regular consumer contracts would include some allowance and then the host's software would automatically send the user's premium traffic (VoIP say) as Conex-enabled. In this case the ISP needs to upgrade its traffic management box to check the consumer stays within their Conex allowance.

- Benefits: The premium service is offered to other CDNs, ISPs and consumers - effectively QoS is controlled by the CDN or end user, so that they choose which of their traffic is within which class of QoS, always bearing in mind that they must stay within the limits that they have paid for.
- Incremental: Conex capability is added a CDN or end user at a time.

5   Enhancing the Framework

One important development in telecoms is virtualisation. Although the basic idea is long-standing, it has recently come to much greater practical importance with the rise of cloud networking. Normally the advantages are explained in terms of storage and processing "in the cloud" at lower cost, due to efficiency gains through better aggregation. However, there is also an interesting advantage from the perspective of deployment. A new application can be deployed "on the cloud" – effectively the end users use a virtualised instance of the new application. Although our adoption framework is still valid, there are now differences in emphasis:

- Roll out of the software should be cheaper, therefore the expected benefits of the deployment can be less.
- There is no need to coordinate end users all having to upgrade. Every user can immediately use the new (virtualised) software, so effectively a large number of users can be enabled simultaneously.
- These factors reduce the deployment risk, especially as it should also be easier to "roll back" if there is some problem with the new software.

Virtualisation is not suitable for all types of software, for instance new transport layer functionality, such as MPTCP and CONEX, needs to be on the actual devices.

There is an analogy with the digitalisation of content, which has greatly lowered the costs of distribution. Virtualisation should similarly lower the cost of distribution – in other words, it eases deployment.

Another aspect is the interaction of a new protocol with existing protocols. It is important that the design minimises negative interactions, and to test for this. For instance the MPTCP is designed to cope if a middlebox removes MPTCP-options.

There will also be cases of positive interactions, where a new protocol suddenly enables an existing protocol to work better. One set of examples is the various IPv4-IPv6 transition mechanisms that try to release the (currently hidden) benefits of IPv6. Another example is a protocol "bundle", for instance telepresence offerings now wrap together several services that separately had less market traction.

6 Conclusions

The main message of this Chapter is that implementation, deployment and adoption need to be thought about carefully during the design of the protocol, as even the best technically designed protocol can fail to get deployed. Initial narrow and subsequent widespread scenarios should be identified and mental experiments performed concerning these scenarios in order to improve the protocol's design. We have presented a framework; by using it we believe a designer improves the chances that their protocol will be deployed and adopted.

We have applied the framework to two emerging protocols which we are developing in the Trilogy project [28]. Multipath TCP (MPTCP) is designed to be incrementally deployable by being compatible with existing applications and existing networks, whilst bringing benefits to MPTCP-enabled end users. For Congestion Exposure (Conex), a reasonable initial deployment scenario is a combined CDN-ISP that offers a premium service using Conex, as it requires only one party to deploy Conex functionality.

Acknowledgments. This research was supported by Trilogy (http://www.trilogy-project.org), a research project (ICT-216372) partially funded by the European Community under its Seventh Framework Programme. The views expressed here are those of the authors only. The European Commission is not liable for any use that may be made of the information in this document. Alexandros Kostopoulos is co-financed by the European Social Fund and National Resources (Greek Ministry of Education – HERAKLEITOS II Programme).

References

1. Rogers, E.: Diffusion of Innovations. Free Press, New York (1983)
2. Thaler, D., Aboba, B.: What Makes for a Successful Protocol? RFC 5218 (2008)

3. Burness, L., Eardley, P., Akhtar, N., Callejo, M.A., Colas, J.A.: Making migration easy: a key requirement for systems beyond 3G. In: VTC 2005-Spring, IEEE 61st Vehicular Technology Conference (2005)
4. Hovav, A., Patnayakuni, R., Schuff, D.: A model of Internet Standards Adoption: the Case of IPv6. Information Systems Journal 14(3), 265–294 (2004)
5. Katz, M., Shapiro, C.: Technology Adoption in the Presence of Network Externalities. Journal of Political Economics 94, 822–841 (1986)
6. Joseph, D., Shetty, N., Chuang, J., Stoica, I.: Modeling the Adoption of New Network Architectures. In: International Conference on Emerging Networking Experiments and Technologies (2007)
7. Dovrolis, C., Streelman, T.: Evolvable network architectures: What can we learn from biology? ACM SIGCOMM Computer Communications Review 40(2) (2010)
8. Kostopoulos, A., Warma, H., Leva, T., Heinrich, B., Ford, A., Eggert, L.: Towards Multipath TCP Adoption: Challenges and Perspectives. In: NGI 2010 - 6th EuroNF Conference on Next Generation Internet, Paris (2010)
9. Kostopoulos, A., Richardson, K., Kanakakis, M.: Investigating the Deployment and Adoption of re-ECN. In: ACM CoNEXT ReArch'10, Philadelphia, USA (2010)
10. Multipath TCP Working Group, IETF. Latest status at http://tools.ietf.org/wg/mptcp (2010)
11. Ford, A., Raiciu, C., Handley, M.: TCP Extensions for Multipath Operation with Multiple Addresses, draft-ford-mptcp-multiaddressed-02.txt, work in progress (2010)
12. Huitema, C.: Multi-homed TCP, draft-huitema-multi-homed-01.txt, work in progress (2005)
13. Hsieh, H.-Y., Sivakumar, R.: pTCP: An End-to-End Transport Layer Protocol for Striped Connections. In: IEEE International Conference on Network Protocols, ICNP (2002), http://www.ece.gatech.edu/research/GNAN/work/ptcp/ptcp.html
14. Rojviboonchai, K., Aida, H.: An Evaluation of Multi-path Transmission Control Protocol (M/TCP) with Robust Acknowledgement Schemes. Internet Conference IC (2002)
15. Zhang, M., Lai, J., Krishnamurthy, A., Peterson, L., Wang, R.: A Transport Layer Approach for Improving End-to-End Performance and Robustness Using Redundant Paths. In: Proc. of the USENIX 2004 Annual Technical Conference (2004)
16. Dong, Y.: Adding concurrent data transfer to transport layer, ProQuest ETD Collection for FIU, Paper AAI3279221 (2007), http://digitalcommons.fiu.edu/dissertations/AAI3279221
17. Sarkar, D.: A Concurrent Multipath TCP and Its Markov Model. In: IEEE International Conference on Communications, ICC (2006)
18. Hasegawa, Y., Yamaguchi, I., Hama, T., Shimonishi, H., Murase, T.: Improved data distribution for multipath TCP communication. In: IEEE GLOBECOM (2005)
19. Kelly, F., Voice, T.: Stability of end-to-end algorithms for joint routing and rate control. Computer Communication Review 35, 2 (2005)
20. Key, P., Massoulie, P., Towsley, D.: Combined Multipath Routing and Congestion Control: a Robust Internet Architecture, no. MSR-TR-2005-111 (2005), http://research.microsoft.com/pubs/70208/tr-2005-111.pdf
21. Honda, M.: Call for contribution to middlebox survey (2010), http://www.ietf.org/mail-archive/web/multipathtcp/current/msg01150.html
22. Becke, M., Dreibholz, T., Iyengar, J., Natarajan, P., Tuexen, M.: Load Sharing for the Stream Control Transmission Protocol (SCTP), draft-tuexen-tsvwg-sctp-multipath-00.txt, work in progress (2010)

23. Singh, V., Karkkainen, T., Ott, J., Ahsan, S., Eggert, L.: Multipath RTP (MPRTP), draft-singh-avt-mprtp, work in progress (2010)
24. Ford, A., Handley, M.: HTTP Extensions for Simultaneous Download from Multiple Mirrors, draft-ford-http-multi-server, work in progress (2009)
25. Raiciu, C., Plunkte, C., Barre, S., Greenhalgh, A., Wishcik, D., Handley, M.: Data center Networking with Multipath TCP. ACM Sigcomm Hotnets (2010)
26. Warma, H., Levä, T., Eggert, L., Hämmäinen, H., Manner, J.: Mobile Internet In Stereo: an End-to-End Scenario. In: 3rd Workshop on Economic Traffic Management, ETM (2010)
27. Congestion Exposure Working Group, IETF. Latest status at http://tools.ietf.org/wg/conex (2010)
28. Trilogy project, http://trilogy-project.org/ (2010)

An Approach to Investigating Socio-economic Tussles Arising from Building the Future Internet

Costas Kalogiros[1], Costas Courcoubetis[1], George D. Stamoulis[1], Michael Boniface[2],
Eric T. Meyer[3], Martin Waldburger[4], Daniel Field[5], and Burkhard Stiller[4]

[1] Athens University of Economics and Business, Greece
ckalog@aueb.gr, courcou@aueb.gr, gstamoul@aueb.gr
[2] University of Southampton IT Innovation, United Kingdom
mjb@it-innovation.soton.ac.uk
[3] Oxford Internet Institute, University of Oxford, United Kingdom
eric.meyer@oii.ox.ac.uk
[4] University of Zürich, Switzerland
waldburger@ifi.uzh.ch, stiller@ifi.uzh.ch
[5] Atos Origin, Spain
daniel.field@atosresearch.eu

Abstract. With the evolution of the Internet from a controlled research network
to a worldwide social and economic platform, the initial assumptions regarding
stakeholder cooperative behavior are no longer valid. Conflicts have emerged in
situations where there are opposing interests. Previous work in the literature has
termed these conflicts tussles. This article presents the research of the SESERV
project, which develops a methodology to investigate such tussles and is carrying out a survey of tussles identified within the research projects funded under
the Future Networks topic of the FP7. Selected tussles covering both social and
economic aspects are analyzed also in this article.

Keywords: Future Internet Socio-Economics, Incentives, Design Principles,
Tussles, Methodology

1   Introduction

The Internet has already long since moved from the original research-driven network
of networks into a highly innovative, highly competitive marketplace for applications,
services, and content. Accordingly, different stakeholders in the Internet space have
developed a wide range of on-line business models to enable sustainable electronic
business. Furthermore, the Internet is increasingly pervading society [3]. Wide-spread
access to the Internet via mobile devices, an ever-growing number of broadband users
world-wide, lower entry barriers for non-technical users to become content and service providers, and trends like the Internet-of-Things or the success of Cloud services,
all provide indicators of the high significance of the Internet today. Hence, social and
economic impacts of innovations in the future Internet space can be reasonably ex-

J. Domingue et al. (Eds.): Future Internet Assembly, LNCS 6656, pp. 145–159, 2011.

pected to increase in importance. Thus, since the future Internet can be expected to be characterized by an ever larger socio-economic impact, a thorough investigation into socio-economic tussle analysis becomes highly critical [9].

The term tussle was introduced by Clark et al. [5] as a process reflecting the competitive behavior of different stakeholders involved in building and using the Internet. That is, a tussle is a process in which each stakeholder has particular self-interests, but which are in conflict with the self-interests of other stakeholders. Following these interests results in actions – and inter-actions between and among stakeholders. When stakeholder interests conflict, inter-actions usually lead to contention. Reasons for tussles to arise are manifold. Overlay traffic management and routing decisions between autonomous systems [11] and mobile network convergence [10] constitute only two representative examples for typical tussle spaces.

The main argument for focusing on tussles in relation to socio-economic impact of the future Internet is in the number of observed stakeholders in the current Internet and their interests. Clark et al. speak of tussles on the Internet as of today. They argue [5] that *"[t]here are, and have been for some time, important and powerful players that make up the Internet milieu with interests directly at odds with each other."* With the ongoing success of the Internet and with the assumption of a future Internet being a competitive marketplace with a growing number of both users and service providers, tussle analysis becomes an important approach to assess the impact of stakeholder behavior.

This paper proposes a generic methodology for identifying and assessing socio-economic tussles in highly-dynamic and large systems, such as the current and future Internet. In order to help an analyst during the tussle identification task, the approach presented here provides several examples of tussles, together with their mappings to four abstract tussle patterns. Furthermore, a survey of tussles is also presented and the way those have been addressed by several FP7 projects.

SSM (Soft Systems Methodology) proposed by Checkland [4] and CRAMM (CCTA Risk Analysis and Management Method) [7] have similar objectives to our methodology. The former, which is extensively used when introducing new information systems into organizations, suggests an iterative approach to studying complex and problematic real-world situations (called systems) and evaluating candidate solutions. The latter approach aims at identifying and quantifying security risks in organizations. The situations analyzed by the aforementioned methodologies are often associated with certain kinds of tussles. However are quite restrictive in the way evaluation of situations is performed, suggesting specific qualitative methods. On the other hand, the proposed tussle analysis methodology provides a higher-level approach allowing and/or complementing the application of a wide range of techniques (both qualitative and quantitative). For example, microeconomic analysis can be applied, which uses mathematical models aiming to understand the behavior of single agents, as part of a community, who selfishly seek to maximize some quantifiable measure of well-being, subject to restrictions imposed by the environment and the actions of others [6]. Similarly, game-theoretic models that aim at finding and evaluating all possible equilibrium outcomes when a set of interdependent decision makers interact with each other is another candidate method. In this way, one can derive what the possible equilibrium points are, under what circumstances these are reached, and compare different protocols and the tussles enabled thereby with respect to a common metric. Such a metric can be social welfare or the "Price of Anarchy" [12], i.e., the ratio of the worst case Nash equilibrium to the social optimum.

The remaining of this article is organized as follows. In Section 2 we describe the proposed methodology for identifying and analyzing socio-economic tussles in the Future Internet. In Section 3 we provide a classification of tussles according to stakeholders' interests into social and economic ones, which can be used as a reference point when applying this methodology. In Section 4 we provide examples of existing and potential tussles being studied by several FP7 research projects and we conclude in Section 5 by outlining our future work.

2   A Methodology for Identifying and Assessing Tussles

The *Design for Tussle* goal is considered to be a normal evolution of Internet design goals to reflect the changes in Internet usage. This paradigm shift should be reflected in new attempts for building the Future Internet. However, identifying both existing and future socio-economic tussles, understanding their relationship, assessing their importance and making informed technical decisions can be very complicated and costly, requiring a multi-disciplinary approach to grasp the potential benefits and consequences.

Providing a systematic approach for this task has received little attention by Future Internet researchers. Such a methodology should be a step-by-step procedure that can be applied to any Internet functionality, acting as a guide for making sure that all important factors are considered when making technology decisions. This would support policy-makers (such as standardization bodies) to prepare their agenda by addressing critical issues first, or protocol designers so that functionality is future-proof. For example the latter could apply this methodology before and after protocol introduction in order to estimate the adoptability and other possible effects, both positive and negative ones, for the Future Internet.

The proposed methodology is composed of three steps and can be executed recursively, allowing for more stakeholders, tussles, etc. to be included in the analysis. It is out of the article's scope to suggest where the borderline for the analysis should be drawn, as this choice depends on subjective factors like the goals of the analysts and their views on the criticality of each research issue. Nevertheless, this requires all steps of the methodology to be performed in a justifiable way; following a code of research ethics (e.g. assumptions should be realistic and agreed by all team members). It is also important to note that for each step of this procedure many techniques could be available for completing this task, but not all of them may be perfectly suitable. A multidisciplinary team, composed of engineers, economists and social scientists, would allow for suggesting candidate techniques and incorporating useful insights from different domains at each step of the methodology.

The proposed methodology is the following:

1. *Identify all primary stakeholders and their properties for the functionality under investigation.*
2. *Identify tussles among identified stakeholders and their relationship.*

3. *For each tussle:*

   a. *Assess the impact to each stakeholder;*

   b. *Identify potential ways to circumvent and resulting spill-overs. For each new circumventing technique, apply the methodology again.*

The first step of the methodology suggests identifying and studying the properties of all important stakeholders affected by a functionality related to a protocol, a service, or an application instance. The outcome of this step is a set of stakeholders and attributes such as their population, social context (age, entity type, etc.), technology literacy and expectations, openness to risk and innovation. Furthermore, it should be studied whether and how these attributes, as well as the relative influence across stakeholders, change over time.

The next step aims at identifying conflicts among the set of stakeholders and their relationship. In performing the first part of this step the analyst could find particularly useful to check whether any tussle pattern described in the next section can be instantiated. After the identification task the analyst should check for potential dependencies among the tussles, which can be useful in understanding how these are interrelated (for example are some of them orthogonal, or have a cause-and-effect relationship?).

The third step of the methodology proposes to estimate the impact of each tussle from the perspective of each stakeholder. In the ideal scenario a tussle outcome will affect all stakeholders in a non-negative way and no one will seek to deviate; thus an equilibrium point has been reached. Usually this is a result of balanced control across stakeholders, which means that the protocols implementing this functionality follow the Design for Choice design principle [5]. Such protocols allow for conflict resolution at run-time, when the technology under investigation is being used. However there will be cases where some – or all – stakeholders are not satisfied by the tussle outcome and have the incentive to take advantage of the functionality provided, or employ other functionalities (protocols/tricks) to increase their socio-economic welfare triggering a tussle spill-over. Tussle spill-overs can have unpredictable effects and are considered to be a sign of flawed architectural design [9]. If a tussle spill-over can occur then another iteration of the methodology should be performed for the enabling functionality, broadening the scope of the analysis.

Of course, one difficult aspect to these approaches is acquiring the empirical evidence from which one can draw inferences about stakeholders and tussles. For all steps of this methodology except for 3a, system modeling by using use-case scenarios and questionnaires would be the most straightforward way to go. However, in complex systems with multiple stakeholders, multiple quantitative and qualitative sources of evidence may be required to better understand the actual and potential tussles. Thus, one can think of the tussle approach outlined here as just one of a set of tools necessary to identify, clarify, and help in resolving existing and emergent tussles. For instance, impact assessment (3a) could be performed by mathematical models for assessing risk or utility, as well as providing benchmarks like the price of anarchy ratio. Ideally a single metric should be used so that results for each tussle are comparable. Note that the assessment of each side-effect (step 3b) is performed in the next iteration.

In the following the methodology above is applied in case of congestion control with TCP, assuming the analyst stops at the third iteration.

In the first iteration, congestion control mainly affects heavy users (HUs), interactive users (IUs) and ISPs. Two tussles have been identified, which are closely related: (a) contention among HUs and IUs for bandwidth on congested links and (b) contention among ISPs and HUs since the aggressive behavior of the latter has a negative effect on IUs and provision of other services. Assuming that the ISP's network remains the same, control in both tussles is considered biased. An IU gets K1 bps by opening a single TCP connection, while an HU opens N TCP connections and gets K2 bps (where K1<<K2), regardless of their utility on instantaneous bandwidth. Similarly, only a HU controls how many TCP connections will be active, since the ISP has no means to correlate connections with applications. In order to assess the impact of the first tussle, an analyst could measure social welfare loss or calculate the price of anarchy ratio, noticing that the latter can be very large due to starvation of IUs. On the other hand, risk assessment techniques seem more relevant for the second tussle since high congestion can have an impact on ISP's plans to offer other real-time services. Identifying possible spill-over effects for the tussle among HUs and IUs it can be mentioned that the possibility for developers of interactive applications or ASPs (Application Service Providers) to adopt more aggressive techniques, resulting in greater contention. In the second tussle, an ISP could employ middle-boxes and perform traffic shaping based on port number, which has a negative impact on QoS-aware applications of third-party ASPs.

In the second iteration the focus laid will be on the network neutrality issue that is considered a side-effect of traffic-shaping (but not the only reason). In this case, the set of stakeholders is extended to include ASPs as well. The new tussle involves ISPs and ASPs (e.g. VoIP providers), since the traffic of the latter is being throttled by middle-boxes (either on purpose or not). Again, control is imbalanced; only ISPs can configure the middle-boxes since there is no API (Application Programming Interface) for ASPs to affect how their traffic will be handled. ASPs and HUs can employ protocol obfuscation techniques and ISPs can reply by more aggressive traffic shaping, resulting in an endless arms' race. Risk assessment techniques could be used in this case, as well as models for estimating social welfare loss. A side-effect of this tussle is innovation discouragement since new applications are harder to become widely known, which may result in regulatory intervention.

In the third iteration it will be assumed that the policy-maker (a new stakeholder) decides to intervene, with the important advantage of proactively seeking the socially optimum solution. The regulator's decision will redistribute control across stakeholders in a balanced way or, in more complex cases, cause new tussles to arise. Since future tussles depend on the regulator's action and the possible set of actions can be large, the regulator should perform an iteration of the methodology for each scenario. Then the policy-maker should select the action with the most favorable properties. In practice, of course, other factors will often intercede and result in actors making less than perfectly rational decisions, but it will be assumed for the sake of argument that the actors are seeking optimum solutions.

3   Taxonomy of Socio-economic Tussles

Many articles have been published building on Clark's work as applied to specific technical domains [12,13]. However, the extensive range of tussles to be addressed and analyzed by SESERV requires a classification framework for tussles based on abstract tussle patterns.

On the left part of Figure 1 we see a general model for a single tussle. In the model, agents have resources that are used to realize their interests. A tussle occurs when two agents have an interest in a resource that cannot be satisfied for both through the utilization of the resource. On the right part of Figure 1 we see that agents acting selfishly can lead to new tussles (spill-over) that may involve new stakeholders as well. For example, the Tussle I among Actor A and Actor B may trigger the Tussle II involving the same stakeholders, or a Tussle III among Actor B and Actor C. This basic model is extended to identify abstract tussle patterns that can be used to identify and analyze a broad range of tussles from the desired topic space. Each tussle pattern identifies agents, their interests in resources and how conflicts of interests emerge between actors. Each tussle pattern has distinctive characteristics that make them difficult to resolve.

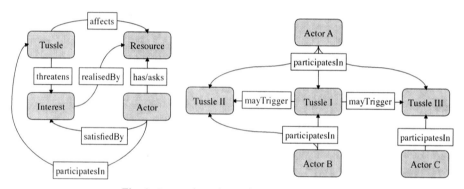

Fig. 1. An ontology for socio-economic tussles

Tussles can be categorized based on their nature as economic and social ones. Economic tussles refer to conflicts between stakeholders, motivated from an expected reward gained (or cost avoided) when using scarce resources rationally, while social tussles refer to conflicts between stakeholders that do not share the same social interests, or that have repercussions into broader society as a result of changes in the technical domain. Tussles related to engineering decisions during design-time are out of article's scope, but we argue that most of them have their roots in economic and social domain. We should note that a single tussle instance could have arisen because a set of stakeholders follow economic objectives and their actions affect the social interests of other stakeholders.

3.1   Tussle Patterns

We have identified an initial set of four tussle patterns that include contention, repurposing, responsibility and control. Figure 2 shows the actors involved in each tussle pattern, and their interests that result in conflict for a set of resources. Dotted arrows represent a conflict among two stakeholders, while a dotted rectangle shows the selected set of resources when at least one stakeholder has the ability to influence the outcome. Based on the context, a reverse tussle pattern may also be present. The characteristics of each pattern can be seen in many current and future Internet scenarios. Each pattern looks at relationships between consumers and suppliers and how conflicts of interest can emerge through technical innovations. The dynamics of a relationship over time is important, as interests, values and technologies change. By classifying tussle patterns we envisage the provision of a reference point in performing the second step of the proposed methodology for identifying and assessing tussles. It is important to note that the roles "consumer" and "provider" are context specific, and an individual stakeholder can be a resource consumer in one tussle, but a provider of a resource in another. For instance, while individual Internet users are typically consumers, when they are creating data that a business would like to sell, with or without their knowledge and consent, they are "providers" of the resource in such a scenario. The initial set of tussle patterns is described below.

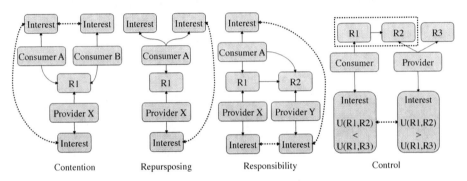

Fig. 2. The Initial Set of Tussle Patterns

The *contention* tussle pattern involves two or more consumers (A and B) using a single resource R1 from a provider X for the same or different interests. The tussle exists either between consumer interests due to the scarcity of resource, or among a consumer and the provider due to the impact on a provider's ability to exploit the resource. The role of the consumer may be played by an end-user or even a provider that receives services at the wholesale level (we refer to this case as the reverse contention tussle). In the reverse case, two providers may compete for a resource owned by a single consumer. Instances of this tussle pattern have their roots in economics and thus are typically resolved through the process of economic equilibrium or through regulation when an interest becomes a citizen's right. Examples include cloud resources utilization like bandwidth of bottleneck links. If pricing schemes for such shared resources are not sensitive to the volume consumed then a "tragedy of the commons" can arise.

In the *repurposing* tussle pattern, a consumer A wants to use a resource R1 from a provider X for an interest not acceptable to X. The tussle exists between consumer and provider if A's new interest utilizes R1 in unforeseen ways that affects X's ability to deliver R1 sustainably, and/or the value A derives from their new interest fair exceeds that gained by X. The situation often results in X restricting the capabilities of resources. Examples of economic tussles include sharing of copy-righted files (e.g. music) and selling of personal information. It is important to note that many innovations in the Internet space have involved repurposing of resources, so identifying this sort of tussle also represents a way to find potential areas of growth and innovation.

In the *responsibility* pattern, a consumer A uses a resource R1 from provider X and resource R2 from provider Y to fulfill an interest that is not acceptable to provider Y. The tussle exists between providers as it is not in X's interest to defend Y's interests. The situation is difficult to resolve as acceptance of responsibility has a cost, which when not aligned with a business objective is difficult to incentivise. Example includes distribution of copy-righted content.

In the *control* tussle pattern, a consumer – or provider – X uses a resource R1 but relies on provider Y in order for the service to be completed. Provider Y can use either resource R2 or R3, but chooses R2 that is different from the one provider X prefers. This tussle pattern arises because each provider makes decisions following different policies and is mostly related to economic objectives. An example of such tussles is attempts by an ISP to restrict how a consumer uses the resource (e.g. performing deep packet inspection and throttling so that quality of other services is acceptable).

3.2   Economic Tussles

Economic tussles refer to conflicts between stakeholders, motivated from an expected reward gained (or cost avoided) when acting rationally. These tussles are realized by taking advantage of imbalanced access to necessary information, or uneven control abilities. The latter case stems from protocol features not designed for being used in that way, or were intentionally left out of scope. Economic tussles are mostly related to the scarcity of certain resources that need to be shared. Furthermore, such tussles can occur between collaborating stakeholders due to different policies or, in economic terms, different valuations of the outcome. Tussles can also appear when a stakeholder is being bypassed.

Contention tussles are usually caused by the existence of scarce resources and can be seen as evidence of misalignment between demand and supply in the provisioning of services. A popular example is bandwidth of bottleneck links and radio frequencies shared between users and wireless devices. In the former case, modern transport control protocols perform congestion control without considering the utility of the sender on instantaneous bandwidth or the number of their active connections. This, together with the prevalence of flat pricing schemes, has led to a contention tussle among user types, which economists identify as a "tragedy of the commons". Similar contention tussles can take place for other cloud resources as well, such as processing and storage capabilities of servers and networking infrastructure. For example, routing table memory of core Internet routers can be considered a "public good" that retail ISPs have an incentive to over-consume by performing prefix de-aggregation with Border Gateway Protocol (BGP). Another type of scarce Internet resources is network identifiers, like IPv4 addresses and especially "Provider Independent" ones that ease network management and avoid ISP lock-in. Sometimes a contention tussle between consumers can have side effects on the owner of the scarce resource, which is an economic entity and must protect its investments. Examples include the deployment of Deep Packet Inspection techniques by ISPs in order to control how bandwidth is allocated across users and services.

The remaining tussle patterns are mostly seen in bilateral or multilateral transactions where one party has an information advantage over others; a situation known as "information asymmetry" in the economic theory literature. This imbalance of power can sometimes lead to "market failures", a case where the final outcome is not preferable by any participant. Two well-known effects of information asymmetry are "adverse selection" and "moral hazard".

Adverse selection arises when several providers offer the same service, with possibly different quality features known only to each seller, and the buyer who seeks a high-quality service is prepared to pay on average less than the price he would be willing to pay if he could infer the quality of all candidates and select the most suitable one. But, this lower price would lead some sellers of higher quality services to stop selling (since they do not cover their costs anymore) and thus, in the long term, only low quality services will be available. Similarly, if a service provider were the less informed party, then setting - for example - a low price would increase his risk of being selected by the least profitable customers. This would increase his costs and trigger a rise in prices, making this service less attractive to a number of profitable customers. Eventually this "adverse selection spiral" might, in theory, lead to the collapse of the market. The *repurposing* tussle pattern described above can be associated with the adverse selection issue.

Moral hazard can occur when one party of the transaction cannot (or it is very costly to) infer the actions of the other one and thus the latter one may have the incentive to behave inappropriately. *Responsibility* tussles are related to moral hazard and usually arise when a service contract term is violated and the consumer has economic transactions with multiple providers. This is the case when a set of providers collaborate during service provision with strict requirements, like long-distance phone conversations taking place over Internet. Each provider has partial private information about the problem and no one is willing to take responsibility and the resulting cost. Furthermore, this type of tussle can occur as a side effect to a contention tussle. In the example of file sharing applications, if an ISP deployed middle-boxes and performed traffic shaping then it may have negative impact on the services, and thus, on the viability of new ASPs, who however cannot safely attribute these effects to the ISP.

Closely related to moral hazard is the "principal-agent problem", where one party - the principal – delegates control to the agent, but their interests are not aligned and thus the latter has the ability and incentive to shirk. The *control* tussle pattern is a principal-agent type of problem, observed when the involved parties have a customer-provider relationship or, in general, when a contract outlines their obligations. For example, control tussles can appear when a pair of entities makes decisions following different policies and conflicting objectives. Examples of such tussles include different policies on routing decisions, for example ISPs selecting the next hop of user traffic while users selecting the traffic source in their requests. In the former case, a provider may seek redundancy and reliability asking for a backup path towards a destination, or prefer avoiding specific upstream ISPs. In the latter case, when multiple candidate servers are available, a consumer may prefer the one offering better QoS, while a provider selects the server that minimizes its cost; e.g., this is possible if the provider operates a local DNS service. However, the control pattern includes also cases where there is no contract between the participants and these may have conflicting interests on the possible outcomes. In this last case, information about each other's preferences, possible actions and ability to observe past actions can have significant effect on the outcome, as is the case with the "prisoner's dilemma" game theoretic problem.

Researchers have proposed several methods in order to deal with the above issues. For example, incentive schemes like reputation systems and penalties to promote effort and care are suggested as a countermeasure for moral hazard issues. Similarly, the proposed way for mitigating the effects of adverse selection is for the less informed party to gather more information (called "signaling") and select candidate transaction partners (called "screening") by using, for example, auctions.

3.3   Social Tussles

What does SESERV mean when discussing social tussles? At the most basic level, these tussles represent issues that arise as a result of a disconnect between the technical affordances of the network and the interests of regulators, business and individuals at the micro level and societal values and social goods at the macro level. SESERV can identify social tussles that arise as a result of how individuals interact with each other and with technology, based on their roles, identities, and psychology.

Repurposing tussles occur in regards to the privacy of user communication data between users, ISPs, service providers and regulators. The users are social actors who have a desire, generally speaking, that networks are trustworthy and private [2]. The privacy of communications is based on democratic ideals, that persons should be secure from unwarranted surveillance. However, the issue turns into a tussle over the very definition of what constitutes unwarranted surveillance, and when surveillance may be warranted in ways that individual users are willing to forego their privacy concerns in the interest of broader societal concerns. Governments frequently argue that in order to protect national security, they must be given access to network communication data. Furthermore, ISPs and other companies such as Google and Amazon have increasingly been able to monetize their user transaction data and personal data. Google is able to feed advertisements based on past searching and browsing habits, and Amazon is able to make recommendations based on viewing and purchasing habits. These applications of user data as marketing tools are largely unregulated. And in many cases, users have proved willing to give up some of their privacy in exchange for the economic benefit of better deals that can come from targeted advertising. However, for users who wish to opt out of such systems, the mechanisms for doing so are often less than clear, since the owners of the system prefer to keep people in, rather than easily let them out.

Responsibility tussles occur with ISPs that often inhabit a middle ground – they are the bodies with direct access to the data, but are simply businesses, trying to make a profit. ISPs, however, are often placed in the uncomfortable position of trying to negotiate a balance between their users' expectations of privacy (which, if breached, could cause them to take their business elsewhere), the potential profits to be made from monitoring and monetizing the communication of their users, and the demands of government bodies to be able to monitor the networks for illegal or unwanted activities.

Control tussles in a social context relate, for instance, to digital citizenship and understanding the balance between individual and corporate rights and responsibilities, and how such a balance can be achieved through accountability and enforceable consequences (e.g. loss of privileges). This is a difficult issue, which must be debated and resolved in the real world by policy makers and legal experts. However, these processes tend to be slow to deal with change, particularly when compared to the speed of change in many technological systems such as the Future Internet. In practice, technology that upsets the balance of control is often released and the debates over control and resulting policy changes follow. In some ways, new technology that unbalances existing systems of control can be the impetus and focus of debates that would otherwise be quite dry and difficult to interest politicians and citizens in. This is a very tough problem and relates to those promoting principles of open society and those wishing to maintain confidential communication.

For instance, is Wikileaks right or wrong to distribute leaked documents containing the details of government and corporate communications? Until Wikileaks started releasing real documents of widespread interest, few people were interested in debating the societal risks and values surrounding a platform that could potentially distribute previously secret documents. However, once Wikileaks began distributing documents, millions of people worldwide began to debate these very issues in the media, in seats of government and power, and at the dinner table. Suddenly, questions regarding whether Wikileaks should do this, whether governments and businesses had the right to censor or attack them for doing so, and what role ISPs and service providers such as PayPal have in supplying services to controversial online bodies come to the forefront. If Wikileaks is wrong, what sanctions would be appropriate, and what technical designs would be appropriate to implement them? Conversely, if Wikileaks is right, what technical designs can protect such sites from being attacked by entities inconvenienced or embarrassed by their revelations? The Internet makes this a particularly contentious issue because with the global nature of the Internet one can't just assume Western values (as if it were possible even within Europe to agree to what that means). Where does national sovereignty fit into all of this? Such a tussle of control would need to be assessed by philosophers and politicians as well as security and trust experts.

4   Survey of Work on Social and Economic Tussles as Highlighted in FP7 Projects

In this section, SESERV looks at specific projects in the FP7 Future Networks project portfolio, and discuss the socio-economic tussles related to them.

The Trilogy project [16] studied extensively the *contention* tussle among users as well as among an ISP and its customers, due to the aggressive behavior of popular file-sharing applications. On the one hand it proposed two protocols and a novel congestion control algorithm that gives the right incentives to users of bandwidth intensive applications. Re-ECN protocol makes senders accountable for the congestion they cause. It requires a sender to inform the network about the congestion that each packet is expected to cause; otherwise the packet will be dropped with high probability before reaching its destination. MPTCP is a new multi-path transport protocol that carefully couples the congestion control of multiple sub-paths so that ISPs' resources are shared between users in a fairer manner. This is achieved by configuring MPTCP so that it acts less aggressively than TCP when the latter flows experience congestion and more aggressively otherwise. Furthermore, the adoption of several protocols (*i.e.* MPTCP, LISP) and pricing schemes (based on traffic volume and congestion volume) has been studied as a control tussle among providers.

The Trilogy project also studied the social tussles surrounding "phishing", the attempt to acquire sensitive personal data of end-users by masquerading as a trustworthy entity, as a reverse *contention* tussle among two website owners (the "consumers"). The tussle is being played out in the routing domain: the fraudulent one advertises more specific BGP prefixes so that ISPs update the entries in their routing tables (the resource) and route end-user requests to the fake website instead of the real one. This situation has been shown to be a real problem due to the incentives of ISPs to increase their revenues by attracting traffic, but no mechanism has been suggested to deal with this security problem and the fears that it raises among end-users. There is a special social concern regarding vulnerable populations such as the elderly, who are often considered to be easy targets for such "phishing" attempts.

The ETICS project (Economics and Technologies for Inter-Carrier Services) [8] studies a *repurposing* tussle arising when an ISP (the "provider") requests a share of an ASP's revenues (the "consumer") due to its higher investment risks and operational costs. ETICS proposes technical solutions and economic mechanisms that will allow network providers to offer inter-domain QoS assurance and obtain higher bargaining power during negotiations for service terms (e.g. pricing). The need for collaboration among ISPs gives rise to a *control* tussle and a *responsibility* tussle in case of contract term violation.

The SmoothIT project (Simple Economic Management Approaches of Overlay Traffic in Heterogeneous Internet Topologies) studies the *control* tussle that arises between ISPs and ASPs with respect to the routing decisions of each party. An ASP or peer-to-peer (P2P) application may employ advanced probing techniques for estimating the performance on each path and select the path (or destination) that maximizes its utility. At the same time an ISP performs traffic engineering without being able to predict how ASPs will react. This results to an endless loop of selfish actions that increases the cost of ISPs and limits performance gains of ASPs. To this end, an incentive-based approach was developed, referred to as the Economic Traffic Management (ETM). ETM offers better coordination among the aforementioned players that is mutually beneficial [15].

The development and investigation of In-Network Management mechanisms was a novel paradigm to manage networks according to the 4WARD project [1]. Since it is based on a lean architecture to operate new services in the Future Internet, the discovery of capabilities and the adaptation of many management operations to current working conditions of a network are major elements in the new approach. Thus, a *control* tussle arises, where embedded capabilities of networking devices and elements see "default-on" management functionality, which consist out of autonomous components interacting with each other in the same device and with components in neighboring devices. This requires device vendors to change their management model and ISPs to enable respective embedded management functionality within their networks.

The MOBITHIN project [13] is related to a *responsibility* tussle between users of wireless services, mobile operators and regulators that has arisen from the social interest to reducing carbon footprint of the ICT sector and the economic incentive to minimize costs. The regulator (who is in charge of allocating and administering how spectrum is being utilized and thus can be seen as "Provider Y" in Figure 2) is trying to place limits on energy consumption of both consumers and providers and may introduce penalty fees to those that don't use efficient technologies. Due to economies of scale the thin-client paradigm, where most applications run on a remote server, is considered to achieving energy savings but to the disadvantage of the server provider. However under some assumptions, WiFi hotspots can consume much less energy than UMTS (Universal Mobile Telecommunications System) networks. Thus, responsibility cannot be easily checked. Furthermore, this situation triggers a *control* tussle between wireless network operators and users of dual-band devices (e.g. WiFi and UMTS) on the technology used to communicate. Next generation networks, where a provider can control which access technology is used by its end-users, could affect the user's ability to derive maximum value from the service.

The SENDORA project [14] identifies a *contention* tussle based on their own ecosystem design for Sensor Network aided Cognitive Radio technology that utilizes wireless sensor networks to support the coexistence of licensed and unlicensed wireless users in an area. In this case, the spectrum is the resource in contention and the "provider" is the regulator, which is not the owner but the administrator of the resource. Existing mobile operators, TV broadcasters and new operators are the "consumers" of the resource in contention. The latter is looking to have a slice of the resource in order to develop business whilst the former two are at once trying to block the entry of new entrants to the market and minimize any impact on their existing business. The solution proposed by SENDORA is to build this tussle into their business ecosystem and to design benefits for the incumbent resource consumers (e.g. mobile operators and TV broadcasters) such as reduced operating costs, superior technology and potentially lucrative spectrum trading. Furthermore, there is a *repurposing* tussle between a regulator for anti-competitive tactics and the provider. The spectrum can be used for providing a service as well as a barrier-to-entry which is in conflict with the regulator's interest for preventing monopolies.

These are just a few examples among the many tussles that exist or potentially exist as a result of technological changes and innovations being researched to advance the Future Internet. One challenge for the technologists designing new hardware, software, systems, and platforms, however, is to be aware that technology is not value-free, since it can have several consequences. To some extent, this message has already been taken on board by many policy makers, computer scientists, and systems designers. The recognition that technology-in-use frequently differs from technology-during-design is growing. Thus technology will have socio-economic consequences when released, and the challenge is to take steps to anticipate those consequences where possible, to identify unanticipated emergent consequences as they arise, and to learn the lessons of previous tussle negotiations and resolutions to smooth the implementation of future designs.

5   Conclusions and Future Work

The SESERV Coordination and Support Action was designed to help fill the gap between socio-economic priorities and the Future Internet research community by offering selected services to FP7 projects in Challenge 1. SESERV provides access to socioeconomic experts investigating the relationship between FI technology, society, and the economy through white papers, workshops, FIA sessions, and research consultancy.

In this paper SESERV proposes a methodology for identifying and assessing tussles that are present in the Internet, or may arise after a protocol or service has been introduced. Although the suitability of such a methodology cannot be easily quantified, we believe it can capture the evolving relationships among stakeholders, and thus tussles, across time. Furthermore, we provide a taxonomy of economic and social tussles linked to a number of identified patterns and give examples of such tussles and how these are studied by several European research projects under FP7. The tussle analysis methodology will be evaluated in the context and work of other FP7 projects during the lifetime of the project, and will be enriched and complemented by other techniques; thus forming a toolbox of approaches to understanding the socio-economic issues inherent in FP7 FI projects. This toolbox will be further enhanced and finalized after the workshops, sessions, and consultations, as the project empirically identifies socio-economic issues arising from FP7 and links those to socio-economic tools and methods for analyzing and resolving these issues.

Even though the SESERV project is at its initial phase SESERV can state preliminary observations on whether and the extent to which socio-economic issues are being addressed by the FP7 Future Network project portfolio. At this early stage, SESERV noticed that a significant number of research projects show a major technical viewpoint. For example, MIMAX, EUWB, FUTON, and WIMAGIC try to design technical solutions that achieve efficient spectrum usage for mobile devices. Following the increasing consensus on benefits of incorporating economic incentive mechanisms in technical solutions, several projects like Trilogy, SmoothIT, ETICS, and PURSUIT follow a techno-economic approach. However, SESERV feels that less focus has been given on the interplay of technical, social, and economic objectives. This could be attributed to the difficulty in setting up such a multi-disciplinary team in order to apply a holistic approach, when making technology decisions and/or the inherent difficulty of addressing socioeconomic issues in the Internet when such challenges still exist in the real world.

References

1. 4WARD, http://www.4ward-project.eu (accessed December 1, 2010)
2. Blackman, C., Brown, I., Cave, J., Forge, S., Guevara, K., Srivastava, L., Tsuchiya, M., Popper, R.: Towards a Future Internet: Interrelation between Technological, Social and Economic Trends, Final Report for DG Information Society and Media, European Commission DG INFSO, Project SMART 2008/0049 (2010)
3. Blazic, B.J.: The Future of the Internet: Tussles and Challenges in the Evolution Path as Identified. In: Fourth International Conference on Digital Society 2010 (ICDS'10), pp. 25–30.
4. Checkland, P.: Systems Thinking, Systems Practice. Wiley, Chichester (1981)
5. Clark, D.D., Wroclawski, J., Sollins, K.R., Braden, R.: Tussle in Cyberspace: Defining Tomorrow's Internet. IEEE/ACM Transactions on Networking 13(3), 462–475 (2005)
6. Courcoubetis, C., Weber, R.: Pricing Communication Networks: Economics, Technology and Modeling. Wiley, Chichester (2003)
7. CRAMM: http://www.cramm.com (accessed December 1, 2010)
8. ETICS: https://www.ict-etics.eu (accessed December 1, 2010)
9. Ford, A., Eardley, P., van Schewick, B.: New Design Principles for the Internet. In: IEEE International Conference on Communications Workshops, June 2009, pp. 1–5 (2009)
10. Herzhoff, J.D., Elaluf-Calderwood, S.M., Sørensen, C.: Convergence, Conflicts, and Control Points: A Systems-Theoretical Analysis of Mobile VoIP in the UK. In: Ninth International Conference on Mobile Business and 2010 Ninth Global Mobility Roundtable (ICMB-GMR), June 2010, pp. 416–424 (2010)
11. Wang, J.H., Chiu, D.M., Lui, J.C.S.: Modeling the Peering and Routing Tussle between ISPs and P2P Applications. In: IWQoS 2006, pp. 51–59 (2006)
12. Koutsoupias, E., Papadimitriou, C.: Worst-case Equilibria. In: 16th Annual Symposium on Theoretical Aspects of Computer Science 1999, pp. 404–413 (1999)
13. MOBITHIN project: D2.5 Business Models, Public Deliverable, http://www.mobithin.eu (accessed December 1, 2010)
14. SENDORA project: D2.2 Business Case and Ecosystem Evaluations, Public Deliverable, http://www.sendora.eu (accessed December 1, 2010)
15. Soursos, S., Rodriguez, M., Pussep, K., Racz, P., Spirou, S., Stamoulis, G.D., Stiller, B.: ETMS: A System for Economic Management of Overlay Traffic. In: Towards the Future Internet - Emerging Trends from European Research, IOS Press, Amsterdam (2010)
16. Trilogy: D10 - Initial Evaluation of Social and Commercial Control Progress, 2009, Public Deliverable, http://trilogy-project.org (accessed December 12, 2010)

Part III:

Future Internet Foundations: Security and Trust

Introduction

If you are asking for the major guiding principles of Future Internet technology and applications, the answer is likely to include "sharing and collaboration". Cloud computing, for instance, is built on shared resources and computing environments, offering virtualized environments to individual tenants or groups of tenants, while executing them on shared physical storage and computation resources. The concept of Platform-as-a-Service provides joint development and execution environments for software and services, with common framework features and easy integration of functionality offered by third parties. The Internet of Services allows the forming of value networks through on-demand service coalitions, built upon service offerings of different provenance and ownership. And, finally, the principle of sharing and collaboration reaches to the applications and business models, ranging from the exchange of data of physical objects for the optimization of business scenarios in, e.g., retail, supply chain management or manufacturing, – the Internet of Things – to social networks.

While it is evident that sharing and collaboration brings the Internet, its technologies, applications and users to the next level of evolution, it also raises security and privacy concerns and introduces additional protection needs. The Future Internet is characterized by deliberate exposure of precious information and resources on one hand and a number of likely previously unknown interacting entities on the other hand, including service and platform providers as well as service brokers and aggregators. Valuable and sensitive information, be it business or personal data, should, however, only be exposed to known and trusted entities and in a controlled way, allowing the owner of the data to decide and control how, when, and where it is going to be used. These are not new requirements in nature, but the Future Internet adds new dimensions of scale and complexity. The number of participating and collaborating entities reaches billions when we consider the inclusion of physical objects, and they need to be identified and trusted when information is passed along. Entities interact spontaneously and form ad-hoc coalitions, asking for trust establishment for short-term relations. Data travel through a multitude of different domains, contexts and locations while being processed by a large number of entities with different ownership. Hence, they need to be traced and treated according to the data owner's policy, in balance with the processing entities' policies.

Increased dynamics, frequently changing relations of entities, distributed infrastructures and shared information, in addition, change the threat model and increase the attack surface. An attack can potentially be launched by a malicious or fake service provider, service consumer, infrastructure provider or service broker or any other Future Internet entity, while distribution and exchange of data serve for additional entry points that can potentially be exploited to penetrate a system. The challenge is to design security and trust solutions that scale to Future Internet complexity and keep the information and resource owner in control, balancing potentially conflicting requirements while still supporting flexibility and adaptation. Explicit specification of protection needs in terms of declarative policies is key, as well as providing assurance about security properties of exposed services and information.

The chapters presented in the Security and Trust section of this volume look at the challenges mentioned above from three different angles. First, Future Internet principles are supported by revised communication paradigms, which address potential security issues from the beginning, but also imply the need for novel solutions like integrity and availability. The chapter, "Security Design for an Inter-domain Publish/Subscribe Architecture" by K. Visala et al. looks into security implications of a data-centric approach for the Future Internet, replacing point-to-point communication by a publish/subscribe approach. This allows the recipient to control the traffic received and, hence, avoids issues of unwanted traffic from the beginning. The authors introduce a security architecture based on self-certifying name schemes and scoping that ensure the availability of data and maintains their integrity. It is a good example of how clean-slate approaches to the Future Internet can support security needs by design, rather than provided as an add-on to an existing approach, as has been the case for the current Internet.

The second group of chapters investigates the provision of assurance of the security properties of services and infrastructures in the Future Internet. The provision of evidence and a systematic approach to ensure that best security practices are applied in the design and operation of Future Internet components are essential to provide the needed level of trustworthiness of these components. Without trustworthy components, the opportunities of sharing and collaboration cannot be leveraged. The chapter "Engineering Secure Future Internet Services" by W. Joosen et al. makes a point for establishing an engineering discipline for secure services, taking the characteristics of the Future Internet into account. Such a discipline is required to particularly emphasize multilateral security requirements, the composability of secure services, the provision of assurance through formal evidence and the consideration of risk and cost arguments in the Secure Development Life Cycle (SDLC). The authors propose security support in programming and execution environments for services, and suggest using rigorous models through all phases of the SDLC, from requirements engineering to model-based penetration testing. Their considerations lead to the identification of Future Internet specific security engineering research strands. One of the major ingredients of this program, the provision of security assurance through formal validation of security properties of services, is investigated in detail in the chapter 'Towards Formal Validation of Trust and Security in the Internet of Services" by R. Carbone et al. They introduce a language to specify the security aspects of services and a validation platform based on model-checking. A number of distinguished features ensure the feasibility of the approach to Future Internet scenarios and the scalability to its complexity: it supports service orchestration and hierarchical reasoning, the language is sufficiently expressive so that translators from commonly used business process modeling languages can be used, and the automated technology is integrated in industrial-scale process modeling and execution environments. The two chapters demonstrate the way towards rigorous security and trust assurance in the Future Internet addressing one of the major obstacles preventing businesses and users to fully exploit the Future Internet opportunities today.

While engineering and validation approaches provide a framework for the secure design of Future Internet artifacts adapted to its characteristics, the third group of chapters looks into specific instances of the information sharing and collaboration principle and introduces novel means to establish their security. The chapter "Trustworthy Clouds underpinning the Future Internet" of R. Glott et al. discusses latest trends in cloud computing and related security issues. The vision of clouds-of-clouds describes collaboration and federation of independent cloud providers to provide seamless access to end users, as if they were working within a single cloud environment. This advanced level of distribution offers increased economic benefits, but also faces new security risks, from the breach of separation between tenants to the compliance challenge in case of distribution over different regulatory domains. The authors discuss these risks and provide an outlook to their mitigation, embedded in a systematic security risk management process. In cloud computing, but also in most other Future Internet scenarios like the Internet of Services, the need for data exchange leads to sensitive data, e.g., personally identifiable information, travelling across a number of processes, components, and domains. All these entities have the means to collect and exploit these data, posing a challenge to the enforcement of the users' protection needs and privacy regulations. This is amplified by the dynamic nature of the Future Internet, which does not allow one to predict by whom data will be processed or stored. To provide transparency and control of data usage, the chapter "Data Usage Control in the Future Internet Cloud" proposes a policy-based framework for expressing data handling conditions and enforcing them. Policies relating events and obligations are coupled with data ("sticky policies") and, hence, cannot get lost in transition. A common policy framework based on tamper-proof event handlers and obligation engines allows for the evaluation of user-defined policies and their execution, leaving control to the user.

With the three groups of chapters, this section of the book provides directions on how security and trust risks emerging from the increased level of sharing and collaboration in the Future Internet can be mitigated, removing a major hurdle for using its exciting opportunities in sensitive scenarios of both the business and societal worlds.

Volkmar Lotz and Frances Cleary

Security Design for an Inter-Domain Publish/Subscribe Architecture

Kari Visala[1], Dmitrij Lagutin[1], and Sasu Tarkoma[2]

[1] Helsinki Institute for Information Technology HIIT /
Aalto University School of Science and Technology, Espoo, Finland
{Kari.Visala, Dmitrij.Lagutin}@hiit.fi
[2] Department of Computer Science, University of Helsinki, Helsinki, Finland
Sasu.Tarkoma@cs.helsinki.fi

Abstract. Several new architectures have been recently proposed to replace the Internet Protocol Suite with a data-centric or publish/subscribe (pub/sub) network layer waist for the Internet. The clean-slate design makes it possible to take into account issues in the current Internet, such as unwanted traffic, from the start. If these new proposals are ever deployed as part of the public Internet as an essential building block of the infrastructure, they must be able to operate in a hostile environment, where a large number of users are assumed to collude against the network and other users. In this paper we present a security design through the network stack for a data-centric pub/sub architecture that achieves availability, information integrity, and allows application-specific security policies while remaining scalable. We analyse the solution and examine the minimal trust assumptions between the stakeholders in the system to guarantee the security properties advertised.

Keywords: Future Internet, publish/subscribe networking, network security

1 Introduction

Data-centric pub/sub as a communication abstraction [2,3,4] reverses the control between the sender and the receiver. Publication in the middle decouples the publisher from the subscriber and there is no direct way of sending a message to a given network, which provides a good starting point against distributed denial of service (DDoS) attacks, where a number of nodes try to flood part of the network with unwanted traffic.

Most pub/sub systems have been overlays on top of IP but our goal is to replace the whole Internet protocol suite with a clean-slate data-centric pub/sub network waist [14]. This enables new ways to secure the architecture in a much more fundamental way compared to overlay solutions, as the underlay cannot anymore be used to launch DDoS attacks against arbitrary nodes in the network. Our architecture comprises of rendezvous, topology, and forwarding functions and the security design presented here covers all these as a whole. In this paper we refine and extend our work in [5] and especially concentrate on the concept of *scope* and how it can be used flexibly to

J. Domingue et al. (Eds.): Future Internet Assembly, LNCS 6656, pp. 167–176, 2011.

support many types of application-specific security policies. Some of the techniques used in our architecture, such as securing of forwarding, delivery tree formation, and rendezvous system interconnection are explained in [5] in more detail.

Our security goals concur with [1] except that confidentiality and privacy are expected to be handled on top of the network layer and are outside the scope of this paper. The security goals are:

- **Availability**, which means that the attackers cannot prevent communication between a legitimate publisher and a subscriber inside a trusted scope.
- **Information integrity**, which guarantees that the binding between the identity of the publication and its content cannot be broken without the subscriber noticing it.
- **Application-specific security policies**, which mean that the architecture can cater for the specialized security policies of different types of applications while partially same resources can be shared by them.

In addition to aforementioned goals, the solution is restricted by the requirements of *scalability* and *efficiency*. For example, it must be assumed that the core routers forward packets at line-speeds of tens of Gigabits per second, which requires expensive, high speed memory for the routing tables. In the inter-domain setting, we have to take into account the various stakeholders such as ISPs, end-users, and governments, and *tussles* [6] between their goals. That is, we cannot enforce certain design choices but keep the architecture flexible and let the balance of power between stakeholders to decide the stable configuration. The design should also adhere to architectural constraints such as the *end-to-end principle* (E2E) [13] by which we mean that the network itself should only have minimal functionality that is required to efficient utilization of the invested resources, and the rest of the features should be implemented on higher layers at the endpoints to be more flexible. The architecture must be incrementally deployable, and minimal in complexity and trust assumptions between stakeholders.

2   Basic Concepts

Data- or content-centric networking can be seen as the inversion of control between the sender and the receiver compared to message passing: Instead of naming sinks to which senders can address messages, the receiver expresses its interest in some identified data that the network then returns when it becomes available taking advantage of multicast and caching [2,3]. We use the term information-centric for this communication pattern to emphasize that the data items can link to other named data and that the data has structure.

An immutable association can be created between a *rendezvous identifier* (Rid) and a data value by a *publisher* and we call this association a *publication*. At some point in time, a *data source* may then publish the publication inside a set of *scopes* that determine the distribution policies such as access control, routing algorithm, reachability, and QoS for the publication and may support transport abstraction specific policies such as replication and persistence for data-centric communication. The scope must be trusted by the communicating nodes to function as promised and much of the security of our architecture is based on this assumption as we explain in [5]. Scopes are identified with a special type of Rid called *scope identifier* (Sid).

Even though the control plane of our architecture, implementing the rendezvous function, operates solely using data-centric pub/sub model, it can be used to set up communication using any kind of transport abstraction on the data plane *fast path*, that is used for the payload communication. The data-centric paradigm is a natural match with the communication of topology information that needs to be distributed typically to multiple parties and the ubiquitous caching considerably reduces the initial latency for the payload communication as popular operations can be completed locally based on cached data.

Below the control plane, t he network is composed of *domains*, that encapsulate resources such as links, storage space, processing power in routers, and information. The concept of domain is here very general, and can refer to abstractions of any granularity, such as software components, individual nodes, or ASes. An *upgraph* of a node is the set of potential resources, that can be represented as a network map of domains and their connectivity, to which a node has an independent access to based on its location and contracts between providers. We have developed our own language called *advanced network description language* (ANDL) for the communication of network topology information in control plane publications, but it is outside the scope of this paper.

The links in the upgraphs represent resources with minimal abstraction and can be limited to carrying only various *transport* abstractions or protocols over them. Each transport protocol implements a specific communication abstraction and every actualized *instance of interaction* or communication event consuming the resources of the network has an associated transport, *topic*, a *graphlet* and a set of *roles*. For example, when IP is seen as a transport protocol in our network architecture, the roles for the endpoints are a source and a destination or for data-centric transport: a *data source* and a *subscriber*. The topic is identified with an Rid and is used to match the end nodes in correct interaction instances by the scope. For example, for data-centric communication, the topic identifies the requested publication.

A graphlet defines the network resources used for the payload communication and it can be anything from the path of an IP packet to private virtual circuits. Some protocols may require an additional phase for the reservation of a graphlet before the payload communication. A graphlet adheres to a set of scopes that are responsible for policy-compliant matching of nodes to interaction instances, collecting the needed information to build end-to-end paths and placing constraints on the chosen resources for the graphlet. A graphlet connects a set of end nodes that each implement a certain role in the transport.

C ommunication always happens inside at least a single scope, but the policies can be divided into aspects of communication handled modularly by different scopes implemented by different entities. Scopes are responsible for combining upgraph information from multiple nodes requesting to participate in an interaction instance inside the scope [15] and the scope selects a subset of the given resource that adhere to its policies. In cases such as CDNs, the scope can bring its own additional resources to be used in the graphlet. It should be noted, that because a scope can act as a neutral 3rd party for the route selection, it can balance the power between endpoints and optimize the path as a whole.

2.1   Identifier Structure

All identifiers in our system have the similar self-certifying, DONA-like structure [4]. An identifier is a *(P,L)* pair where P is the public-key of the *namespace owner* of the identifier and L is a variable length label of binary data. Only fixed length hash of the identifier is used in-network, for example in caches, to identify a publication, but variable length names are needed for dynamically generated content, where the data source uses the label as an argument to produce the publication on the fly. This could have been emulated by using only fixed length identifiers, but it would have required additional round-trips in some cases. Because each namespace is managed by its owner, we do not need a special hierarcy or centralized entity managing the labels.

The self-certification is achieved by the namespace owner authorizing a publisher to publish a certain set of labels in the namespace. The actual content segments are then signed by the publisher and the certificate from the namespace owner is included. Together these form a trust chain starting from the namespace owner that can be verified by the subscriber. Here the security model only guarantees the integrity of the association between an identifier and its content. It is assumed that the subscriber knows that she has the correct identifier and trusts the scope enough for availability. For the cases where the content of the publication is known beforehand and no human-readable labels are required, we support a stronger model by allowing the hash of the content to be stored in the label part of the Rid. Confidentiality of publications can be achieved by encryption of the content and/or the labels.

Fig. 1 depicts a simplified example of "My movie edit meta-data" publication that has Rid *(P_N, "My Robin Hood video edit")*, where P_N is the public key of the user's own namespace. The contents of this publication point to another "movie frame data" publication indirectly using a so called *application level identifier* (Aid) of the referred publication. Alternatively, a network level Rid could have been used directly, but they are not assumed to have a long life-time as the security mechanism is coupled with the identifier. We assume that multiple application level schemes for identifiers with different properties will be developed and these can be translated into Rids by various means. DNS is one such orthogonal mechanism that could be used to map long-term human-readable names to Rid namespace public keys.

As can be seen in Fig. 1, the publication contents are split into segments, that each have their own signature chain starting from the P_N verifying the contents of the segment. This makes it possible to check segments independently, for example, before caching. Because it is not feasible to use traditional cryptographic solutions like RSA on a per-segment basis in the payload communication, we use *packet level authentication* (PLA) [25] that uses elliptic curve cryptography (ECC) [23]. An FPGA based hardware accelerator has been developed for PLA [24] accelerating cryptographic operations.

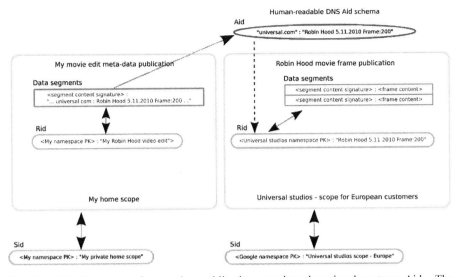

Fig. 1. Publications can refer to other publications persistently using long-term Aids. The scopes, where publications are made available are orthogonal to the structure of the data.

In Fig. 1, the publication on the left is published inside "My home scope" that is fully controlled by the local user. On the other hand, the movie frame publication on the right is stored inside movies studio's localized scope, which can, for example, limit the access to the scope only to the customers of the company. In this example, it is easy to see that the logical structure of the data, e. g. the link between the two publications, is orthogonal to the scoping of the data that determines the communication aspects for each publication.

2.2   Interdomain Structure

Each node has an access to a set of network resources. In the current Internet, most policy compliant paths have the so-called *valley-free* property [16], which means that, on the AS business relationship level, packets first follow 0-n logical customer-provider "up-hill" links, then 0-1 peer-peer links, and finally 0-n provider-customer „down-hill" links. The relationships between ASes are typically based on simple bilateral contracts and we assume that this remains to be the case for the future.

In NIRA [17], it was discovered that almost all policy-compliant paths can be constructed by joining the upgraphs of the communicating endpoints. An upgraph contains transitively all ASes reachable from the node following only customer-provider links and possibly a single peer-peer link as the last hop of the path. The upgraphs contain typically little less than 300 ASes without peering links and around 1500 ASes when taking the peering links into account. The total number of ASes is about 30000, which means that the "two-sided" approach can reduce 50-fold the number of links to consider. NIRA cannot express all BGP policies as the upgraphs will always be the same independent of the packet destination, but our ANDL can be also used to model BGP.

3   Architecture

A central component in our architecture is the rendezvous system, which implements a data-centric pub/sub primitive as a recursive, hierarchical structure, which first joins node local rendezvous implementations into *rendezvous networks* (RN) and then RNs into a global *rendezvous interconnect* (RI) using a hierarchical Chord DHT [18,19] as shown in Fig. 2. The RNs can be implemented, for example, using DONA [4] implementations. In another dimension, the rendezvous system is split into common *rendezvous core* and scope-specific implementations of *scope home* nodes that implement the functionality for a set of scopes.

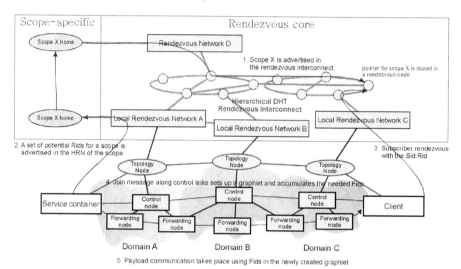

Fig. 2. A client and a service rendezvous on the control plane at the scope home of the scope in which the communication takes place. The scope joins the upgraphs and produces an end-to-end path between the service container (e.g. a data source) and the client (e.g. a subscriber) and returns the information to the client that can then use this information to join a graphlet (e.g. a delivery tree) that can then be used for the fast-path payload communication.

At every level of the hierarchy, the rendezvous core provides an anycast routing to the approximately closest scope home that hosts a given scope. A *subscription message* contains the *(Sid,Rid)* tuple of the publication of which contents the client wishes to receive. Each node in the rendezvous core can cache results and immediately route the answer back along the reverse route to the client. The rendezvous message is first hierarchically routed towards a *scope pointer* based on the hash of the Sid and then the pointer redirects the request towards the approximately closest scope home. Also scope pointers can be cached. *Publication messages* are routed similarly and they contain also the contents of the publication in addition to the *(Sid,Rid)* pair. The scope then stores the contents of the publication so that it can serve the subscribers requesting it. In accordance with the *fate-sharing principle* [20], both the publication and subscription messages need to be periodically repeated in order to keep the publication data or pending subscription alive. This pub/sub primitive is the only functionality implemented by the rendezvous core. We refer to our work in [5] for a detailed description of the rendezvous security mechanisms.

Scopes, however, can have varying implementations. When a cached result cannot be found in the rendezvous core, the subscription reaches the scope, which can then dynamically generate the response if it has enough information available. It is possible to avoid caching by including version information in the Rid. Each scope implementation may be scaled up by adding more scope home nodes and implementing their own coordination protocol internally. We note that the slow-path control plane rendezvous is not meant for transfer of large publications, but this type of applications should be supported by adding a data-centric transport to the data plane as we did in [2].

Topology manager (TM) is another function that is implemented by each independently managed domain. Its task is to crawl and collect neighbourhood ANDL map and metadata publications of network components using the rendezvous. From this information TM builds a complete view of the local network and publishes this information back into the rendezvous for other nodes to listen to. TM may also hide parts of the network and setup higher-level links in the network map. TM also subscribes to route advertisements from neighbouring domains and updates its upgraph publication accordingly. TM can also act as the local scope home and control its policies. Information about other networks can be found because the publications are named in a standard fashion based on the cryptographic identity of each domain and scope. Basically, a domain X is assumed to have a cryptographic identity DK_X that can be carried in network description attributes. Each domain has a scope with Sid $(DK_X, "external scope")$ and another scope called "local scope", which is reachable only locally. Each scope also publishes a meta-data publication inside itself named $(DK_X, "scope meta-data")$ describing which transports the scope supports, among others. It should be noted that the upgraph combination based routing does not require any type of central entity to manage addresses but the internetwork can freely evolve based on trust between neighbouring domains.

4   Phases of Communication

Each node wishing to communicate first requests the description of end-to-end fast-path resources used for the graphlet from the scope by subscribing to a publication labeled "$<Topic\_Rid>, <U_N>, t$" where U_N is the *(Sid,Rid)* pair naming the ANDL upgraph map of the node and t is the current time. If the scope performing the upgraph joining is access controlled, then the subscribed label also includes the node identity. The upgraph data itself is published by the provider domain of the node. Because many nodes share the same upgraph, the data-centric rendezvous system caches them orthogonally close to the *scope homes* that are nodes implementing the scope in question. Similarly, the result of the rendezvous is automatically cached and reused if another node in the same domain requests the same service. ANDL maps can also link to other maps and a complete map can be built recursively from smaller publications, which minimizes the amount of communication as only relevant information needs to be transfered. Multiple scopes can be used together by performing the rendezvous independently for each of the scopes and then taking the logical *AND* of the resulting policy-compliant resources at the end nodes.

A typical sequence of operations is shown in Fig. 2. After a successful rendezvous in the scope, the client is returned information about end-to-end resources that it can use for the graphlet formation. If the transport in question is multicast data dissemination, then a separate resource allocation protocol could be coupled with the protocol as we did in [2]. The client side implementation of the transport would then take the resource description from rendezvous as an input and exchange packets with the selected fast-path domains that would reserve the graphlet. The resource allocation layer could produce, for example, a set of *forwarding identifiers* (Fid), such as zFilter [21], that could be used as an opaque capability to access the graphlet securely without affecting the rest of the network. For this we use the zFormation technique described in [22].

We note that, for example, *onion routing* could be used for the resource allocation, which would hide the destination of the communication from the transit domains and thus guarantee some level of network neutrality. Also pseudonyms for domains can be used to hide the location of the service from its users. We refer to our work in [2] for a more detailed example of graphlet formation in an intra-domain architecture where the dedicated nodes handling a transport can be scattered in the network. The TM of the domain just routes the transports through compatible nodes while balancing the load to each node. This means that the transport functionality does not even have to operate at backbone line speeds because of demultiplexing the flows to multiple nodes. Thus, we claim that the deployment of new transport functionality in the network to be run at branching points of graphlets can be done scalably.

5   Related Work

This section covers related work for publish/subscribe systems and network layer security solutions. A data-oriented network architecture DONA [4] replaces a traditional DNS-based namespace with self-certifying flat labels, which are derived from cryptographic public keys. DONA names are expressed as a P:L pair, where P is a hash of a principal's public key which owns the data and L is a label. DONA utilizes an IP header extension mechanism to add a DONA header to the IP header, and separate resolution handlers (RHs) are used to resolve P:L pairs in topological locations.

In content-centric networking (CCN) [3] every packet has an unique human-readable name. CCN uses two types of packets. Consumers of data send interest packets to the network, and a nodes possessing the data reply with the corresponding data packet. Since packets are independently named, a separate interest packet must be sent for each required data packet. In CCN data packets are signed by the original publisher allowing independent verification, however interest packet's are not always protected by signatures.

Security issues of the content-based pub/sub system have been explored in [7]. The work proposes secure event types, where the publication's user friendly name is tied to the publisher's cryptographic key.

5.1   Security Mechanisms

Most of existing network layer security proposals utilize hash chains or Merkle trees [8]. Examples of hash chain based solutions include TESLA [9], which is time-based hash chain scheme, and ALPHA [10] that relies on interaction between the sender and receiver. While hash chain approaches are very lightweight, they have several downsides, such as path dependency and complex signaling. Merkle tree based solutions have high bandwidth overhead for large trees, and their performance suffers if packets arrive in out-of-order.

Accountable Internet Protocol (AIP) [11] aims to improve security by providing accountability on the network layer. AIP uses globally self-certifying unique endpoint identifiers (EID) to identify and address the source and the destination of the connection, in addition to normal IP addresses. EIDs contain hashes of host's public keys that are communicating within the network. AIP aims to prevent source address spoofing in the following way: If the router receives a packet from the unknown EID, the router will send a verification message back and the node will reply with a message signed by its private key. Since EID is hash of node's public key, this proves that the node owns a corresponding private key and thus has a right to use the EID.

Identity-based encryption and signature scheme (IBE) [12] allows a label, e.g., the user's e-mail address to be used as user's public key, simplifying the key distribution problem. However, IBE relies on a centralized entity called private key generator (PKG), which knows all private keys of its users.

6   Conclusion and Future Work

In this paper we introduced a data-centric inter-domain pub/sub architecture addressing availability and data integrity. We used the concept of scope to separate the logical structure of linked data from the orthogonal distribution strategies used to determine how the data is communicated in the network. This is still ongoing work and, for example, the ANDL language and quantitative analysis will be covered in our future work.

References

1. Wang, C., Carzaniga, A., Evans, D., Wolf, A.L.: Security issues and requirements for Internet-scale publish-subscribe systems. In: HICSS '02, Hawaii, USA (2002)
2. Visala, K., Lagutin, D., Tarkoma, S.: LANES: An Inter-Domain Data-Oriented Routing Architecture. In: ReArch'09, Rome, Italy (2009)
3. Jacobson, V., Smetters, D.K., Thornton, J.D., Plass, M., Briggs, N., Braynard, R.L.: Networking named content. In: ACM CoNEXT 2009, Rome, Italy (2009)
4. Koponen, T., Chawla, M., Chun, B.-G., Ermolinskiy, A., Kim, K.H., Shenker, S., Stoica, I.: A Data-Oriented (and Beyond) Network Architecture. In: ACM SIGCOMM 2007, Kyoto, Japan (2007)

5. Lagutin, D., Visala, K., Zahemszky, A., Burbridge, T., Marias, G.: Roles and Security in a Publish/Subscribe Network Architecture. In: ISCC'10, Riccione, Italy (2010)

6. Clark, D., Wroclawski, J., Sollins, K., Braden, R.: Tussle in Cyberspace: Defining Tomorrow's Internet. IEEE/ACM Transactions on Networking 13(3), 462–475 (2005)

7. Pesonen, L.I., Bacon, J.: Secure event types in contentbased, multi-domain publish/subscribe systems. In: 5th international workshop on Software engineering and middleware, pp. 98–105 (2005)

8. Merkle, R.: Secrecy, authentication, and public key systems. Ph.D. dissertation, Department of Electrical Engineering, Stanford University (1979)

9. Perrig, A., Canetti, R., Tygar, J.D., Song, D.: The Tesla broadcast authentication protocol. Cryptobytes 5(2), 2–13 (2002)

10. Heer, T., Götz, S., Morchon, O.G., Wehrle, K.: Alpha: An adaptive and lightweight protocol for hopbyhop authentication. In: Proceedings of ACM CoNEXT (2008)

11. Andersen, D.G., Balakrishnan, H., Feamster, N., Koponen, T., Moon, D., Shenker, S.: Accountable internet protocol (AIP). In: Proceedings of the ACM SIGCOMM 2008, pp. 339–350 (2007)

12. Shamir, A.: Identity-based cryptosystems and signature schemes. In: Brauer, W. (ed.) CRYPTO 1980. LNCS, vol. 84, pp. 47–53. Springer, Heidelberg (1980)

13. Saltzer, J., Reed, D., Clark, D.: End-to-end arguments in system design. ACM Transactions on Computer Systems 2(4), 277–288 (1984)

14. Lagutin, D., Visala, K., Tarkoma, S.: Publish/Subscribe for Internet: PSIRP Perspective. Valencia FIA book (2010)

15. Tarkoma, S., Antikainen, M.: Canopy: Publish/Subscribe with Upgraph Combination. In: 13th IEEE Global Internet Symposium 2010 (2010)

16. Gao, L.: On Inferring Autonomous System Relationships in the Internet. IEEE/ACM Transactions on Networking 9(6), 733–745 (2001)

17. Yang, X., Clark, D., Berger, A.W.: NIRA: A New Inter-Domain Routing Architecture. IEEE/ACM Trans. Netw. 15(4), 775–788 (2007)

18. Rajahalme, J., Särelä, M., Visala, K., Riihijärvi, J.: Inter-Domain Rendezvous Service Architecture. PSIRP Technical Report TR09-003 (2009)

19. Ganesan, P., Gummadi, K., Garcia-Molina, H.: Canon in G Major: Designing DHTs with Hierarchical Structure. In: ICDCS'04, pp. 263–272. IEEE Computer Society Press, Los Alamitos (2004)

20. Carpenter, B.: rfc1958: Architectural Principles of the Internet. IETF (June 1996)

21. Jokela, P., Zahemszky, A., Esteve, C., Arianfar, S., Nikander, P.: LIPSIN: Line speed Publish/Subscribe Inter-Networking. In: SIGCOMM'09 (2009)

22. Esteve, C., Nikander, P., Särelä, M., Ylitalo, J.: Self-routing Denial-of-Service Resistant Capabilities using In-packet Bloom Filters. In: European Conference on Computer Network Defence, EC2ND (2009)

23. Miller, V.S.: Use of elliptic curves in cryptography. In: Williams, H.C. (ed.) CRYPTO 1985. LNCS, vol. 218, pp. 417–426. Springer, Heidelberg (1986)

24. Forsten, J., Järvinen, K., and Skyttä, J.: Packet level authentication: Hardware subtask final report. Helsinki University of Technology, Tech. Rep (2008), http://www.tcs.hut.fi/Software/PLA/new/doc/PLA_HW_final_report.pdf

25. Lagutin, D.: Securing the Internet with Digital Signatures. Doctoral dissertation, Department of Computer Science and Engineering, Aalto University, School of Science and Technology (2010)

Engineering Secure Future Internet Services

Wouter Joosen[1], Javier Lopez[2], Fabio Martinelli[3], and Fabio Massacci[4]

[1] Katholieke Universiteit Leuven
wouter.joosen@cs.kuleuven.be
[2] University of Malaga
jlm@lcc.uma.es
[3] National Research Council of Italy
Fabio.Martinelli@iit.cnr.it
[4] University of Trento
massacci@dit.unitn.it

Abstract. In this paper we analyze the need and the opportunity for establishing a discipline for engineering secure Future Internet Services, typically based on research in the areas of software engineering, of service engineering and security engineering. Generic solutions that ignore the characteristics of Future Internet services will fail, yet it seems obvious to build on best practices and results that have emerged from various research communities.

The paper sketches various lines of research and strands within each line to illustrate the needs and to sketch a community wide research plan. It will be essential to integrate various activities that need to be addressed in the scope of secure service engineering into comprehensive software and service life cycle support. Such a life cycle support must deliver assurance to the stakeholders and enable risk and cost management for the business stakeholders in particular. The paper should be considered a call for contribution to any researcher in the related sub domains in order to jointly enable the security and trustworthiness of Future Internet services.

1 Introduction

1.1 Future Internet Services

The concept named Future Internet (FI) aggregates many facets of technology and its practical use, often illustrated by a set of usage scenarios and typical applications. The Future Internet may evolve to use new infrastructures, network technologies and protocols in support of a growing scale and a converging world, especially in light of smaller, portable, ubiquitous and pervasive devices. Besides such a network-level evolution, the Future Internet will manifest itself to the broad mass of end users through a new generation of services (e.g. a hybrid aggregation of content and functionality), service factories (e.g., personal and enterprise mash-ups), and service warehouses (e.g., platform as a service). One specific service instance may thus be created by multiple service development organizations, it may be hosted and deployed by multiple providers, and may

J. Domingue et al. (Eds.): Future Internet Assembly, LNCS 6656, pp. 177–191, 2011.

be operated and used by a virtual consortium of business stakeholders. While the creative space of services composition is in principle unlimited, so is the fragmentation of ownership of both services and content, as well as the complexity of implicit and explicit relations among participants in each business value chain that is generated. In addition, the user community of such FI services evolves and widens rapidly, including masses of typical end users in the role of *prosumers*(producing and consuming services). This phenomenon increases the scale, the heterogeneity and the performance challenges that come with FI service systems.

This evolution obviously puts the focus on the **trustworthiness** of services. Multiparty service systems are not entirely new, yet the Future Internet stretches the present know how on building secure software services and systems: more stakeholders with different trust levels are involved in a typical service composition and a variety of potentially harmful content sources are leveraged to provide value to the end user. This is attractive in terms of degrees of freedom in the creation of service offerings and businesses. Yet this also creates more vulnerabilities and risks as the number of trust domains in an application gets multiplied, the size of attack surfaces grows and so does the number of threats. Furthermore, the Future Internet will be an intrinsically dynamic and evolving paradigm where, for instance, end users are more and more empowered and therefore decide (often on the spot) on how content and services are shared and composed. This adds an extra level of complexity, as both risks and assumptions are hard to anticipate. Moreover, both risks and assumptions may evolve; thus they must be monitored and reassessed continuously.

1.2   The Need for Engineering Secure Software Services

The need to organize, integrate and optimize the research on engineering secure software services to deal effectively with this increased challenge is pertinent and well recognized by the research community and by the industrial one. Indeed, there is also a growth of successful attacks on ICT-service systems, both in terms of impact and variety. This obviously harms the economic impact of Future Internet services and causes significant monetary losses in recovering from those attacks. In addition, this induces users at several levels to lose confidence in the adoption of ICT-services.

From a business perspective, however, we are now witnessing the emergence of new and unprecedented models for service-oriented computing for the Future Internet: Infrastructure as a Service (IaaS), Platform as a Service (PaaS) and Software as a Service (SaaS). These models have the potential to better adhere to an economy of scale and have already shown their commercial value fostered by key players in the field. Nevertheless, those new models present change of control on the applications that will run on an infrastructure not under the direct control of the business service provider. For business critical applications this could be difficult to be accepted, when not appropriately managed and secured. These issues are of an urgent practical relevance, not only for academia, but also for industry and governmental organizations. New Internet services will have to be provided in the near future, and security breaches in these services may lead to large financial loss and damaged reputation.

1.3   Research Focus on Developing Secure FI Services

Our focus is on the creation and correct execution of a set of methodologies, processes and tools for secure software development. This typically covers requirements engineering, architecture creation, design and implementation techniques. However this is not enough! We need to enable **assurance**: approving that the developed software is secure. Assurance must be based on **justifiable evidence**, and the whole process designed for assurance. This would allow the uptake of new ICT-services according to the latest Future Internet paradigms, where services are composed by simpler services (provided by separate administrative domains) integrated using third parties infrastructures and platforms. The need of managing the intrinsic **modularity and compose-ability** of these architectures, traditionally shared with commercial off the shelf components (COTS), should drive the development of corresponding methodologies. Clearly industry needs to prioritize its efforts in order to improve their return of investments (ROI). Thus, embedding **risk/cost analysis** in the SDLC is currently one of the key research directions in order to link security concerns with business needs and thus supporting a business case for security matters.

Our research addresses the early phases of the development process of services, bearing in mind that the discovery and remediation of vulnerabilities during the early development stages saves resources. Thus our joint research activities fall in six areas: (1) *security requirements* for FI services, (2) *creating secure service architectures and secure service design*, (3) supporting *programming environments for secure and compose-able services*, (4) *enabling security assurance*, integrating the former results in *(5) a risk-aware and cost-aware software development life-cycle (SDLC)*, and (6) the delivery of case studies of future internet application scenarios.

The first three activities represent major and traditional stages of (secure) software development: from requirements over architecture and design to the composition and/or programming of working solutions. These three activities interact to ensure the integration between the methods and techniques that are proposed and evaluated. This is a first element that drives to *research integration*.

In addition, the research programme adds two horizontal activities that span the service creation process. Both the security assurance programme and the programme on Risk and Cost aware SDLC will interact with each of the initial three activities, drive the requirements of these activities and leverage upon, even integrate their outcome. This is a second element that drives to research integration.

Finally, notice that all 5 research activities mentioned above will be inspired and evaluated by their application in specific FI application scenarios. This third element complements the overall research programme that leads to integrated research and intensive research collaboration in the area.

In the sequel of this paper we elaborate on the relevant sub domains and techniques that we consider useful for engineering secure Future internet services.

2   Security Requirements Engineering

The main focus of this research strand is to enable the modeling of high-level requirements that can be expressed in terms of high-level concepts such as compliance, privacy, trust, and so on. These can be subsequently mapped into more specific requirements that refer to devices and to specific services. A key challenge is to support dealing with an unprecedented multitude of autonomous stakeholders and devices – probably one of the most distinguishing characteristics of the FI.

The need for assurance in the Future Internet demands a set of novel engineering methodologies to guarantee secure system behavior and provide credible evidence that the identified security requirements have been met from the point of view of all stakeholders. The security requirements of Future Internet applications will differ considerably from those of traditional applications. The reason is that Future Internet applications will not only be distributed geographically, as are traditional applications, but they will also involve multiple autonomous stakeholders, and may involve an array of physical devices such as smart cards, phones, RFID sensors and so on that are perpetually connected and transmit a variety of information including identity, bank accounts, location, and so on. Some of these transactions might even happen transparently to the user; for example, a person's identity could be seamlessly communicated by a personal device to the store she is entering to do the shopping. Addressing concerns about identity theft, unauthorized credit card usage, unauthorized transmission of information by third-party devices, trust, privacy, and so on are critical to the successful adoption of FI applications.

Service-orientation and the fragmentation of services (both key characteristics of FI applications) imply that a multitude of stakeholders will be involved in a service composition and each one will have his own security requirements. Hence, eliciting, reconciling, and modeling all the stakeholders' security requirements become a major challenge [5]. Multilateral Security Requirements Analysis techniques have been advocated in the state of the art [14] but substantial research is still needed. In this respect, agent-oriented and goal-oriented approaches such as Secure Tropos [12] and KAOS [8] are currently well recognized as means to explicitly take the stakeholders' perspective into account. These approaches will represent a promising starting point but need to be uplifted in order to be able to cope with the level of complexity put forward by FI applications. Furthermore, it is important that security requirements are addressed from a higher level perspective, e.g., in terms of the actors' relationships with each other. Unfortunately, most current requirements engineering approaches consider security only at the technological level. In other words, current approaches provide modeling and reasoning support for encryption, authentication, access control, non-repudiation and similar requirements. However, they fail to capture the high-level requirements of trust, privacy, compliance, and so on.

This picture is further complicated by the vast number and the geographical spread of smart devices stakeholders would deploy to meet their requirements. Sensor networks, RFID tags, smart appliances that communicate not only with the user but with their manufacturers, are examples of such devices. Such deployments inherit security risks from the classical Internet and, at the same time, create new and more complex security challenges. Examples include illicit tracking of RFID tags (privacy violation) and cloning of data on RFID tags (identity theft). Applications that involve such deployments typically cross organization boundaries.

In light of the challenges and principles highlighted above, we identify the following detailed objectives:

- The definition of techniques for the identification of all stakeholders (including attackers), the elicitation of high-level security goals for all stakeholders, and the identification and resolution of conflicts among different stakeholder security goals;
- The refinement of security goals into more detailed security requirements for specific services and devices;
- The identification and resolution of conflicts between security requirements and other requirements (functional and other quality requirements);
- The transformation of a consolidated set of security requirements into security specifications.

The four objectives listed above obviously remain generic by nature, one should bear in mind though that the forthcoming techniques and results will be applied to a versatile set of services, devices and stakeholder concerns.

3   Secure Service Architecture and Design

FI applications entail scenarios in which there exist a huge amount of heterogeneous users and a high level of composition and adaptation is required. These factors increase the complexity of applications and make it necessary to leverage existing mechanisms and methodologies for software construction as well as researching about new ways to take this complexity into account in a holistic manner. These applications enable pervasive, ubiquitous scenarios where multiple users, devices, third-party components interact continuously and seamlessly, so security enforcement mechanisms are indispensable. The design phase of the software service and/or system is a timely moment to enforce and reason about these security mechanisms, since by that phase one must have already grasped a thorough understanding of the application domain and of the requirements to be fulfilled. Furthermore, at design-time a preliminary version of the application architecture has been produced.

The software architecture encompasses the more relevant elements of the application, providing either a static or/and a dynamic view of the application. A more comprehensive definition can be found in [2], where it is defined as "the structure or structures of the system, which comprise software elements, the externally visible properties of those elements, and the relationships among them".

The security architecture for the system must enforce the visible security properties of components and the relationships between them. All this information makes it feasible to enforce, assess and reason about security mechanisms at an early phase in the software development cycle.

The research topics one must focus on in this subarea relate to model-driven architecture and security, the compositionality of design models and the study of design patterns for FI services and applications. The three share the common ambition to maximize reuse and automation while designing secure FI services and systems.

As for the first element the aim is to support methodologies that utilize several easy-to-understand models to represent the application. According to the model-driven approach, these models may be manipulated and converted automatically into other models, in such a way that they all preserve certain properties. So, it would be possible to specify a first high-level model with some high-level security policies. Then, by automation, this model could be converted into another more specific model, in which the security policies become more detailed, closer to the enforcement mechanisms that will fulfil them. This process should be applied until a basic version of the application architecture can be released.

The integration of security aspects into this paradigm is the so-called model-driven security [6], leading to a design for assurance methodology in which every step of the design process is performed taking security as a primary goal. A way of carrying out this integration includes first decomposing security concerns, so that the application architecture and its security architecture is decoupled. This makes possible for architects to assess more easily tradeoffs among different security mechanisms, simulate security policies and test security protocols before the implementation phase, where changes are typically far more expensive.

In order to achieve this, it is first needed to convert the security requirements models into a security architecture by means of automatic model transformations. These transformations are interesting, since whilst requirements belong to the problem-domain, the architecture and design models are within the solution-domain, so there is an important gap to address. In the context of security modeling, it is extremely relevant to incept ways to model usage control (e.g., see [21,22,18]), which encompasses traditional access control, trust management and digital rights management and goes beyond these building blocks in terms of definition and scope. Finally, by means of transformation patterns, it is required to research on new ways to map the high-level policies established at requirements stage into low-level, enforceable policies at run-time. Furthermore note that FI scenarios include Cloud and GRID services and although some work has already been made in the area [23], further research is necessary to find out what kind of security architecture is required in the context and how to carry out the decomposition of such fairly novel architectures.

Until this point in the software and service development process, different concerns – security among them – of the whole application have been separated into different models, each model representing different functional and non-functional concerns that different stakeholder may have about it. However, in order to grasp a comprehensive understanding of the application as a whole, it is required to integrate all these views into a unified one. This process is called composition [25,11] and, as a recent work suggests [20], it is possible to perform it at run-time, adding a new level of flexibility and adaptation for FI applications. Regarding composition, several topics will be studied. First, it is desirable to define contracts for model composition, in such a way that only correct compositions are allowed, limiting the propagation of design flaws through the models. Second, given that different sub-architectures may exist, each addressing different concerns – even different security sub-architectures for different security requirements – it is required to assure that the composition of all these architectures is accomplished and that all the requirements are met in this composition. Finally, adaptation of composite services is a key area of interest. FI scenarios are very dynamic, so threats in the environment may change along the time and some reconfiguration may be required to adapt to that changes.

The last research focus is on design patterns and on reusable architectural know-how. A design pattern is a general repeatable solution to a commonly occurring problem in software design. Design patterns, once identified, allow reuse of design solutions that have proved to be effective in the past, reducing costs and risks usually arisen by uncertainty, leveraging a risk and cost-aware . There are large catalogues and surveys on security patterns available [26,13], but the FI applications yet to come and the new scenarios enabled by FI need to extend and tailor these catalogues. In this context, the first step is studying the patterns currently available and, what is more important, to analyze the relationships amongst them [17], identifying those which may be useful for FI scenarios. Finally, how to bridge the gap among problem patterns – such as problem frames or KAOS refinement patterns and solution patterns – architectural patterns – must be analyzed, both from a general perspective and from a security perspective for security-critical software systems.

4   Security Support in Programming Environments

Security Support in Programming Environments is not new; still it remains a grand challenge, especially in the context of Future Internet (FI) Services. Securing Future Internet Service is inherently a matter of secure software and systems. The context of the future internet services sets the scene in the sense that (1) specific service architectures will be used, that (2) new types of environments will be exploited, ranging from small embedded devices ("things") to service infrastructures and platform in the cloud, and (3) a broad range of programming technologies will be used to develop the actual software and systems.

The search for security support in programming environments has to take this context in account. The requirements and architectural blueprints that will be produced in earlier stages of the software engineering process cannot deliver the expected security value unless the programs (code) respect these security artefacts that have been produced in the preceding stages. This sets the stage for model driven security in which transformations of architecture and design artefacts is essential, as well as the verification of code compliance with various properties. Some of these properties have been embedded in the security specific elements of the software design; other may simply be high priority security requirements that have articulated – such as the appropriate treatment of concurrency control and the avoidance of race conditions – in the code, as a typical FI service in the cloud may be deployed with extreme concurrency in mind.

Supporting security requirements in the programming – code – level requires a comprehensive approach. The service *creation* means must be improved and extended to deal with security needs. Service creation means both aggregating and *composing* services from pre-existing building blocks (services and more traditional components), as well as *programming* new services from scratch using a state-of-the-art programming language. The service creation context will typically aim for techniques and technologies that support compile and build-time feedback. One could argue that security support for service creation must focus on and enable better static verification. The service *execution* support must be enhanced to deal with hooks and building blocks that facilitate effective security enforcement at run-time. Dependent on the needs and the state-of-the-art this may lead to interception and enforcement techniques that "simply" ensure that the application logic consistently interacts with underpinning security mechanisms such as authentication or audit services. Otherwise, the provisioning of the underpinning security mechanisms and services (e.g. supporting mutual non repudiation, attribute based authorization in a cloud platform etc.) will be required as well for many of the typical FI service environments. Next we further elaborate on the needs and the objectives of community wide research activities.

4.1   Secure Service Composition

Future Internet services and applications will be composed of several services (created and hosted by various organizations and providers), each with its own security characteristics. The business compositions are very dynamic in nature, and span multiple trust domains, resulting in a fragmentation of ownership of both services and content, and a complexity of implicit and explicit relations among the participants.

Service Composition Languages. One of the challenges for the secure service composition is the need for new formalisms to specify service requests (properties of service compositions) and service capabilities, including their security policies, and tools to generate code for service compositions that are able to fulfil these requirements based on the available services. In addition to complying with the requested functional and quality-of-service-related characteristics, composition languages must support means to preserve at least the security policy of those services being composed. The research community needs to consider the cases where only partial or inadequate information on the services is available, so that the composition will have to find compliant candidates or uncover the underspecified functionality.

Middleware Aspects. The research community should re-investigate service-oriented middleware for the Future Internet, with a special emphasis on enabling deployment, access, discovery and composition of pervasive services offered by resource-constrained nodes.

4.2 Secure Service Programming

Many security vulnerabilities arise from programming errors that allow an exploit. Future Internet will further reinforce the prominence of highly distributed and concurrent applications, making it important to develop methodologies that ensure that no security hole arises from implementations that exploit the computational infrastructure of the Future Internet. The research community must further investigate advances over state-of-the-art in fine-grained concurrency to enable highly concurrent services of the Future Internet, and will improve analysis and verification techniques to verify, among others, adherence to programming principles and best-practices [10].

Verifiable Concurrency. Lock-free wait-free algorithms for common software abstractions (queues, bags, etc.) are one of the most effective approaches to exploit multi-core parallelism. These algorithms are hard to design and prove correct, error-prone to program, and challenging to debug. Their correctness is crucial to the correct behaviour of client programs. Research should now focus on build *independently* checkable proofs of the absence of common errors, including deadlock, race conditions, and non-serialize-ability [16].

Adherence to Programming Principles and Best-Practices. Programming support must include methods to ensure the adherence of a particular program to well-known programming principles or best-practices in secure software development. Emphasis will be put on language extensions that guarantee adherence to best-practices, and verified design patterns that can be used during development. The research community might investigate and re-visit methods from language-based security, in particular type systems, to enforce best-practises currently used in order to prevent cross-site scripting attacks and similar vulnerabilities associated with web-based distributed applications. Obviously, the logical rationales underlying such best-practises must be articulated, enabling he development of type systems enforcing these practises directly – thus allowing users to deviate from rigid best-practices while still maintaining security.

4.3 Platform Support for Security Enforcement

Future Internet applications span multiple trust domains, and the hybrid aggregation of content and functionality from different trust domains requires complex cross-domain security policies to be enforced, such as end-to-end information flow, cross-domain interactions and usage control. In effect, the security enforcement techniques that are triggered by built-in security services and by realistic in the FI setting, must address the challenge of *complex interactions* and of *finely grained control* [15]. Research should therefore focus on the enforcing cross-domain barriers in the interaction among different cross-domains, and on the enforcement of fine-grained security policies via execution monitoring.

Secure Cross-Domain Interactions. Web technology inherently embeds the concept of cross-domain references, and applications are isolated via the Same-Origin-Policy (SOP) in the browser. From a functional perspective, the SOP puts limitations on compose-ability and cooperation of different applications, and from a security perspective, the SOP is not strong enough to achieve the appropriate application isolation.

Finely Grained Execution Monitoring. Trustworthy applications need run-time execution monitors that can provably enforce advanced security policies [19,3] including fined-grained access control policies usage control policies and information flow policies [24].

Supporting Security Assurance for FI Services. Assurance will play a central role in the development of software based services to provide confidence about the desired security level. Assurance must be treated in a holistic manner as an integral constituent of the development process, seamlessly informing and giving feedback at each stage of the software life cycle by checking that the related models and artefacts satisfy their functional and security requirements and constraints. Obviously the security support in programming environments that must be delivered will be essential to incept a transverse methodology that enables to manage assurance throughout the software and service development life cycle (SDLC). The next section clarifies these issues.

5   Embedding Security Assurance and Risk Management during SDLC

Engineering secure Future Internet services demands for at least two traversal issues, security assurance and risk and cost management during SDLC.

5.1   Security Assurance

The main objective is to enable assurance in the development of software based services to ensure confidence about their trustworthiness. Our core goal is to incept a transverse methodology that enables to manage assurance throughout the software development life cycle (SDLC). The methodology is based on two strands: A first sub-domain covers early assurance at the level of requirements, architecture and design. A second sub-domain includes the more conventional and complementary assurance techniques based on implementation.

Assurance during the Early Stages of SDLC. Early detection of security failures in Future Internet applications reduces development costs and improves assurance in the final system. This first strand aims at developing and applying assurance methods and techniques for early security verification. These methods are applied to abstract models that are developed from requirements to detailed designs.

One main area of research is step-wise refinement of security, by developing refinement strategies, from policies down to mechanisms, for more complex secure protocols, services, and systems. This involves the definition of suitable service and component abstractions (e.g., secure channels) and the setup of the corresponding reasoning infrastructure (e.g., facts about such channels). Moreover, we need to extend the refinement framework with *compositional* techniques for model-based secure service development. Model decomposition supports a divide-and-conquer approach, where functional and security-related design aspects can be refined independently. Model composition must preserve the refinement relation and component properties. Our aim is to offer developers support for smoothly integrating security aspects into the system development process at any step of the development.

Enabling rigorous and formal analysis processes. There is an increasing demand of models and techniques to allow the formal analysis of secure services. The objective is to develop methodologies, based on formal mappings from the constraint languages, to other formalisms for which theorem proving and/or (semi-)decision procedures are available, to support formal (and, when possible, automated) reasoning about the security policies models.

The methodologies must be supported by automatic protocol verification tools, such as the AVISPA [1] tool set and the Scyther tool [7], for the verification of Future Internet protocols. The planned extensions require not only significant efficiency improvements, but also the ability to deal with more complex primitives and security properties. Moreover, the Dolev-Yao attacker model [9] used by these tools needs to be extended to include new attack possibilities such as adaptive corruptions, XSS attacks, XML injection, and guessing attacks on weak passwords. In addition, for assurance, there is the need to extend model checking methods to enable automatic generation of protocol correctness proofs that can be independently verified by automated theorem proving.

Security Assurance in Implementation. Several assurance techniques are available to ensure the security at the level of an implementation. Security policies can be implemented correctly by construction through a rigorous secure programming discipline. Internet applications can be validated through testing. In that case, it is possible to develop test data generation that specifically targets the integration of services, access control policies or specific attacks. Moreover, implementations can be monitored at run-time to ensure that they satisfy the required security properties.

Complementing activities are related to **secure programming**. This strand addresses a comprehensive solution for program verification, while adding a particular focus on session management in concurrent and distributed service compositions.

In addition, an important set of research activities must address **testing**. This strand covers the testing activities which complement programming and coding. We can consider three aspects, that although not comprehensive, present characteristic for service-oriented applications in the future Internet: penetration testing that leverages on the high-level models that are generated in early stages of the software life cycle, automated generation in XML-based input data to maximize the efficiency in the security testing process, and testing of policies that are the typical high-level front end of a complex service composition. The latter part will focus on access control policies. i

Finally, an important set of activities relates to **run-time verification**. This strand concludes the trilogy of implementation-level testing: run-time *verification* must complement programming-level verification and testing in order to provide the final assurance that the latter cannot deliver, be it for scientific and technological reasons, be it for reasons of organizational complexity. The latter may frequently occur in a multi-organizational context, typical for service compositions in Future Internet. We will study approaches for run-time monitoring of data flow, as well as technologies for privacy-preserving usage control.

Towards a Traverse Methodology. Security concerns are specified at the business-level but have to be implemented in complex distributed and adaptable systems of FI services. We need comprehensive assurance techniques in order to guarantee that security concerns are correctly taken into account through the whole SDLC. A chain of techniques and tools crossing the above areas is planned.

Security Metrics. Measurements are essential for objective analysis of security systems. Metrics can be used directly for computing risks (e.g., probability of threat occurrence) or indirectly (e.g., time between antivirus updates). Security metrics in the Future Internet applications become increasingly important. Service-oriented architectures demand for assurance indicators that can explicitly indicate the quality of protection of a service, and hence indicate the effective level of trustworthiness. These metrics should be assessed and communicable to third parties. Clients want to be sure that their data outsourced to other domains, which the clients cannot control, are well protected. We need to define formal metrics and measurements that can be practically calculated. Compositional calculation approaches will be studied in this context. Many of the proposed metrics will be linked to and determined by the various techniques in the Engineering process.

5.2   Risk and Cost Aware SDLC

There is the need of the creation of a methodology that delivers a risk and cost aware SDLC for secure FI services. Such a life cycle model aims to ensure the stakeholders' return of investment when implementing security measures during various stages of the SDLC. We can envision several aspects of this kind of SDLC support (see also [4]).

Process: The methodology for risk and cost aware SDLC should be based on an *incremental and iterative process* that is accommodated to an incremental software development process. While the software development proceeds through incremental phases, the risk and cost analysis will undergo new iterations for each phase. As such the results of the initial risk and cost analyses will propagate through the software development phases and become more refined. In order to support the propagation of analysis results through the phases of the SDLC one needs to develop methods and techniques for the refinement of risk analysis documentation. Such refinement can be obtained both by refining the risk models, e.g. by detailing the description of relevant threats and vulnerabilities, and by accordingly refining the system and service models.

Aggregation: In order to accommodate to a modular software development process, as well as effectively handling the heterogeneous and compositional nature of Future Internet services, one needs to focus on a modular approach to the analysis of risks and costs. In a compositional setting, also risks become compositional and should be analysed and understood as such. This requires, however, methods for aggregating the global risk level through risk composition which will be investigated.

Evolution: The setting of dynamic and evolving systems furthermore implies that risk models and sets of chosen mitigations are dynamic and evolving. Thus, in order to maintain risk and cost awareness, there is a need to continuously reassess risks and identify cost-efficient means for risk mitigation as a response to service or component substitution, evolving environments, evolving security requirements, etc., both during system development and operation. Based on the modular approach to risk and cost analysis one needs methods to manage the dynamics of risks. In particular, the process for risk and cost analysis is highly iterative by supporting updates of global analysis results through the analysis of only the relevant parts of the system as a response to local changes and evolvements.

Interaction: The methodology of this strand spans the orthogonal activities of security requirement engineering, secure architecture and design, secure programming as well as assurance and the relation to each of these ingredients must be investigated. During security requirements engineering risk analysis facilitates the identification of relevant requirements. Furthermore, methods for risk and cost analysis offer support for the prioritization and selection among requirements through e.g. the evaluation of trade-off between alternatives or the impact of priority changes on the overall level of risks and cost. In the identification of security mechanisms intended to fulfil the security requirements, risk and cost analysis can be utilized in selecting the most cost efficient mechanisms. The following architecture and design phase incorporates the security requirements into the system design. The risk and cost models resulting from the previous development phase can at this point be refined and elaborated to support the management of risks and costs in the design decisions. Moreover, applying cost metrics to design models and architecture descriptions allows early validation of cost estimates. Such cost metrics may also be used in combination with security metrics for the optimization of the balance between risk and cost. The assurance techniques can therefore be utilized in providing input to risk and cost analysis, and in supporting the identification of means for risk mitigation based on security metrics.

6 Conclusion

We have advocated in this paper the need and the opportunity for firmly establishing a discipline for engineering secure Future Internet Services, typically based on research in the areas of software engineering, security engineering and of service engineering. We have clarified why generic solutions that ignore the characteristics of Future Internet services will fail: the peculiarities of FI services must be reflected upon and be addressed in the proposed and validated solution.

The various lines of research and the strands within each of research line have been articulated while founding the NESSoS Network of Excellence (www.nessos-project.eu). Clearly, the needs and challenges sketched in this paper reach beyond the scope and capacity of a closed consortium. The topics listed above should and will be shared and tackled by an entire and open research community.

Acknowledgments. We would like to thank the anonymous reviewers for the helpful comments. Work partially supported by EU FP7-ICT project NESSoS (Network of Excellence on Engineering Secure Future Internet Software Services and Systems) under the grant agreement n.256980.

References

1. Armando, A., Basin, D., Boichut, Y., Chevalier, Y., Compagna, L., Cuellar, J., Drielsma, P.H., Heám, P.C., Kouchnarenko, O., Mantovani, J., Mödersheim, S., von Oheimb, D., Rusinowitch, M., Santiago, J., Turuani, M., Viganò, L., Vigneron, L.: The AVISPA tool for the automated validation of internet security protocols and applications. In: Etessami, K., Rajamani, S.K. (eds.) CAV 2005. LNCS, vol. 3576, pp. 281–285. Springer, Heidelberg (2005)
2. Bass, L., Clements, P., Kazman, R.: Software Architecture in Practice, 2nd edn. Addison-Wesley, Boston (2003)
3. Bauer, L., Ligatti, J., Walker, D.: Composing security policies with polymer. SIG-PLAN Not. 40, 305–314 (2005)
4. Braber, F., Hogganvik, I., Lund, M.S., Stølen, K., Vraalsen, F.: Model-based security analysis in seven steps — a guided tour to the coras method. BT Technology Journal 25, 101–117 (2007)
5. Bresciani, P., Perini, A., Giorgini, P., Giunchiglia, F., Mylopoulos, J.: Tropos: An agent-oriented software development methodology. Autonomous Agents and Multi-Agent Systems 8, 203–236 (2004)
6. Clavel, M., da Silva, V., de O. Braga, C., Egea, M.: Model-driven security in practice: An industrial experience. In: Schieferdecker, I., Hartman, A. (eds.) ECMDA-FA 2008. LNCS, vol. 5095, pp. 326–337. Springer, Heidelberg (2008)
7. Cremers, C.J.: The scyther tool: Verification, falsification, and analysis of security protocols. In: Gupta, A., Malik, S. (eds.) CAV 2008. LNCS, vol. 5123, pp. 414–418. Springer, Heidelberg (2008)

8. Dardenne, A., van Lamsweerde, A., Fickas, S.: Goal-directed requirements acquisition. Sci. Comput. Program. 20, 3–50 (1993)
9. Dolev, D., Yao, A.C.: On the security of public key protocols. In: Proceedings of the 22nd Annual Symposium on Foundations of Computer Science, Washington, DC, USA, pp. 350–357. IEEE Computer Society Press, Los Alamitos (1981), doi:10.1109/SFCS.1981.32
10. Erlingsson, U., Schneider, F.B.: Irm enforcement of java stack inspection. In: Proceedings of the 2000 IEEE Symposium on Security and Privacy, Washington, DC, USA, pp. 246–255. IEEE Computer Society Press, Los Alamitos (2000)
11. France, R., Fleurey, F., Reddy, R., Baudry, B., Ghosh, S.: Providing support for model composition in metamodels. In: Proceedings of the 11th IEEE International Enterprise Distributed Object Computing Conference, Washington, DC, USA, p. 253. IEEE Computer Society Press, Los Alamitos (2007)
12. Giorgini, P., Mouratidis, H., Zannone, N.: Modelling security and trust with secure tropos. In: Integrating Security and Software Engineering: Advances and Future Vision, IDEA (2006)
13. Group, O.: Security design pattern technical guide, http://www.opengroup.org/security/gsp.htm
14. Gürses, S.F., Berendt, B., Santen, T.: Multilateral security requirements analysis for preserving privacy in ubiquitous environments. In: Proc. of the Workshop on Ubiquitous Knowledge Discovery for Users at ECML/PKDD, pp. 51–64 (2006)
15. Hamlen, K.W., Morrisett, G., Schneider, F.B.: Computability classes for enforcement mechanisms. ACM Trans. Program. Lang. Syst. 28, 175–205 (2006), doi:10.1145/1111596.1111601
16. Jacobs, B., Piessens, F., Smans, J., Leino, K.R.M., Schulte, W.: A programming model for concurrent object-oriented programs. ACM Trans. Program. Lang. Syst. 31, 1–1 (2008), doi:10.1145/1452044.1452045
17. Kubo, A., Washizaki, H., Fukazawa, Y.: Extracting relations among security patterns. In: SPAQu'08 (Int. Workshop on Software Patterns and Quality) (2008)
18. Lazouski, A., Martinelli, F., Mori, P.: Usage control in computer security: A survey. Computer Science Review 4(2), 81–99 (2010)
19. Le Guernic, G., Banerjee, A., Jensen, T., Schmidt, D.A.: Automata-based confidentiality monitoring. In: Okada, M., Satoh, I. (eds.) ASIAN 2006. LNCS, vol. 4435, pp. 75–89. Springer, Heidelberg (2008)
20. Morin, B., Fleurey, F., Bencomo, N., Jézéquel, J.-M., Solberg, A., Dehlen, V., Blair, G.S.: An aspect-oriented and model-driven approach for managing dynamic variability. In: Czarnecki, K., Ober, I., Bruel, J.-M., Uhl, A., Völter, M. (eds.) MODELS 2008. LNCS, vol. 5301, pp. 782–796. Springer, Heidelberg (2008)
21. Park, J., Sandhu, R.S.: The ucon$_{abc}$ usage control model. ACM Trans. Inf. Syst. Secur. 7(1), 128–174 (2004)
22. Pretschner, A., Hilty, M., Basin, D.A.: Distributed usage control. Commun. ACM 49(9), 39–44 (2006)
23. Rosado, D.G., Fernandez-Medina, E., Lopez, J.: Security services architecture for secure mobile grid systems. Journal of Systems Architecture. In Press (2010)
24. Sabelfeld, A., Myers, A.C.: Language-based information-flow security. IEEE Journal on Selected Areas in Communications 21(1), 2003 (2003)
25. Whittle, J., Moreira, A., Araújo, J., Jayaraman, P., Elkhodary, A.M., Rabbi, R.: An expressive aspect composition language for UML state diagrams. In: Engels, G., Opdyke, B., Schmidt, D.C., Weil, F. (eds.) MODELS 2007. LNCS, vol. 4735, pp. 514–528. Springer, Heidelberg (2007)
26. Yoshioka, N., Washizaki, H., Maruyama, K.: A survey on security patterns. Progress in Informatics 5, 35–47 (2008)

Towards Formal Validation of Trust and Security in the Internet of Services

Roberto Carbone[1], Marius Minea[2], Sebastian Alexander Mödersheim[3],
Serena Elisa Ponta[4,5], Mathieu Turuani[6], and Luca Viganò[7]

[1] Security & Trust Unit, FBK, Trento, Italy
[2] Institute e-Austria, Timişoara, Romania
[3] DTU, Lyngby, Denmark
[4] SAP Research, Mougins, France
[5] DIST, Università di Genova, Italy
[6] LORIA & INRIA Nancy Grand Est, France
[7] Dipartimento di Informatica, Università di Verona, Italy

Abstract. Service designers and developers, while striving to meet the requirements posed by application scenarios, have a hard time to assess the trust and security impact of an option, a minor change, a combination of functionalities, etc., due to the subtle and unforeseeable situations and behaviors that can arise from this panoply of choices. This often results in the release of flawed products to end-users. This issue can be significantly mitigated by empowering designers and developers with tools that offer easy to use graphical interfaces and notations, while employing established verification techniques to efficiently tackle industrial-size problems. The formal verification of trust and security of the Internet of Services will significantly boost its development and public acceptance.

1 Introduction

The vision of the *Internet of Services (IoS)* entails a major paradigm shift in the way ICT systems and applications are designed, implemented, deployed and consumed: they are no longer the result of programming components in the traditional meaning but are built by composing *services* that are distributed over the network and aggregated and consumed at run-time in a demand-driven, flexible way. In the IoS, services are business functionalities that are designed and implemented by producers, deployed by providers, aggregated by intermediaries and used by consumers. However, the new opportunities opened by the IoS will only materialize if concepts, techniques and tools are provided to ensure security. Deploying services in future network infrastructures entails a wide range of trust and security issues, but solving them is extremely hard since making the service components trustworthy is not sufficient: composing services leads to new, subtle and dangerous, vulnerabilities due to interference between component services and policies, the shared communication layer, and application functionality. Thus, one needs validation of both the service components and their composition into secure service architectures.

J. Domingue et al. (Eds.): Future Internet Assembly, LNCS 6656, pp. 193–207, 2011.

Standard validation technologies, however, do not provide automated support for the discovery of important vulnerabilities and associated exploits that are already plaguing complex web-based security-sensitive applications, and thus severely affect the development of the future internet. Moreover, security validation should be carried out at all phases of the service development process, in particular during the design phase by the service designers themselves or by security analysts that support them in their complex tasks, so as to prevent the production and consumption of already flawed services.

Fortunately, a new generation of analyzers for automated security validation at design time has been recently put forth; this is important not just for the results these analyzers provide, but also because they represent a stepping stone for the development of similar tools for validation at service provision and consumption time, thereby significantly improving the all-round security of the IoS. In this chapter, we give a brief overview of the main scientific and industrial challenges for such verification tools, and the solutions they provide; we also discuss some concrete case studies and success stories, which provide proof of concept. As an actual example, we discuss the main ideas and results of one such rigorous technology: the *AVANTSSAR Validation Platform* (or *AVANTSSAR Platform* for short) is an integrated toolset that has been developed in the context of the AVANTSSAR project (www.avantssar.eu, [4]) for the formal specification and automated validation of trust and security of *service-oriented architectures (SOAs)*. This technology, which involves the design of a suitable specification language and is based on a variety of complementary techniques[8], has been tuned and proven on a number of relevant industrial case studies. We also report on our activities in migrating project results to industry and disseminating them to standardization bodies, which will ultimately speed up the development of new network and service infrastructures, enhance their security and robustness, and thus increase the development and public acceptance of the IoS.

We proceed as follows. In Sections 2 and 3, we discuss, respectively, some of the main features of specification languages and automated validation techniques that have been developed for the verification of trust and security of services. In Section 4, we present the AVANTSSAR Platform and the AVANTSSAR Library, and then, in Section 5, we present some case studies and validation success stories, and the migration of results into industrial practice. Section 6 concludes the chapter.

2   Specification Languages

Modeling and reasoning about trust and security of SOAs is complex due to three main characteristics of service orientation.

First, SOAs are *heterogeneous*: their components are built using different technology and run in different environments, yet interact and may interfere with each other.

---

[8] Such as model checking with constraints, approaches based on SAT (i.e., satisfiability) or SMT (i.e., satisfiability modulo testing), or abstract interpretation.

Second, SOAs are also *distributed* systems, with functionality and resources distributed over several machines or processes. The resulting exponential state-space complexity makes their design and efficient validation difficult, even more so in hostile situations perhaps unforeseen at design time.

Third, SOAs and their security requirements are *continuously evolving*: services may be composed at runtime, agents may join or leave, and client credentials are affected by dynamic changes in security policies (e.g., for incidents or emergencies). Hence, security policies must be regarded as part of the service specification and as first-class objects exchanged and processed by services.

The security properties of SOAs are, moreover, very diverse. The classical data security requirements include confidentiality and authentication/integrity of the communicated data. More elaborate goals are structural properties (which can sometimes be reduced to confidentiality and authentication goals) such as authorization (with respect to a policy), separation or binding of duty, and accountability or non-repudiation. Some applications may also have domain-specific goals (e.g., correct processing of orders). Finally, one may consider liveness properties (under certain fairness conditions), e.g., for a given web service for online shopping one may require that every order will eventually be processed if the intruder cannot block the communication indefinitely. This diversity of goals cannot be formulated with a fixed repertoire of generic properties (like authentication); instead, it suggests the need for specification of properties in an expressive logic.

Various languages have been proposed to model trust and security of SOAs, e.g., BPEL [24], π calculus [19], F# [5], to name a few. Each of them, however, focuses only on some aspects of SOAs, and cannot cover all previously described features, except perhaps in an artificial way. One needs a language fully dedicated to specifying trust and security aspects of services, their composition, the properties that they should satisfy and the policies they manipulate and abide by. Moreover, the language must go beyond static service structure: a key challenge is to integrate policies that are dynamic (e.g., changing with the workflow context) with services that can be added and composed dynamically themselves.

As a concrete solution, in the AVANTSSAR project, we have defined a language, the *AVANTSSAR Specification Language ASLan*, that is both expressive enough that many high-level languages, such as BPEL, can be translated to it, and amenable to formal analysis.[9] A key feature of ASLan is the integration of *Horn clauses* that are used to describe policies in a clear, logical way, with a *transition system* that expresses the dynamics of the system, e.g., agents can become members of a group or leave it, with immediate consequences for their access rights.

---

[9] The AVANTSSAR Platform allows users also to input their services by specifying them using the high-level formal specification language ASLan++, which we have defined to be close to specification languages for security protocols/services and to procedural and object-oriented programming languages. The semantics of ASLan++ is formally defined by translation to ASLan.

As a simple, general (i.e., not AVANTSSAR/ASLan specific) example, consider the policies that a user U has access to a file F if U belongs to a group G that is the owner of F, or U is the deputy of a user that has access to F:

$$access(U, F) \leftarrow member(U, G) \wedge owner(G, F)$$
$$access(U, F) \leftarrow deputy(U, U') \wedge access(U', F)$$

Policies are dynamic, since facts like *member*, *owner*, and *deputy* can change over time, which in turn affects *access* rights. For instance, if user *Alice* changes to another group within the organization, she will immediately obtain all access rights to files of the new group, but lose access rights to files of her old group, except for those that she maintains due to her being a deputy for other users.

We consider transition systems in which a state is a set of facts like *member*, *owner*, etc.; they can be used to describe service workflows and steps in security protocols. For instance, an employee (*Alice*) changing group membership at the command of her manager (*Peter*) can be formalized as:

$$member(Alice, g_1) \wedge isManager(Peter, Alice) \wedge canAssign(Peter, g_3) \Rightarrow$$
$$member(Alice, g_3) \wedge isManager(Peter, Alice) \wedge canAssign(Peter, g_3)$$

The above transition is applicable in a state that includes all facts on the left hand side. When the transition is applied, *Alice*'s membership to g_1 is removed, she is added as member to g_3, and other facts are preserved.

We can now integrate policies with the dynamic aspects of transition systems by defining: the set of facts that hold true in a state is the least closure of the state under all Horn clauses. For example, if a state contains the facts

$$member(a, g_1), member(b, g_2), owner(g_1, f_1), owner(g_2, f_2), deputy(a, b)$$

then the policy implies the following access rights:

$$access(a, f_1), access(b, f_2), access(a, f_2)$$

The least closure represents a "default deny": if the policies do not imply the access right, it is false. The main difference between policies expressed via Horn clauses and transitions expressed via rewrite rules is that the effects of the policy are inferred in the same state (repeatedly, after each transition), while the effects of a transition are inferred in the next state. Thus, the use of Horn clauses enables a deduction chain which is performed for every state of the transition system. Integrating Horn clauses with transition systems is, of course, a broader concept: next to policies, we can model other consequences of facts that become true.

Finally, we need to model the *security properties*. While this can be done by using different languages, in ASLan we have chosen to employ a variant of linear temporal logic (LTL, e.g. [18,25]), with backwards operators and ASLan facts as propositions. This logic gives us the desired flexibility for the specification of complex goals. As an example, in a system employing the policy described above, we may require a *separation of duty* property, namely that for privacy purposes, no agent can access both files f_1 and f_2. This can be expressed in LTL as follows:

$$G(\ (access(U, f_1) \rightarrow \neg F(access(U, f_2)))\ \wedge\ (access(U, f_2) \rightarrow \neg F(access(U, f_1)))\ )$$

3   Automated Validation Techniques

Due to the inherent complexity (heterogeneity, distribution and dynamicity) of the Internet of Services, the challenge of validating services and service-oriented applications cannot be addressed simply by scaling up the current generation of formal analysis approaches and tools. Rather, novel and different validation techniques are required to automatically reason about services, their composition, their required security properties and associated policies. In particular, one has to consider the various ways in which component services can be coordinated, and develop new techniques, such as model checking, that allow for compositional validation reflecting this modularity, as well as cope with the complexity problem. Moreover, for the practical use and take-up by industry and standardisation organisations, it is essential that any such verification technique provides a high degree of automation.

An important solution to overcome the complexity of SOAs and the heterogeneous security contexts is to integrate different technologies into a single analysis tool, in such way that they can interact and benefit from each other's features. For instance, the AVANTSSAR Platform comprises three validation backends (CL-AtSe [27], OFMC [23], and SATMC [1]), which are based on different automated deduction techniques operating on the same ASLan input, and which thus provide complementary strengths.

In the following subsections, we discuss these four points — orchestration, model checking of SOAs, compositional reasoning, and abstraction-based validation techniques — in more detail and describe how they have been implemented in the AVANTSSAR Platform.

3.1   Orchestration

Composability, one of the basic principles and design-objectives of SOAs, expresses the need for providing simple scenarios where already available services can be reused to derive new added-value services. In their SOAP incarnation, based on XML messaging and relying on a rich stack of related standards, SOAs provide a flexible yet highly inter-operable solution to describe and implement a variety of e-business scenarios possibly bound to complex security policies.

When security constraints are to be respected, it can be very complex to discover or even to describe composition scenarios. This motivates the introduction of automated solutions to scalable services composition. Two key approaches for composing web services have been considered, which differ by their architecture: *orchestration* is centralized and all traffic is routed through a *mediator*, whereas *choreography* is distributed and all web services can communicate directly.

Several "orchestration" notions have been advocated (see, e.g., [20]). However, in inter-organizational business processes it is crucial to protect sensitive data of each organization; and our main motivation is to take into account the security policies while computing an orchestration. The AVANTSSAR Platform, for example, implements an idea presented in [11] to automatically generate a *mediator*. We specify a web service profile from its *XML Schema* and *WS-SecurityPolicy* using first-order terms (including cryptographic functions). The *mediator* is able to use cryptography to produce new messages, and is constructed with respect to security goals using the techniques we developed for the verification of security protocols.

3.2   Model Checking of SOAs

Model checking [13] is a powerful and automatic technique for verifying concurrent systems. It has been applied widely and successfully in practice to verify digital sequential circuit designs, and, more recently, important results have been obtained for the analysis of security protocols. In the context of SOAs, a model-checking problem is the problem of determining whether a given model — representing the execution of the service under scrutiny in a hostile environment — enjoys the security properties specified by a given formula. As mentioned in Section 2, these security properties can be complex, requiring an expressive logic.

Most model-checking techniques in this context make a number of simplifying assumptions on the service and/or on its execution environment that prevent their applicability in some important cases. For instance, most techniques assume that communication between honest principals is controlled by a Dolev-Yao intruder [17], i.e. a malicious agent capable to overhear, divert, and fake messages. Yet we might be interested in establishing the security of a service that relies on a less insecure channel. In fact, services often rely on transport protocols enjoying some given security properties (e.g. TLS is often used as a unilateral or a bilateral communication authentic and/or confidential channel), and it is thus important to develop model-checking techniques that support reasoning about communication channels enjoying security-relevant properties, such as authenticity, confidentiality, and resilience.

Among general model-checking techniques, *bounded model checking*, by supporting reasoning about LTL formulae, allows one to reason about complex trace-based security properties. In particular, the AVANTSSAR Platform integrates a bounded model-checking technique for SOAs [1] that allows one to express complex security goals that services are expected to meet as well as assumptions on the security offered by the communication channels.

3.3   Channels and Compositional Reasoning

A common feature of SOAs is an organization in *layers*: we may have a layer that provides a secure communication infrastructure between participants, e.g. a virtual private network or a TLS [26] channel, and run applications on top of it, as if the participants were directly connected via tamper-proof lines. It is, of course, undesirable to verify the entire system as a whole: this can easily be too complex for automated methods, and lacks generality and reuse. In fact, an application that requires a secure connection should not depend on the details of the realization of the secure connection and, vice-versa, a system that establishes a secure channel should be able to run arbitrary protocols over it. Thus, a compositionality result is desired: if components are safe in isolation and satisfy certain properties, then they can be composed into a larger system that is also safe.

Progress has been made in this direction for the parallel (and sequential) composition of protocols [12,15,16], i.e., independently using several protocols over the same communication medium. Moreover, there are first results for the layered compositional reasoning needed for SOAs, namely running an application over a channel [22]. This means verifying that *(i)* a protocol such as TLS indeed provides an authentic and confidential channel, *(ii)* an application system is safe if its communication is routed over a secure channel, and *(iii)* both satisfy certain sufficient conditions (their message formats do not interfere). Here, *(i)* and *(ii)* are verification tasks that the tools can check in isolation, and *(iii)* is a format property that can be checked statically. If *(i)*, *(ii)*, and *(iii)* hold, then we can conclude that running the application over the established channel is also safe.

3.4 Abstract Interpretation

The complexity of SOAs is a major challenge for classical model-checking methods. To cope with this, formal validation approaches often strictly bound all aspects of a system, e.g., the number of service runs that honest agents (and a dishonest agent) can perform. However, one would rather verify a system without such limitations, e.g., no matter how many agents use it in parallel. Hence, methods based on abstract interpretation have recently become increasingly popular [6,7,8,9,14,28]. For instance, TulaFale [6], a tool by Microsoft Research based on ProVerif [7], exploits abstract interpretation for verification of web services that use SOAP messaging, using logical predicates to relate the concrete SOAP messages to a less technical representation that is easier to reason about.

There are two basic ideas here: *(i)* partition the infinite set of constants (representing, for instance, agents, request numbers, or cryptographic keys) into finitely many equivalence classes and to compute on those equivalence classes instead; *(ii)* avoid reasoning about a transition system and rather compute an over-approximation of everything that will ever hold true. This allows for the use of classical automated first-order reasoning techniques, in particular resolution or fixed-point computations of static analysis. Thanks to the over-approximation, these systems completely avoid the state-explosion problems of model checking and can analyze systems without bounds on the number of runs that agents participate in. On the downside, the over-approximation can introduce false positives, i.e., attacks introduced by the over-approximation while the actual system is safe.

The ability of abstraction methods to avoid exploration of concrete transition systems has been the reason for their success, but on the other hand also implies a serious limitation for the verification of complex SOAs. Since there is no notion of time, we cannot model that at some time-point, a key, certificate, access right, or membership is revoked. The *set-based abstraction method* [21] can overcome this limitation while preserving the benefits of abstract interpretation. The idea is to organize data by means of sets and to abstract data by set membership. In the example of Section 2, we may consider the set U_g of users that are currently members of a group g. We can then identify all users that belong to the same set of groups as one abstract equivalence class. The difficult part is how to handle the change of the set memberships, e.g., if a user changes from one group to another. Here, the set-abstraction method defines a mechanism to reason about how facts about one class imply facts about a new class, e.g., roughly, a dishonest user would not delete any information that he has learned in the old group, but he cannot read any new information that the old group produces. Thus, revocation of facts can be modeled without the need to directly express sequencing in time.

4   The AVANTSSAR Platform and Library

We have implemented the AVANTSSAR Platform as a service-oriented architecture. As shown in Fig. 1, the platform includes a *connectors layer*, i.e., a layer of software modules that carry out the translation from application-level specification languages (such as BPMN and BPEL, as well as our own AnB and ASLan++) into ASLan, and vice versa for the platform output. The platform then takes as concrete input a policy stating the functional and security requirements of a goal service and a description of the available services (including a specification of their security-relevant behavior, possibly including the local policies they satisfy) and applies automated reasoning techniques in order to build an orchestration of the available services that meets the security requirements stated in the policy. More specifically, the platform comprises of two main components:

- The *Orchestrator* tries to build an orchestration, i.e., a composition, of the available services in a way that is expected (but not yet guaranteed) to satisfy the input policy. It takes as input an ASLan file with a specification of the available services and either a specification of the client or a partial specification of the goal, and it produces as output an ASLan file with the specification of the available services, a full specification of the goal, and a specification of the client (a putative one, if it was not given as input).
- The *Validator* takes as input an orchestration and a security goal formally specified in ASLan, and automatically checks whether the orchestration meets the security goal. If this is the case, then the ASLan specification of the validated orchestration is given as output, otherwise a counterexample is sent back to the Orchestrator (where a failed validation means the existence of vulnerabilities that need to be fixed).

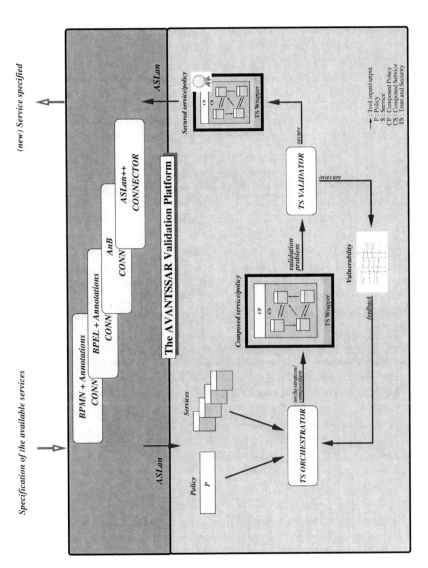

Fig. 1. The AVANTSSAR Validation Platform and its usage towards Enterprise SOA.

As proof of concept, we have applied the AVANTSSAR Platform to the case studies in the *AVANTSSAR Library*, which comprises of the formalization of 10 application scenarios and 94 problem cases of service-oriented architectures from the e-Business, e-Government and e-Health application areas. In this way, we have been able to detect a considerable number of attacks in the considered services and provide the required corrections. Moreover, the formal modeling of case studies has allowed us to consolidate our specification languages and has driven the evolution of the validation platform, both in terms of support for the new language and modeling features, as well as in efficiency improvements needed for the validation of significantly more complex models. We expect that the library will provide a useful test-suite for similar validation technologies.

5   Case Studies, Success Stories, and Industry Migration

The landscape of services that require validation of their security is very broad. The validation is made more difficult by the tension between the need for flexibility, adaptability, and reconfigurability, and the need for simple, understandable, coherent, declarative policies. These must contain all relevant information required to determine the access to private data and to the meta-policies that control them. For example, e-Health practitioners are under pressure to reduce redundant and inefficient behaviors in daily workflows and methods. They must establish repeatable, standards-based solutions that promote a "plug-and-play" approach to context-based information access, making clinical and non-clinical data available anywhere and anytime in a health care organization, while lowering infrastructure costs. Clearly, privacy requirements will be much more difficult to implement and assess in such environments. To ease the analysis, it is necessary to factor out the access control policies and meta-policies from the possible workflow, and to understand and validate the authorization conditions and the security mechanisms that implement them independently of their use in particular workflows. There is thus a clear advantage in having a language allowing the specification of policies via clauses (e.g., Horn clauses) next to the transition system defining the workflow as put forward in Section 2.

Within the AVANTSSAR project, services from a wide variety of application areas have been modeled: banking (loan origination), electronic commerce (anonymous shopping), e-Government (citizen and service portals, public bidding, digital contract signing), and e-Health (electronic health records). Classes of properties that have been verified include authorization policies, accountability, trust management, workflow security, federation and privacy.

A highlight of the effectiveness of the AVANTSSAR methods and tools is the detection of a serious flaw in the SAML-based SSO solution for Google Apps [3]. Though well specified and thoroughly documented, the OASIS SAML security standard is written in natural language that is often subject to interpretation. Since the many configuration options, profiles, protocols, bindings, exceptions, and recommendations are laid out in different, interconnected documents, it is hard to establish which message fields are mandatory in a given profile and which are not. Moreover, SAML-based solution providers have internal requirements that may result in small deviations from the standard. For instance, internal requirements (or denial-of-service considerations) may lead the service provider to avoid checking the match between the ID field in the AuthResp and in the previously sent AuthReq. What are the consequences of such a choice? The technical overview document provided by OASIS SAML as a non-official addendum increases the clarity in this respect. Still, when Google developed their SAML-based SSO solution for Google Apps they released a flawed product, which allowed a dishonest service provider to impersonate the victim user on Google Apps, granting unauthorized access to private data and services (email, docs, etc.). The vulnerability was detected by the SATMC backend of the AVANTSSAR Platform and the attack was reproduced in an actual deployment of SAML-based SSO for Google Apps. Google and the US Computer Emergency Readiness Team (US-CERT) were informed and the vulnerability was kept confidential until Google developed a new version of the authentication service and Google's customers updated their applications accordingly. The severity of the vulnerability has been rated High in a note issued by the National Institute of Standard and Technology (NIST).

Moreover, as shown in [2], the SATMC backend of the AVANTSSAR Platform also allowed us to detect that the prototypical SAML SSO use case (as described in the SAML technical overview) suffers from an authentication flaw that, under some conditions, allows a malicious service provider to hijack a client authentication attempt and force the latter to access a resource without its consent or intention. It also allows an attacker to launch Cross-Site Scripting (XSS) and Cross-Site Request Forgery attacks (XSRF). This last type of attack is even more pernicious than classic XSS, because XSRF requires the client to have an active session with the service provider, whereas in this case, the session is created automatically hijacking the client's authentication attempt. This may have serious consequences, as witnessed by the new XSS attack identified in the SAML-based SSO for Google Apps and that could have allowed a malicious web server to impersonate a user on any Google application. In [2], solutions that can be used to mitigate and even solve the problem are described. These possible solutions are being discussed with OASIS.

In another notable validation success story, the tool Tookan [10], which is based on SATMC, has automatically found vulnerabilities in PKCS#11-based products by Aladdin, Bull, Gemalto, RSA, and Siemens among others. PKCS# 11 specifies an API for performing cryptographic operations such as encryption and signature using cryptographic tokens (e.g., USB tokens or smart cards). Sensitive cryptographic keys, stored inside the token, should not be revealed to the outside and it should be impossible for an attacker to change those keys. The attacks found show that in many implementations this is not the case: the compromise of a key allows an attacker to clone the token and, more generally, to perform the same security-critical operations as the legitimate token user.

Formal validation of trust and security will become a reality in the Internet of Services only if and when the available technologies will have migrated to industry, as well as to standardization bodies (which are mostly driven by industry and influence the future of industrial development). Such an industry migration has to face the gap between advanced *formal methods (FM)* techniques and their real exploitation within industry and standardisation bodies. Though the use of FM would promote a more secure development environment, a variety of practical and cultural reasons lead the industrial world to perceive FM approaches as being expensive in terms of time and effort in comparison to the benefits they provide, and difficult to be integrated within industrial processes. In order to ease their adoption, several obstacles have to be overcome, such as: *(i)* the lack of automated FM technology, *(ii)* the gap between the problem case that needs to be solved in industry and the abstract specification provided by FM, and *(iii)* the differences between formal languages and models and those used in industrial design and development environments (e.g., BPMN, Java, ABAP).

The problem is how to make new, efficient methodologies and technologies accessible and readily exploitable, benefitting industry designers and developers. This amounts to migrating the research outcomes of the logical level into the application level by providing a push-button technology so that industry and standardization bodies could check more rapidly the correctness of the proposed solutions without having a strong mathematical background. In particular, industrially suited specification languages (model-driven languages), equipped with easy-to-use GUIs and translators to and from the core formal models should be devised and migrated to the selected development environments.

A concrete example is the industry migration of the AVANTSSAR Platform to the SAP environment. Two valuable migration activities have been carried out by building contacts with core business units. First, in the trail of the successful analysis of Google's SAML-based SSO, an internal project has been run to migrate AVANTSSAR results within SAP NetWeaver Security and Identity Management (SAP NW SIM) with the objective of exploiting the AVANTSSAR technology to initiate a deep formal analysis of the SAP NetWeaver SAML Next Generation Single Sign-On (NW-NGSSO) to formally establish its soundness, i.e., to have formal evidence that the employed service providers and identity provider services fulfill the expected security desiderata in the considered SAP relevant scenarios. This has included the evaluation of those configurations of the highly configurable SAML SSO standard that are relevant for SAP as well as design and development decisions SAP could have taken to fulfill internal customer requirements. More than 50 formal specifications capturing these scenarios, the variety of configuration options, and SAP internal design and implementation choices have been specified. By means of the AVANTSSAR Platform, safe and unsafe configurations for NW-NGSSO for several SAML profiles relevant for SAP have been identified. All discovered risks and flaws in the SAML protocol have been addressed in NW-NGSSO implementation and counter-measures have been taken. The results have been collected in tables that can be used by SAP in setting-up the NW-NGSSO services on customer production systems.

Besides this, the results triggered very valuable discussions in the steering committee that was supervising this internal project. For instance, the authentication flaw in the SAML standard helped SAP business units to get major insights in the SAML standard than the security considerations described in there and helped SAP Research to better understand the vulnerability itself and to consolidate the results.

The AVANTSSAR technology has been also integrated into the SAP Net-Weaver Business Process Management (NW BPM) product to formally validate security-critical aspects of business processes. An eclipse plug-in extension for NW BPM was proposed through the design and development of a security validation plug-in that enables a business process modeler to easily specify the security goals one wishes to validate such as least privilege which can be accomplished by means of the Need-to-Know principle (giving to the users enough rights to perform their job, but no more than that). It also proposes to control the access over automated tasks through the restriction on the invocation and consumption of remote services. A scalability study has also been conducted on a loan origination process case study with a few security goals and on a more complex aviation maintenance process (designed with 70 human activities). The performance analysis helped us to devise a number of optimizations (up to three orders of magnitude).

These results show that the AVANTSSAR technology can provide a high level of assurance within industrial BPM systems, as it allows for validating all the potential execution paths of the BP under-design against the expected security desiderata. In particular, the migration activity succeeded in overcoming obstacles for the adoption of model-checking techniques to validate security desiderata in industry systems by providing an automatic generation of the formal model on which to run the analysis, as well as highlighting the model-checking results as a comprehensive feedback to a business analyst who is neither a model-checking practitioner nor a security expert. As a successful result, the security validation plug-in is currently listed in the productization road-map of SAP products for business process management.

6   Conclusions and Outlook

As exemplified by these case studies and success stories, formal validation technologies can have a decisive impact for the trust and security of the IoS. The research innovation put forth by AVANTSSAR aims at ensuring global security of dynamically composed services and their integration into complex SOAs by developing an integrated platform of automated reasoning techniques and tools. Similar technologies are being developed by other research teams. Together, all these research efforts will result in a new generation of tools for automated security validation at design time, which is a stepping stone for the development of similar tools for validation at service provision and consumption time. These advances will significantly improve the all-round security of the IoS, and thus boost its development and public acceptance.

References

1. Armando, A., Carbone, R., Compagna, L.: LTL Model Checking for Security Protocols. Journal of Applied Non-Classical Logics, special issue on Logic and Information Security, 403–429 (2009)
2. Armando, A., Carbone, R., Compagna, L., Cuéllar, J., Pellegrino, G., Sorniotti, A.: From Multiple Credentials to Browser-based Single Sign-On: Are We More Secure? In: Proceedings of IFIP SEC 2011 (to appear)
3. Armando, A., Carbone, R., Compagna, L., Cuellar, J., Tobarra Abad, L.: Formal Analysis of SAML 2.0 Web Browser Single Sign-On: Breaking the SAML-based Single Sign-On for Google Apps. In: Proceedings of the 6th ACM Workshop on Formal Methods in Security Engineering (FMSE 2008), pp. 1–10. ACM Press, New York (2008)
4. AVANTSSAR: Automated Validation of Trust and Security of Service-Oriented Architectures. FP7-ICT-2007-1, Project No. 216471, http://www.avantssar.eu, 01.01.2008–31.12.2010
5. Bhargavan, K., Fournet, C., Gordon, A.D.: Verified Reference Implementations of WS-Security Protocols. In: Bravetti, M., Núñez, M., Zavattaro, G. (eds.) WS-FM 2006. LNCS, vol. 4184, pp. 88–106. Springer, Heidelberg (2006)
6. Bhargavan, K., Fournet, C., Gordon, A.D., Pucella, R.: Tulafale: A security tool for web services. In: de Boer, F.S., Bonsangue, M.M., Graf, S., de Roever, W.-P. (eds.) FMCO 2003. LNCS, vol. 3188, pp. 197–222. Springer, Heidelberg (2004)
7. Blanchet, B.: An efficient cryptographic protocol verifier based on Prolog rules. In: Proceedings of the 14th IEEE Computer Security Foundations Workshop, pp. 82–96. IEEE Computer Society Press, Los Alamitos (2001)
8. Bodei, C., Buchholtz, M., Degano, P., Nielson, F., Nielson, H.R.: Static validation of security protocols. Journal of Computer Security 13(3), 347–390 (2005)
9. Boichut, Y., Héam, P.-C., Kouchnarenko, O.: TA4SP (2004), http://www.univ-orleans.fr/lifo/Members/Yohan.Boichut/ta4sp.html
10. Bortolozzo, M., Centenaro, M., Focardi, R., Steel, G.: Attacking and Fixing PKCS#11 Security Tokens. In: Proceedings of the 17th ACM conference on Computer and Communications Security (CCS 2010), pp. 260–269. ACM Press, New York (2010)
11. Chevalier, Y., Mekki, M.A., Rusinowitch, M.: Automatic Composition of Services with Security Policies. In: Proceedings of Web Service Composition and Adaptation Workshop (held in conjunction with SCC/SERVICES-2008), pp. 529–537. IEEE Computer Society Press, Los Alamitos (2008)
12. Ciobâca, S., Cortier, V.: Protocol composition for arbitrary primitives. In: Proceedings of 23rd IEEE Computer Security Foundations Symposium, pp. 322–336. IEEE Computer Society Press, Los Alamitos (2010)
13. Clarke, E.M., Grumberg, O., Peled, D.A.: Model Checking. MIT Press, Cambridge (1999)
14. Comon-Lundh, H., Cortier, V.: New decidability results for fragments of first-order logic and application to cryptographic protocols. Technical Report LSV-03-3, Laboratoire Specification and Verification, ENS de Cachan, France (2003)
15. Cortier, V., Delaune, S.: Safely composing security protocols. Formal Methods in System Design 34(1), 1–36 (2009)
16. Datta, A., Derek, A., Mitchell, J., Pavlovic, D.: Secure protocol composition. In: Proceedings of the 19th MFPS, ENTCS 83, Elsevier, Amsterdam (2004)

17. Dolev, D., Yao, A.: On the Security of Public-Key Protocols. IEEE Transactions on Information Theory 2(29) (1983)
18. Hodkinson, I., Reynolds, M.: Temporal Logic. In: Blackburn, P., van Benthem, J., Wolter, F. (eds.) Handbook of Modal Logic, pp. 655–720. Elsevier, Amsterdam (2006)
19. Lucchi, R., Mazzara, M.: A pi-calculus based semantics for WS-BPEL. Journal of Logic and Algebraic Programming 70(1), 96–118 (2007)
20. Marconi, A., Pistore, M.: Synthesis and Composition of Web Services. In: Bernardo, M., Padovani, L., Zavattaro, G. (eds.) SFM 2009. LNCS, vol. 5569, pp. 89–157. Springer, Heidelberg (2009)
21. Mödersheim, S.: Abstraction by Set-Membership — Verifying Security Protocols and Web Services with Databases. In: Proceedings of 17th ACM conference on Computer and Communications Security (CCS 2010), pp. 351–360. ACM Press, New York (2010)
22. Mödersheim, S., Viganò, L.: Secure Pseudonymous Channels. In: Backes, M., Ning, P. (eds.) ESORICS 2009. LNCS, vol. 5789, pp. 337–354. Springer, Heidelberg (2009)
23. Mödersheim, S., Viganò, L.: The Open-Source Fixed-Point Model Checker for Symbolic Analysis of Security Protocols. In: Aldini, A., Barthe, G., Gorrieri, R. (eds.) FOSAD 2007/2008/2009. LNCS, vol. 5705, pp. 166–194. Springer, Heidelberg (2009)
24. Oasis Consortium. Web Services Business Process Execution Language vers. 2.0 (2007), http://docs.oasis-open.org/wsbpel/2.0/OS/wsbpel-v2.0-OS.pdf
25. Pnueli, A.: The Temporal Logic of Programs. In: Proceedings of the 18th IEEE Symposium on Foundations of Computer Science, pp. 46–57. IEEE Computer Society Press, Los Alamitos (1977)
26. T. Dierks and E. Rescorla. The Transport Layer Security (TLS) Protocol, Version 1.2. IETF RFC 5246 (Aug. 2008)
27. Turuani, M.: The CL-Atse Protocol Analyser. In: Pfenning, F. (ed.) RTA 2006. LNCS, vol. 4098, pp. 277–286. Springer, Heidelberg (2006)
28. Weidenbach, C., Afshordel, B., Brahm, U., Cohrs, C., Engel, T., Keen, E., Theobalt, C., Topic, D.: System Description: Version 1.0.0. In: Ganzinger, H. (ed.) CADE 1999. LNCS (LNAI), vol. 1632, pp. 378–382. Springer, Heidelberg (1999)

Trustworthy Clouds Underpinning the Future Internet

Rüdiger Glott[1], Elmar Husmann[2], Ahmad-Reza Sadeghi[3], and
Matthias Schunter[2]

[1] Maastricht University, The Netherlands
glott.ruediger@gmail.com
[2] IBM Research – Zürich, Rüschlikon, Switzerland
huselmar@de.ibm.com, mts@zurich.ibm.com
[3] TU Darmstadt, Germany
ahmad.sadeghi@trust.rub.de

Abstract. Cloud computing is a new service delivery paradigm that
aims to provide standardized services with self-service, pay-per-use, and
seemingly unlimited scalability. This paradigm can be implemented on
multiple service levels (infrastructures, run-time platform, or actual Software as a Service). They are are expected to be an important component
in the future Internet.
This article introduces upcoming security challenges for cloud services
such as multi-tenancy, transparency and establishing trust into correct
operation, and security interoperability. For each of these challenges, we
introduce existing concepts to mitigate these risks and survey related
research in these areas.

1 Cloud Computing and the Future Internet

Cloud computing is expected to become a backbone technology of the Future
Internet that provides Internet-scale and service-oriented access to virtualized
computing, data storage and network resources as well as higher level services.
In contrast to the current cloud market that is mainly characterized by isolated
providers, cloud computing in the Future Internet is expected to be characterized by a seamless cloud capacity federation of independent providers - similar
to the network peering and IP transit purchasing of ISPs in today's Internet.
For an end-user this means that via interacting with one cloud provider, resources and services provided by multiple similar providers are seamlessly accessed. Cloud computing goes beyond technological infrastructure that derives
from the convergence of computer server power, storage and network bandwidth.
It is a new business and distribution model for computing that establishes a new
relationship between the end user and the data center, which "...gives the user
'programmatic control' over a part of the data center" [1, pp. 8-9].

For this cloud-of-clouds vision[4] this article will investigate the related challenges for trust and security architectures and mechanisms.

---

[4] For which the Internet pioneer Vint Cerf has recently suggested the term "Intercloud"

J. Domingue et al. (Eds.): Future Internet Assembly, LNCS 6656, pp. 209–221, 2011.

FIA projects like RESERVOIR or VISION are conducting research on core technological foundations of the cloud-of-clouds such as federation technologies, interoperability standards or placement policies for virtual images or data across providers. Many of these developments can be expected to be transferred into the Future Internet Core Platform project that will launch in 2011. This goes along with increased collaboration on open cloud standards under developments by groups such as the DMTF Open Clouds Standards Incubator, the SNIA Cloud Storage Technical Working Group or the OGF Open Clouds Computing Interface Working Group.

Trust and security are often regarded as an afterthought in this context, but they may ultimately present major inhibitors for the cloud-of-clouds vision. An important property of this emerging infrastructure will be the need to respect global legal requirements. Today, since the current legal systems are not prepared for the challenges that result from the complexity and pervasiveness of cloud computing, data protection and privacy issues as well as liability and compliance problems may hinder to tap the full potential of cloud computing [22,8,26]. By clouds becoming regulation-aware, in the sense that it will ensure that data mobility is limited to ensure compliance with a wide range of different national legislation including privacy legislation such as the EU Data Protection Directive 95/46/EC.

As of today, cloud computing is facing significant acceptance hurdles when it comes to hosting important business applications or critical infrastructures such as those of the usage domains addressed by FIA. This article will illustrate the reasons for this, and discuss the complex trust and security requirements. Furthermore, we survey existing components to overcome these security and privacy risks. We will explain the state-of-the-art in addressing these requirements and give an overview of related ongoing international, and particularly EU research activities as well as derive future directions of technology development.

2   Trust and Security Limitations of Global Cloud Infrastructures

2.1   Cloud Security Offerings Today

According to the analyst enterprise Forrester Research and their study "Security and the Cloud" [17] the cloud security market is expected to grow to 1.5 billion $ by 2015 and to approach 5 % of overall IT security spending. Whereas today identity management and encryption solutions represent the largest share of this market, particular growth can be expected in three directions:

1. securing commercial clouds to meet the requirements of specific market segments
2. bespoke highly secure private clouds
3. a new range of providers offering cloud security services to add external security to public clouds

An example for the first category is the Google gov.app cloud launched in September 2009 that offers a completely segregated cloud targeted exclusively at US government customers. Similarly, IBM has launched a FISMA compliant Federal Community Cloud in 2010.

Other cloud providers also adapt basic service security to the needs of specific markets and communities. Following its software-plus-services strategy announced in 2007, Microsoft has developed in the past years several SaaS cloud services such as the Business Productivity Online Suite (BPOS). While all of them may be delivered from a multi-tenant public cloud for the entry level user, Microsoft offers dedicated private cloud hosting and supports third-party or customer-site hosting. This allows tailor made solutions to specific security concerns - in particular in view of the needs of larger customers. In the same way, the base security of Microsoft public cloud services is adapted to the targeted market. Whereas Microsoft uses, e.g., for the Office Live Workspace - in analogy to what Google does with Gmail - unencrypted data transfer between the cloud and the user, cloud services for more sensitive markets (such as Microsoft Health Vault) use SSL encryption by default.

On the other hand commodity public cloud services such as the Amazon EC2 are still growing even though they offer only limited base security and largely transfer responsibility for security to the customer. Therefore in parallel to the differentiated security offerings via bespoke private or community clouds, there is also a growing complementary service market to enable enhanced security for public clouds. Here a prime target is the small to mid-size enterprise market. Examples for supplementary services are threat surveillance (e.g,. AlertLogic), access- and identity management (e.g., Novell, IBM), virtual private networking (e.g., Amazon Virtual Private Cloud), encryption (e.g., Amazon managed encryption services) and web traffic filtering services (e.g., Zscaler, ScanSafe).

2.2   Today's Datacenters as the Benchmark for the Cloud

Using technology always constitutes a certain risk. If the IT of any given business failed, the consequences for most of today's enterprises would be severe. Even if multiple lines of defense are used (e.g., firewalls, intrusion defense, and protection of each host), all systems usually contain errors that can be found and exploited. While off-line systems are harder to attack, exchanging media such as USB sticks allows transfer into systems that are not connected to the Internet [5].

Cloudsourcing [15] follows more or less the same economic rationale as traditional IT-outsourcing but provides more benefits, inter alia with regard to upgrades and patches, quick procurement services, avoidance of vendor lock-ins, and legacy modernization [18]. Many cloudsourcers offer bundles of consulting services, application development, migration, and management [14]. A problem that remains with this new stage of IT-outsourcing strategies is that the client still has to find trustworthy service providers. However, this problem has been solved in earlier forms of IT outsourcing, therefore it is not very likely that the emergence of new business opportunities and business models will fail on this point. Rather than that, cloud computing might be significantly hindered by the legal problems that remain to be solved.

For the security objectives when adopting clouds for hosting critical systems we believe that today's datacenters are the benchmark for new cloud deployments. Overall, the benefits need to outweigh the potential disadvantages and risks. While the cost and flexibility benefits of using clouds are easy to quantify, potential disadvantages and risks are harder to qualitatively assess or even quantitatively measure. An important aspect for this equation is the perceived level of uncertainty: For instance, a low but contractually guaranteed availability (such as 98% availability) will allow enterprises to pick workloads that do not require higher guarantees. Today, uncertainty about the actual availability does not allow enterprises to make such risk-management decisions and thus will only allow hosting of uncritical workloads on the cloud.

For security this argument leads to two requirements for cloud adoption by enterprises: The first is that with respect to security and trust, new solutions such as the cloud or cloud-of-clouds will be compared and benchmarked against existing solutions such as enterprise or outsourced datacenters. The second is that in order to allow migration of critical workloads to the cloud, cloud providers must enable enterprises to integrate cloud infrastructures into their overall risk management. We will use these requirements in our subsequent arguments.

3   New Security and Privacy Risks and Emerging Security Controls

Cloud computing being a novel technology introduces new security risks [7] that need to be mitigated. As a consequence, cautious monitoring and management of security risks [13] is essential (see Figure 1 for a sketch following [12]).

We now survey selected security and privacy risks where importance has been increased by the cloud and identify potential security controls for mitigating those risks.

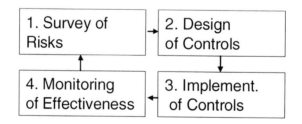

Fig. 1. Simplified Process for Managing Security Risks [12])

3.1   Isolation Breach between Multiple Customers

Cloud environments aim at efficiencies of scale by increased sharing resources between multiple customers. As a consequence, data leakage and service disruptions gain importance and may propagate through such shared resources. An important requirement is that data cannot leak between customers and that malfunction or misbehavior by one customer must not lead to violations of the service-level agreement of other customers.

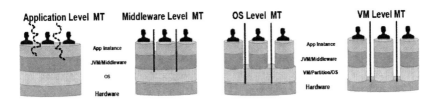

Fig. 2. Multi-tenancy at Multiple Levels [25].

Traditional enterprise outsourcing ensures the so-called "multi-tenant isolation" through dedicated infrastructure for each individual customer and data wiping before re-use. Sharing of resources and multi-tenant isolation can be implemented on different levels of abstraction (see Figure 2). Coarse-grained mechanisms such as shared datacenters, hosts, and networks are well-understood and technologies such as virtual machines, vLANs, or SANs provide isolation. Sharing resources such as operating systems, middleware, or actual software requires a case-by-case design of isolation mechanisms. In particular the last example of Software-as-a-Service requires that each data instance is assigned to a customer and that these instances cannot be accessed by other customers. Note that in practice, these mechanisms are often mixed: While an enterprise customer may own a virtual machine (Machine-level isolation), this machine may use a database server (Middleware isolation) and provide services to multiple individual departments (Application isolation).

In order to mitigate this risk in a cloud computing environment, multi-tenant isolation ensures customer isolation. A principle to structure isolation management is One way to implement such isolation is labeling and flow control:

Labeling: By default all resources are assigned to a customer and labeled with a corresponding label.

Flow control: Shared resources must moderate potential data flow and ensure that no unauthorized data flow occurs between customers. To limit flow control, mechanisms such as access control that ensures that machines and applications of one customer cannot access data or resources from other customers can be used.

Actual systems then need to implement this principle for all shared resources [4] (see, e.g., [2,3] for network isolation). An important challenge in practice is to identify and moderate all undesired information flows [19].

3.2   Insider Attacks by Cloud Administrators

A second important security risk is the accidental or malicious misbehavior of insiders that increased due to global operations and a focus on low cost. Examples may include a network administrator impacting database operations or administrators stealing and disclosing data. This risk is hard to mitigate since security controls need to strike a balance between the power needed to administrate and the security of the administrated systems.

A practical approach to minimize this risk is to adhere to a least-privilege approach for designing cloud management systems. This means that cloud management systems should provide a fine-grained role hierarchy with clearly defined separation of duty constraints. The goal is to ensure that each administrator only holds minimized privileges to perform the job at hand. While today, operators often have god-like privileges, by implementing a least privilege approach, the following objectives can be met:

- Infrastructure administrators can modify their infrastructure (network, disks, and machines) but can no longer access the stored or transported data.
- Security administrators can design and define policies but cannot play any other roles.
- Customer employees can access their respective data and systems (or parts thereof) but cannot access infrastructure or data owned by different customers.

This so-called privileged identity management system is starting to be implemented today and should be mandated for cloud deployments. In today's outsourced datacenters where management tasks are often off-shored, it ensures that the negative impact of remote administrators is limited and that their actions are closely monitored. Such privileged identity management systems usually follow an approach using the following steps:

1. Initially, roles are defined that define the maximum privileges obtained by individual administrators holding these roles. For instance, a database administrator may only obtain administrative privileges over the tables owned by its employer.
2. For a given task at hand, an administrator "checks out" the required privileges while documenting the task. For instance., a database administrator asks for privileges to modify a given database schema.
3. The administrator performs the desired task.
4. The administrator returns the privileges.

Due to the corresponding logging, the security auditors can later determine which employee has held what privileges at any given point in time. Furthermore, for each privilege, the system documents for what task these privileges were requested.

In the long run, these practical approaches may be complemented with stronger protection by, e.g., trusted computing [21] or computations on outsourced data [20].

3.3 Failures of the Cloud Management Systems

Due to the highly automated nature of the cloud management systems and the high complexity of the managed systems, software quality plays an important role in avoiding disruptions and service outages: Clouds gain efficiency by industrializing the production of IT services through complete end-to-end automation. This means that once errors occur in such complex and automated systems, manual intervention for detecting and fixing faults may lead to even more errors. It is furthermore likely that due to the global scale, errors will be replicated globally and thus can only be fixed through automation.

Another source of failure stems from the fact that large-scale computing clouds are often built using low-cost commodity hardware that fails (relatively) often. This leads to frequent failures of machines that may also include a subset of the management infrastructure.

The consequence of these facts is that automated fault tolerance, problem-determination, and (self-)repair mechanisms will be commonly needed in the cloud environment or recover from software and hardware failures.

For building such resilient systems, important tools are data replication, atomic updates of replicated management data, and integrity checking of all data received (see, e.g., [24]). In the longer run, usage of multiple clouds may further improve resiliency (e.g., as pursued by the TClouds project www.tclouds-pro ject.eu or proposed in [11]).

3.4 Lack of Transparency and Guarantees

While the proposed mechanisms to mitigate the identified risks are important, security incidents are largely invisible to a customer: Data corruption may not be detected for a long time. Data leakage by skilled insiders is unlikely to be detected. Furthermore, the operational state and potential problems are usually not communicated to the customer except after an outage has occurred.

An important requirement in a cloud setting is to move away from today's "black-box" approach to cloud computing where customers cannot obtain insight on or evidence of correct cloud operations. A related challenge is how to best foster trust of customers into correct operation of the cloud infrastructure. While partial solutions exist as outlined below, there exists no well-accepted best practice.

The existing approaches range from superficial to academic. The prevailing approach is the so-called best effort approach where operators promise "to do their best" but do not give any guarantees. This is common for free services today. An improvement to this approach is third-party audits. This approach is common to today's outsourcing: (Cloud) service centers are validated by an independent organization to satisfy well-defined standards such as ISO27001 or SAS70. Customers can then be sure that the organization followed these standards at the time of certification. This approach is common best practice today but still only ensures compliance at a point of time and due to it's spot-check approach may miss areas of non-compliance that by accident were not checked.

In the mid-term, it is important that cloud provider provide automated interfaces for observation and incident handling [10]. This will allow customers to automatically identify incidents and to analyze and react to such incidents.

In the long run, the ideal transparency mechanisms would *guarantee* that processes are implemented such that the agreed upon procedures are followed, the functional and non-functional requirements are met, and no data is corrupted or leaked. In practice, these problems are largely unsolved. Cryptographers have designed schemes such as homomorphic encryption [9] that allow verifiable computation on encrypted data. However, the proposed schemes are too inefficient and do not meet the complete range of privacy requirements [23]. A more practical solution is to use Trusted Computing to verify correct policy enforcement [6]. Trusted computing instantiation as proposed by the Trusted Computing Group (TCG) uses secure hardware to allow a stakeholder to perform attestation, i.e., to obtain proof of the executables and configuration that were loaded at boot-time . However, run-time attestation solution still remains an open and challenging problem.

3.5   What about Privacy Risks?

To enable trusted cloud computing, privacy protection is an essential requirement [26]. In simple terms, data privacy aims at protecting personally identifiable data (PID). In Europe, Article 8 of the European Convention on Human Rights (ECHR) provides a right to respect for ones "private and family life, his home and his correspondence". The European Court of Human Rights states in several decisions that this article also safeguards the protection of an individual's PID. Furthermore, the European Data Protection Directive (Directive 95/46/EC) substantiates this right in order to establish a comprehensive data protection system throughout Europe. This directive takes into account the OECD privacy principles [16] which mandate several principles such as, e.g., limited collection of data, the authorization to collect data either by law or by informed consent of the individual whose data are processed ("data subject"), the right to correction and deletion as well as the necessity of reasonable security safeguards for the collected data.

Since cloud computing often means outsourcing data processing, the user as well as the data subject might face risks of data loss, corruption or wiretapping due to the transfer to an external cloud provider. Related to these de-facto obstructions in regard to the legal requirements, there are three particular challenges that need to be addressed by all cloud solutions: Transparency, technical and organizational security safeguards and contractual commitments (e.g., Service Level Agreements, Binding Corporate Rules).

According to European law, the user who processes PID in the cloud or elsewhere remains responsible for the compliance with the aforementioned principles of data privacy. Outsourcing data processing does not absolve the user from his responsibilities and liabilities concerning the data. This means that the user must be able to control and comprehend what happens to the data in the cloud and which security measures are deployed. Therefore, the utmost transparency regarding the processes within the cloud is required to enable the user to carry out his legal obligations. This might be technically realized by, e.g., installing informative event and access logs which enable the user to retrace in detail what happens to his data, where they are stored and who accesses them. Also, the cloud service provider could prove to have an appropriate level of security measurements by undergoing acknowledged auditing and certification processes on a regular basis. Legally, the compliance of the cloud service providers with the European law may be ensured by a commitment to Binding Corporate Rules (BCR). Another method is the implementation of Service Level Agreements (SLAs) into the contracts, which guarantee the adherence to the spelled out privacy requirements. These SLAs could, for example, stipulate an enforcement of privacy via contractual penalties in case of the breach of the agreement.

This applies all the more in cases of cross-border cloud computing with various subcontracting cloud service providers. Subcontracts are already commonly practiced in the cloud computing field. Cloud services commonly rely on each other, since their structures may be consecutively based upon each other. Hence, a computing cloud may use the services of a storage cloud. Unlike local data centers residing in a single country, such cloud infrastructures often extend over multiple legislation and countries. Therefore, the question of applicable law and safeguarding the user's responsibilities regarding data privacy in cross-border cloud scenarios is a matter of consequences for the use of these cloud services. So to avoid unwanted disclosure of data, sufficient protection mechanisms need to be established. These may also extend to the level of technical solutions, such as encryption, data minimization or enforcement of processing according to predefined policies.

4   Open Research Challenges

Today's technology for outsourcing and large-scale systems management laid the foundation for cloud computing. Nevertheless, due to its global scale and the need for full automation, there are still open research challenges that need to be resolved in order to enable hosting of enterprise-class and critical systems on a cloud.

Customer Isolation and Information Flow. For customer isolation, specific challenges are how to reliably manage isolation across various abstraction layers. A single notion of customers needs to be implemented across different systems. Furthermore, data generated by systems need to be assigned to one or more customers to enable access to critical data such as logs and monitoring data. A particularly hard challenge will be to reduce the amount of covert and side channels. Today, such channels are often frozen in hardware and thus cannot easily be reduced.

Insider Attacks. The second area of research are practical and cost-efficient schemes to mitigate the risk of insider fraud. The goal is to minimize the set of trusted employees for each customer through implementing a rigorous least privilege approach as well as corresponding controls to validate employee behavior. Furthermore, a practical scheme needs to support overseas management to reduce cost while still enabling compliance with privacy and other regulations.

Security Integration and Transparency. The third challenge is to allow customers to continue operating a secure environment. This means that security infrastructure and systems within the cloud such as intrusion detection, event handling and logging, virus scans, and access control need to be integrated into an overall security landscape for each individual customers. Depending on the type of systems, this can be achieved by providing more transparency (e.g., visibility of log-files) but may also require security technology within the cloud. One example is intrusion detection: In order to allow customers to 'see' intrusions on the network within the cloud and correlate these intrusions with patterns in the corporate network, the cloud provider either needs to allow the customer to run intrusion detection systems within the cloud (which would raise privacy issues) or else provide generic intrusion detection capabilities that each customer can configure.

Multi-Compliance Clouds. The fourth challenge is how to build clouds that are able to comply with multiple regulations at the same time. One example is the health care sector: A health care cloud would need to satisfy various national or regional privacy and health care regulations. Since manual implementation for each customer will not be cost efficient, an automated way to enforce different (hopefully non-conflicting) regulations would be needed.

One particular challenges in this are is to make regulations and the cloud compatible. Today, regulations often mandate that data needs to be processed in a particular country. This does not align well with today's cloud architectures and will result in higher cost. An alternative could be to define required protections and then leave it to the cloud provider to find a certifiable way to provide sufficient protection.

Federation and Secure Composition The final area of research that we see is cloud federation and secure composition: In order to further reduce the dependency on an individual cloud, services will be obtained from and load balanced over multiple clouds. If this is done properly, services will no longer depend on the availability of any individual cloud.

From a security perspective, this will raise new challenges. Customers need to provide a consistent security state over multiple clouds and provide means to securely fail-over across multiple clouds. Similarly, services will be composed from underlying services from other clouds. Without an accepted way to compose services securely, such compositions would require validation of each individual service based on fixed sub-services.

5   Outlook — The Path Ahead

Cloud computing is not new – it constitutes a new outsourcing delivery model that aims to be closer to the vision of true utility computing. As such, it can rely on security and privacy mechanisms that were developed for service-oriented architectures and outsourcing. Unlike outsourcing, clouds are deployed on a global scale where many customers share one cloud and multiple clouds are networked and layered on top of each other. We surveyed security risks that gain importance in this setting and surveyed potential solutions.

Today, demand for cloud security has increased but the offered security is still limited. We expect this to change and clouds with stronger security guarantees will appear in the market. Initially, they will focus on security mechanisms like isolation, confidentiality through encryption, and data integrity through authentication. However, we expect that they will then move on to the harder problems such as providing verifiable transparency, to integrate with security management systems of the customers, and to limit the risks imposed by misbehaving cloud providers and their employees.

Acknowledgments. We thank Ninja Marnau and Eva Schlehahn from the Independent Centre for Privacy Protection Schleswig-Holstein for substantial and very helpful input to our chapter on privacy risks. We thank the reviewer for helpful comments that enabled us to improve this chapter.

This research has been partially supported by the TClouds project http://www.tclouds-project.eu funded by the European Union's Seventh Framework Programme (FP7/2007-2013) under grant agreement number ICT-257243.

References

1. Babcock, C.: Management Strategies for the Cloud Revolution. McGraw-Hill, New York (2010)
2. Basak, D., Toshniwal, R., Maskalik, S., Sequeira, A.: Virtualizing networking and security in the cloud. SIGOPS Oper. Syst. Rev. 44, 86–94 (2010), doi:10.1145/1899928.1899939
3. Brassil, J.: Physical layer network isolation in multi-tenant clouds. In: Proceedings of the 2010 IEEE 30th International Conference on Distributed Computing Systems Workshops, Washington, DC, USA. ICDCSW '10, pp. 77–81. IEEE Computer Society Press, Los Alamitos (2010), doi:10.1109/ICDCSW.2010.39
4. Cabuk, S., Dalton, C.I., Eriksson, K., Kuhlmann, D., Ramasamy, H.V., Ramunno, G., Sadeghi, A.-R., Schunter, M., Stüble, C.: Towards automated security policy enforcement in multi-tenant virtual data centers. J. Comput. Secur. 18, 89–121 (2010)

5. Chien, E.: W32.Stuxnet dossier. retrieved 2010-13-03, (Sep 2010), From `http://www.symantec.com/connect/blogs/w32stuxnet-dossier`

6. Chow, R., Golle, P., Jakobsson, M., Shi, E., Staddon, J., Masuoka, R., Molina, J.: Controlling data in the cloud: outsourcing computation without outsourcing control. In: ACM Workshop on Cloud Computing Security (CCSW'09), pp. 85–90. ACM Press, New York (2009)

7. Cloud Security Alliance (CSA): Top threats to cloud computing, version 1.0. (March 2010), `http://www.cloudsecurityalliance.org/topthreats/csathreats.v1.0.pdf`

8. Computer and Communication Industry Association (CCIA): Cloud computing (2009), `http://www.ccianet.org/CCIA/files/ccLibraryFiles/Filename/000000000151/Cloud_Computing.pdf`

9. Gentry, C.: Fully homomorphic encryption using ideal lattices. In: Proceedings of the 41st annual ACM symposium on Theory of computing, Bethesda, MD, USA. STOC '09, pp. 169–178. ACM Press, New York (2009), doi:10.1145/1536414.1536440

10. Grobauer, B., Schreck, T.: Towards incident handling in the cloud: challenges and approaches. In: Proceedings of the 2010 ACM workshop on Cloud computing security workshop, Chicago, Illinois, USA. CCSW '10, pp. 77–86. ACM Press, New York (2010), doi:10.1145/1866835.1866850

11. Guerraoui, R., Yabandeh, M.: Independent faults in the cloud. In: Proceedings of the 4th International Workshop on Large Scale Distributed Systems and Middleware, Zürich, Switzerland. LADIS '10, pp. 12–17. ACM Press, New York (2010), doi:10.1145/1859184.1859188

12. International Organization for Standardization (ISO): ISO27001: Information security management system (ISMS) standard (Oct 2005), `http://www.27000.org/iso-27001.htm`

13. Kaliski, Jr., B.S., Pauley, W.: Toward risk assessment as a service in cloud environments. In: Proceedings of the 2nd USENIX conference on Hot topics in cloud computing. pp. 13–13. HotCloud'10, USENIX Association, Berkeley, CA, USA (2010), `http://portal.acm.org/citation.cfm?id=1863103.1863116`

14. Marko, K.: Cloudsourcing - the cloud sparks a new generation of consultants & service brokers (2010), `http://www.processor.com/editorial/article.asp?article=articles%2Fp3203%2F39p03%2F39p03.asp`

15. Oclassen, G.: Why not cloudsourcing for enterprise app user adoption/ training? (2009), `http://velocitymg.com/explorations/why-not-cloudsourcing-for-enterprise-app-user-adoptiontraining/`

16. Organization for Economic Co-Operation and Development (OECD): Guidelines on the protection of privacy and transborder flows of personal data. From `http://www.oecd.org/document/18/0,2340,en_2649_34255_1815186_1_1_1_1,00.html` (last modified January 5 1999), the OECD Privacy Principles

17. Penn, J.: Security and the cloud : Looking at the opportunity beyond the obstacle. Forrester Research (October 2010)

18. Rajan, S.S.: Cloudsourcing vs outsourcing (2010), `http://cloudcomputing.sys-con.com/node/1611752`

19. Ristenpart, T., Tromer, E., Shacham, H., Savage, S.: Hey, you, get off of my cloud: exploring information leakage in third-party compute clouds. In: Proceedings of the 16th ACM conference on Computer and communications security, Chicago, Illinois, USA. CCS '09, pp. 199–212. ACM Press, New York (2009), doi:10.1145/1653662.1653687

20. Sadeghi, A.-R., Schneider, T., Winandy, M.: Token-Based Cloud Computing Secure Outsourcing of Data and Arbitrary Computations with Lower Latency. In: Acquisti, A., Smith, S., Sadeghi, A.-R. (eds.) Proceedings of the 3rd international conference on Trust and trustworthy computing, Berlin, Germany, June 21-23, 2010. LNCS, vol. 6101, pp. 417–429. Springer, Heidelberg (2010)

21. Santos, N., Gummadi, K.P., Rodrigues, R.: Towards trusted cloud computing. In: Proceedings of the 2009 conference on Hot topics in cloud computing. pp. 3–3. HotCloud'09, USENIX Association, Berkeley, CA, USA (2009), http://portal.acm.org/citation.cfm?id=1855533.1855536

22. Sotto, L.J., Treacy, B.C., McLellan, M.L.: Privacy and data security risks in cloud computing. Electronic Commerce & Law Report 15, 186 (2010)

23. Van Dijk, M., Juels, A.: On the Impossibility of Cryptography Alone for Privacy-Preserving Cloud Computing. IACR ePrint 305 (2010)

24. Vukolić, M.: The byzantine empire in the intercloud. SIGACT News 41, 105–111 (2010), doi:10.1145/1855118.1855137

25. Waidner, M.: Cloud computing and security. Lecture Univ. Stuttgart (November 2009)

26. Weichert, T.: Cloud Computing und Datenschutz (2009), http://www.datenschutzzentrum.de/cloud-computing/

Data Usage Control in the Future Internet Cloud

Michele Bezzi and Slim Trabelsi

SAP Labs,
06253, Mougins, France

Abstract. The increasing collection of private information from individuals is becoming a very sensitive issue for citizens, organizations, and regulators. Laws and regulations are evolving and new ones are continuously cropping up in order to try to control the terms of usage of these collected data, but generally not providing a real efficient solution. Technical solutions are missing to help and support the legislator, the data owners and the data collectors to verify the compliance of the data usage conditions with the regulations. Recent studies address these issues by proposing a policy-based framework to express data handling conditions and enforce the restrictions and obligations related to the data usage. In this paper, we first review recent research findings in this area, outlining the current challenges. In the second part of the paper, we propose a new perspective on how the users can control and visualize the use of their data stored in a remote server or in the cloud. We introduce a trusted event handler and a trusted obligation engine, which monitors and informs the user on the compliance with a previously agreed privacy policy.

Keywords: Privacy, Usage control, Privacy Policy

1   Introduction

The vision of the Future Internet heralds a new environment where users, services and devices transparently and seamlessly exchange and combine information, giving rise to new capabilities. In order for it to materialize, this vision needs a mix of adaptation of existing technologies and business models, such as flexible infrastructures and service compositions, distributed ownerships, and large-scale collaborations. The *cloud* is one of the first instantiations of these paradigms. In the cloud users and businesses can buy computing resources (e.g., servers, services, applications) provided by the cloud, that are rapidly provisioned with a minimal management effort and pay-per-use. In the cloud, data may flow around the world, ignoring borders, across multiple services, all in total transparency for the user.

However, this ideal cloud world raises concerns about privacy for individuals, organizations, and society in general. In fact, when data cross borders, they have to comply with privacy laws in every jurisdiction, and every jurisdiction has its own data protection laws. In addition, the risk, for personal data to travel across boundaries and business domains, is that the usage conditions agreed

J. Domingue et al. (Eds.): Future Internet Assembly, LNCS 6656, pp. 223–231, 2011.

upon collection are lost, and, as a consequence, users cannot control their personal information any more, as well as, honest businesses may lose confidence in handling data, when usage conditions are uncertain.

To face these challenges, the concept of *sticky policy* has been introduced [5]. Personal information is associated with a machine-readable policy (*sticky policy*), which stipulates the ways and means to treat that information (for example, expressing that the data should be used for specific purposes only, or the retention period should not exceed 6 months, or the obligation to send a notification to the user when data are transfered to a third party). The sticky policy is propagated with the information throughout its lifetime, and data processors along the supply chain of the cloud have to handle the data in accordance with their attached policies.

The concept of sticky privacy policy represents a powerful instrument to address many privacy requirements. However, its application requires that several problems be solved:

- Expressing privacy policy in a machine-readable language. Although various policy languages have been introduced so far [7,1,2], there is no single language able to completely address the most important privacy scenarios, such as setting and comparing user preferences with server privacy policies, expressing conditions on complex secondary usage cases, specifying obligations and integrating access control policies.
- Providing the data owner with a user-friendly way to express their preferences, as well as to verify the privacy policy the data are collected with.
- Develop mechanisms to enforce these sticky policies in ways that can be verified and audited.

In this paper, we present recent results obtained by the European ICT project PrimeLife which (partly) addresses these problems, introducing a novel policy language to express complex privacy conditions, and the corresponding policy engine able to process these policies. In particular, in Sect. 2 we introduce the PrimeLife Policy Language (PPL), which combines access and data handling policies; we then describe the corresponding policy engine, enabling the deployment, interpretation and enforcement of PPL policies. Although the proposed solution can address the main requirements to manage privacy policies in the cloud, there are still important open problems to address (see Section 3). In particular, the current framework lacks mechanisms to provide the data owner with the guarantee that policy and obligations are actually enforced. In Sect. 4, we present our initial thoughts on how to implement a trusted system for policy enforcement. Conclusions are drawn in the last section.

2   Primelife Privacy Framework

In many web applications, users are asked to provide various kinds of personal information, starting from basic contact information (addresses, telephone, email) to more complex data such as preferences, friends' list, photos. Service providers

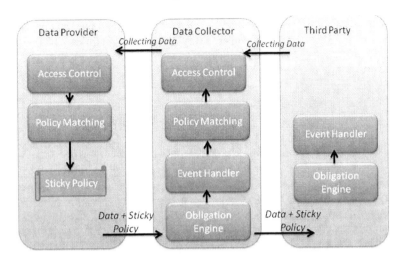

Fig. 1. PPL high level architecture.

describe how the users' data are handled using privacy policy, which is, more or less explicitly, presented to users during the data collection phase. Privacy policies are typically composed of a long text written in legal terms that are rarely fully understood, or even read, by the users. As a result, most of the users creating accounts on web 2.0 applications are not aware of the conditions under which their data are handled.

Therefore, there is need to support the user in this process, providing an as-automatic-as-possible means to handle privacy policies. In this context, the European FP7 project PrimeLife[1] developed a novel privacy policy framework able to express and automatically process privacy policies in web interactions. This approach enables applications, like web browsers, to automate the interpretation of the content of a privacy policy and to compare the service privacy policy with user privacy preferences.

The Primelife project introduced the PrimeLife Policy Language (PPL, herein) [10,4], which allows to describe in an XML machine-readable format the conditions of access and usage of the data. A PPL policy can be used by a service provider to describe his privacy policies (how the data collected will be treated and with whom they will be shared), or by a user to specify his preferences about the use of his data (who can use it and how it should be treated). Before disclosing his personal information, the user can automatically match his preferences with the privacy policy of the website and the result of the matching generates an agreed policy, which is bound to the data (sticky policy) and travels with them. In fact, this sticky policy will be sent to the server and follow the data in all their lifecycle to specify the usage conditions.

The PPL sticky policy defines the following conditions:

---

[1] www.primelife.eu

- Access control: PPL inherits from the XACML [8] language the access control capabilities that express how access to which resource under which condition can be achieved.
- Data Handling: the data handling part of the language defines two conditions:
  - Purpose: expressing the purpose of usage of the data. Purpose can be for example marketing, research, payment, delivery, etc.
  - Downstream usage: supporting a multi-level nested policy describing the data handling conditions that are applicable for any third party collecting the data from the server. This nested policy is applicable when a server storing personal data decides to share the data with a third party
- Obligations: Obligations in sticky policies specify the actions that should be carried out after collecting or storing a data. For example, notification to the user whenever his data are shared with a third party, or deleting the credit card number after the payment transaction is finished, etc..

Introducing PPL policies requires the design of a new framework for the processing of such privacy rules. In particular, it is important to stress that during the lifecyle of personal data, the same actor may play the role of both data collector and data provider. For this reason, PrimeLife proposed the PPL engine based on a symmetric architecture, where any data collector can become a data provider if a third party requests some data (see Figure 1). According to the role played by an entity (data provider or data collector) the engine behaves differently by invoking the appropriate modules.

In more detail, on the data provider side (user) the modules invoked are:

- The access control engine: it checks if there is any access restriction for the data before sending it to any server. For example, we can define black or white lists for websites with whom we do not want to exchange our personal information.
- Policy matching engine: after verifying that a data collector is in the white list, a data provider recovers the server's privacy policy in order to compare it to its preferences and verify whether they are compatible in terms of data handling and obligation conditions. The result of this matching may be displayed through a graphical interface, where a user can clearly understand how the information is handled if he accepts to continue the transaction with the data collector. The result of the matching conditions, as agreed by the user, is transformed into a sticky policy.

On the data collector side, after recovering the personal information with its sticky policy the invoked modules are:

- Event handler: it monitors all the events related to the usage of the collected data. These event notifications are handled by the obligation engine in order to check if there is any trigger that is related to an event. For example, if a sticky policy provides for the logging of any information related to the usage of a data, the event handler will notify the obligation engine whenever an

access (read, write, modification, deletion etc.) to data is detected in order to keep track of this access.
- Obligation engine: it triggers all the obligations required by the sticky policy.

If a third party requests some data from the server, the latter becomes a data provider and acts as a user-side engine invoking access control and matching modules, and the third party plays the role of data collector invoking the obligation engine and the event handler

3   Open Challenges

Although the PPL framework represents an important advancement in fulfilling many privacy requirements of the cloud scenario, there are still some issues, which are not addressed by the PPL framework.

Firstly, in the current PPL framework, the data owner has no guarantee of actual enforcement of the data handling policies and obligations. Indeed, the data collector may implement the PPL framework, thus having the technical capacity of processing the data according to the attached policies, but it could always tamper with this system, which controls, or simply access directly the data without using the PPL engine. In practice, the data owner should trust the data collector to behave honestly.

A second problem relates to the scalability of the sticky policy approach. Clearly, the policy processing adds a relevant computational overhead. Its applicability to realistic scenarios, where large amounts of data have to be transmitted and processed, has to be investigated.

A last issue relates to the privacy business model. The main question is: What should motivate the data collectors/processors to implement such technology? Actually, in many cases, their business model relies on the as-less-restricted-as-possible use of private data. On the user side, a related question is, are the data owners ready to pay for privacy [9]? Both questions are difficult to address, especially when dealing with such a loosely defined concept as privacy. Although studies exist (see [11,3], and references therein), mainly in the context of the web 2.0, we should notice that the advent of cloud changes the business relevance of privacy. In fact, in a typical web 2.0 application the user is disclosing his own data, balancing the value of his personal data with the services obtained. As a matter of fact, users have difficulties to monetize the value of their personal information, and they tend to disclose their data quite easily. In the cloud world, organizations store the data they have collected (under specific restrictions) with the cloud provider. These data have a clear business value, and typically companies can evaluate the amount of money they are risking if such data are lost or made public. For these reasons, it is likely that they are ready to pay for a stronger privacy protection.

All these issues need further research work to be addressed. In the next section, we present our initial thoughts on how we may extend the Primelife framework to address the first problem we mentioned above, i.e., how to provide a secure enforcement for privacy policy.

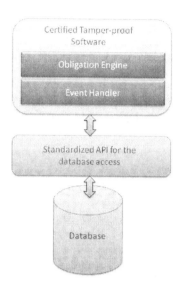

Fig. 2. The key elements of the extension of the PPL framework to guarantee the enforcement of privacy policy.

4 Towards Privacy Policy Enforcement in the Cloud

In the current PPL framework, there is no guarantee of enforcement of the data handling policies and obligations. In other words, we suppose that the server enforces correctly the sticky policies, but, actually, nothing prevents him from creating a back door in his database in order to get unauthorized access to the collected information.

For this reason, we propose in the rest of the paper a secure architecture for the enforcement of the sticky policies and facilitating the task of external auditors to verify the compliance with the privacy requirements, as well as giving the user control on the released data. The main idea is to introduce tamper-proof [6] obligation engine and event handler, certified by a trusted third party, which mediate the communication and the handling of private data in the cloud platform. The schedule of the events, as well as the logs of these components can also be (partly) accessed by the users to monitor the handling of their personal information. Lastly, the trusted-third party can ensure the auditing of the whole system.

Let us sketch how our proposal can work in a simple cloud scenario. Let us consider a cloud platform provider, which hosts one or more services/applications provided by external parties that deal with personal data (e.g., a human resource management application, a remote storage service). Say, these services handle personal data using a PPL framework (as described in Sect. 2). In order to guarantee enforcement of the privacy policies and corresponding obligations by the service, we replace the service provider obligation engine and event handler with a tamper-proof event handler and a tamper-proof obligation engine certified by a trusted third party (e.g., governmental office), see Fig. 2. For instance, the cloud provider may provide these certified components as premium service.

In fact, trust is an essential part of the cloud paradigm. If the data owner has the guarantee from a trusted authority (governmental office, EU commission, etc.) that the application hosted in the cloud is compliant with his privacy requirements, he will tend to transfer his data to the certified host. In order to certify the compliance of an application, the trusted authority has, first, to certify the secure privacy components in charge of enforcing sticky policies, then to perform audits to check if the stored data are handled correctly.

The difficulty comes for the access to the database by the service provider. One solution would be to use a specific tamper-proof database, but this can be technically complex, and impact the business efficiency of the service provider. A possible solution is to specify an API to access the database that is compatible with the event handler. This API should be defined as a standard interface to communicate with the event handler and access to the database. The service has to exclusively use an interface compatible with the standardized API, and this should be subject to audit by an external trust authority (which could be the same or not certifying the tamper proof components).

Data URI	Data Type	Sticky Policy	Pending Obligations	Related Events			Admin actions
				Type	Date	Time	
#12345	e-mail	SP#12345	Delete in 5 minutes	Read	02/10	11:34	Del/Move

Fig. 3. A sketch of data track administration console

The particularity of this API is that all the methods to access the data can be detected by the event handler. For example, if the service adds a new element (data and sticky policy) this action should be detected, managed and logged by the event handler. If there is any method (like table dump) to access the database that cannot be recognized by the event handler, the service will not be certified by the trusted authority.

Using a tamper proof event handler and obligation engine also gives the possibility of providing a monitoring console. The monitoring can be accessible by any data owner, who, once authenticated, can list all the data (or set of data) with their related events and pending or enforced obligations. The data owner can at any time control how his data are handled, under which conditions the information is accessed, and compare them with the corresponding stored sticky policy. Fig. 3 shows a very simple example of how the remote administrative console could be structured, this monitoring console could of course be more complex. The remote monitoring console adds more transparency and more control to the data hosted within the cloud. It also allows the user to detect any improper usage of his data, and, in this case, notify the host or the trusted authority.

The advantages of the proposed solution are twofold. First, from the data owner perspective, there is a guarantee that actual enforcement has taken place, and that he can monitor the status of his data and corresponding policies. Second, from the auditors' point of view, it limits the perimeter of their analysis, since the confidence zone provided by the tamper proof elements and the standardized API facilitate the distinction between authorized and non authorized actions.

5   Conclusions

Cloud computing and the SOA paradigm are fundamental building blocks for the Future Internet, enabling the seamless combination of services across platforms, geographies, businesses and transparently from the user point of view. However, these new capabilities may entail privacy risks. From the user perspective, the risk is that of losing control of his personal information once they are released in the cloud. In particular, when personal data are consumed by multiple services, possibly owned by different entities in different locations, the conditions of the data usage, agreed upon collection, may be lost in the lifecycle of the personal data. From the data consumer point of view, businesses and organizations seek to ensure compliance with the plethora of data protection regulations, and minimize the risk of violating the agreed privacy policy.

The concept of sticky policy may be used to address some of the privacy requirements of the cloud scenario. In this paper we reviewed the recently introduced PPL framework, which provides a flexible language to express privacy policy as well as the necessary mechanisms to process and compare sticky policies. The current PPL framework presents some limitations; it notably requires a high level of trust in the data collector/processor. We presented some initial thoughts about how this problem can be mitigated through the usage of a tamper proof implementation of the architecture. This solution may increase the trust on the cloud, but it still needs further studies to verify its applicability in real life business scenarios.

Acknowledgements. The research leading to these results has received funding from the European Community's Seventh Framework Programme (FP7/2007-2013) under grant agreement no. 216483.

References

1. Ardagna, C.A., Cremonini, M., De Capitani di Vimercati, S., Samarati, P.: A privacy-aware access control system. J. Comput. Secur. 16, 369–397 (2008)
2. Ashley, P., Hada, S., Karjoth, G., Powers, C., Schunter, M.: Enterprise privacy authorization language (EPAL 1.1). IBM Research Report (2003)

3. Bonneau, J., Preibusch, S.: The privacy jungle:on the market for data protection in social networks. In: Moore, T., Pym, D., Ioannidis, C. (eds.) Economics of Information Security and Privacy, pp. 121–167. Springer, New York (2010)
4. Bussard, L., Neven, G., Preiss, F.S.: Downstream usage control. In: IEEE International Workshop on Policies for Distributed Systems and Networks, pp. 22–29 (2010)
5. Karjoth, G., Schunter, M., Waidner, M.: Platform for enterprise privacy practices: Privacy-enabled management of customer data. In: Dingledine, R., Syverson, P.F. (eds.) PET 2002. LNCS, vol. 2482, pp. 69–84. Springer, Heidelberg (2003)
6. Naedele, M., Koch, T.E.: Trust and tamper-proof software delivery. In: Proceedings of the 2006 international workshop on Software engineering for secure systems. SESS '06, New York, NY, USA, pp. 51–58. ACM Press, New York (2006), doi:10.1145/1137627.1137636
7. Reagle, J., Cranor, L.F.: The platform for privacy preferences. Commun. ACM 42, 48–55 (1999), doi:10.1145/293411.293455
8. Rissanen, E.: extensible access control markup language (xacml) version 3.0, extensible access control markup language (xacml) version 3.0, oasis (August 2008)
9. Shostack, A., Syverson, P.: What price privacy? In: Camp, L., Lewis, S. (eds.) Economics of Information Security, Advances in Information Security, vol. 12, pp. 129–142. Springer, New York (2004)
10. Trabelsi, S., Njeh, A., Bussard, L., Neven, G.: The ppl engine: A symmetric architecture for privacy policy handling. W3C Workshop on Privacy and data usage control p. 5 (October 2010), http://www.w3.org/2010/policy-ws/
11. Tsai, J.Y., Egelman, S., Cranor, L., Acquisti, A.: The Effect of Online Privacy Information on Purchasing Behavior: An Experimental Study. In: ICIS 2007 Proceedings, p. 20 (2007)

Part IV:

**Future Internet Foundations:
Experiments and Experimental Design**

Introduction

Research into new paradigms and the comprehensive test facilities upon which the ideas are experimented upon together build a key resource for driving European research into future networks and services. This environment enables both incremental and disruptive approaches, supports multi-disciplinary research that goes beyond network layers, scholastic dogmas and public-private discussions. It provides a core infrastructure, and also a playground for future discoveries and innovations, combining research with experimentation.

The heterogeneous and modular field of Future Internet Research and Experimentation with its national and international stakeholder groups requires community and cohesion building, information sharing, and a single point of contact to co-ordinate and promote a common approach with respect to the following main requirements:

- Testbeds and experimental facilities need synchronisation, resource optimisation, and common efforts in order to offer customers the best possible service and ensure their sustainability beyond project lifetimes.
- Researchers need correct and timely knowledge about the available resources, easy access, high usability and appropriate tools to run and monitor their experiments.

Federation of testbeds aims at creating a physical and logical interconnection of several independent testbeds and experimental facilities to provide a larger-scale, more diverse and higher performance platform for accomplishing tests and experiments. It aims to provide flexibility and preserve autonomy and character for the individual entities. In that sense, high-level federation allows resource sharing and collaboration towards establishing a sustainable customer-friendly facility.

In this context the contribution by Tranoris et al. entitled "A Use-Case on Testing Adaptive Admission Control and Resource Allocation Algorithms on the Federated Environment of Panlab" reports on experiments needing to directly interact with the environment during runtime, and introduced requirements and solutions for a significant upgrade of the federated testbed environment that was used. The chapter by Zseby et al. entitled "Multipath Routing Experiments in Federated Testbeds" demonstrates the practical usefulness of federation and virtualisation in heterogeneous testbeds.

These multipath routing slice experiments were performed over multiple federated testbeds offered by the G-Lab, PlanetLab Europe (PLE), and VINI infrastructures, and they would be not have been possible without the ability to create environments across multiple administrative domains using the concepts of federation, in particular their advanced measurement technologies. Finally the chapter Kousaridas et al. entitled "Testing End-to-End Self-Management in a Wireless Future Internet Environment" reports on the network management protocol test that exploited the availability of different administrative domains in federated testbeds and provides evidence for the benefits of using an experimentally-driven research methodology.

This methodology was used to research and develop a self-management solution for the selection of the appropriate network or service level adaptation to improve end-to-end behaviour and QoS features in wireless networks.

Anastasius Gavras

A Use-Case on Testing Adaptive Admission Control and Resource Allocation Algorithms on the Federated Environment of Panlab

Christos Tranoris, Pierpaolo Giacomin, and Spyros Denazis

Electrical and Computer Engineering department, University of Patras,
Rio, Patras 26500, Greece
tranoris@ece.upatras.gr, yrz@anche.no, sdena@upatras.gr

Abstract. Panlab is a Future Internet initiative which integrates distributed facilities in a federated manner. Panlab framework provides the infrastructure and architectural components that enable testing applications near production environments over a heterogeneous pool of resources. This paper presents a use case where an adaptive resource allocation algorithm was tested utilizing Panlab's infrastructure. Implementation details are given in terms of building a RUBiS testbed that provides all the required resources. Moreover, this experiment needs to directly request, monitor and manage resources that it uses during the experiment. As a result of this use case a new feature for Panlab was developed called Federation Computing Interface (FCI) API which enables applications to access resources during an experiment.

Keywords: Panlab, experimental testing, resource federation, Future Internet

1 Introduction

Future Internet research results in new experimental infrastructures for supporting approaches that exploit extend or redesign current Internet architecture and protocols. The Pan-European laboratory [1], Panlab, is a FIRE[2] initiative and builds on a federation of interconnected and distributed facilities allowing third parties to access a wide variety of resources like platforms, networks, and services for broad testing and experimentation purposes. In this context, Panlab defines a provisioning framework and a meta-architecture that give rise to a number of Federation Mechanisms and Architecture Elements to be used for experimentation in the Future Internet.

The Panlab infrastructure manages interconnections of different geographically distributed testbeds to provide services to customers for various kinds of testing scenarios which in Panlab terminology are called Virtual Customer Testbeds or simply VCTs. A VCT is a specification of required (heterogeneous) resources along with their configurations, offered by a diverse pool of organizations in order to form new richer infrastructures. These VCTs represent customer needs such as i) evaluation and testing specifications of new technologies, products, services, ii) execution of network

J. Domingue et al. (Eds.): Future Internet Assembly, LNCS 6656, pp. 237–245, 2011.

and application layer experiments, or even iii) complete commercial applications that are executed by the federation's infrastructure in a cost-effective way.

Panlab's architecture introduces components for integrating testbeds that belong to various administrative domains, in order to become available to participate in testing scenarios. A Web Portal is available where customers and providers can access services, a visual Creation Environment which is called "Virtual Customer Testbed (VCT) tool" where a customer can define requested services, a repository which keeps all persistent information like resources, partners, defined VCTs, etc. Experimenters can browse through the resource registry content and can select, configure, deploy and access reserved resources. Finally, an Orchestration Engine is responsible for orchestrating the provisioning of the requested services. The above components interact with each other in order to offer a service called "Teagle". Part of Teagle is also the Teagle Gateway, the component that is responsible for transferring provisioning and configuration commands to selected resources lying in various administrative domains. The functionality of Panlab office is complemented by a Policy engine. All components communicate via an HTTP-based (REpresentation State Transfer) RESTful interface. A per domain central controller is the Panlab Testbed Manager (PTM). PTM is responsible for accepting RESTful commands from the Teagle Gateway in order to configure the domain's resources. PTM implements the so called Resource Adaptation Layer where Panlab partners "plug-in" their Resource Adapters (RA). A Resource Adapter (a concept similar to device drivers) wraps a domain's resource API in order to create a homogeneous API defined by Panlab. Details and specifications of Panlab's components can be found at [1].

This paper describes an experiment made utilizing the Panlab's framework and available infrastructure. The challenge here were twofold: i) to run the experiment by moving a designed algorithm from a simulating environment to near production best-effort environment and ii) to exploit the framework in such a way that will allow the system under test to directly request or release resources that it uses. The latter indicates that the experiment needs access to the whole framework after the provisioning during the operational phase. As a result to accomplish the needs of this experiment was the development of a new feature of Panlab's framework called Federation Computing Interface (FCI) API. FCI enables the access of provisioned/reserved resources during the execution of an experiment.

The rest of this paper is organized as follows: The first section describes the use case requirements and the needed infrastructure. The second section describes the implementation of the infrastructure and the deployment of the testbed. The third section discusses the execution of the experiment and how Panlab framework is able by means of Federation Computing Interface API to managed resource. We finally conclude this paper.

2   Use Case Description

In order for one to test an adaptive admission control and resource allocation algorithm, it is necessary to set up an appropriate testbed of a distributed web application like RUBiS benchmark [3], an auction site prototype modeled after eBay.com. It provides a virtualized distributed application that consists of three components, a web server, an application server, a database and a workload generator, which produces the appropriate requests. Furthermore it can be deployed in a virtualized environment using Xen server technology, which allows regulating system resources such as CPU usage and memory, and provides also a monitoring tool, Ganglia, that measures network metrics, such as round trip time and other statistics, and resource usage in virtual machines.

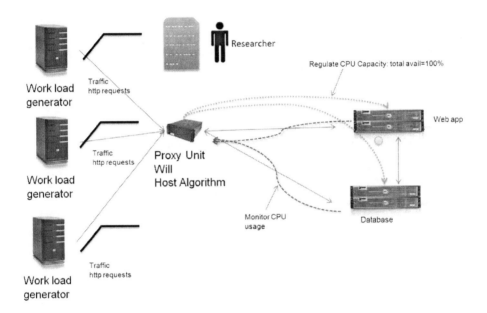

Fig. 1. The setup for testing the algorithm

The adaptive admission control and resource allocation algorithm is applied to succeed in specific target of network metrics, like round trip time and throughput. This will be done by deploying a proxy-like control component for admission control and using Xen server technology to regulate CPU usage. During this scenario the adaptive admission control and resource allocation algorithm is tested against network metrics, like round trip time and throughput. RUBiS clients will produce requests so that push RUBiS components to their limits, so that resource like CPU usage and network throughput get high values.

During the setup, the researcher wants to test http proxy software written in C programming language that implements an admission algorithm. Figure 1 displays the setup for the discussed scenario. The setup consists of 3 work load http traffic generators, making requests through a hosting unit. The algorithm, which is located at the proxy unit, needs to monitor the CPU usage of the Web application and Database machines. Then the algorithm should be able to set new CPU capacity limits on both resources. Additionally the algorithm should be able to start and stop the work load generators on demand.

3   Technical Environment, Testbed Implementation and Deployment

From the requirements of the use case, it is evident that it would benefit from a testbed offering RUBiS resources. Moreover, the experiment needs to manage and monitor resources within the C algorithm. So the resources need to provide monitoring and provisioning mechanisms.

To support such an experiment and similar ones, a required infrastructure needed to be built. The equipment used is as follows:

— Linux machines for the RUBiS based work load generators
— A Linux machine for the hosting the algorithm unit, capable of compiling C and Java software
— Linux machines for running XEN server where on top will run the RUBiS Web app and database

The final user needs to provide the algorithm under test. He will just login to the Proxy Unit, compile the software and execute it. The user will not have access to the RUBiS resources (i.e. cannot login) so there is a need to encapsulate the monitoring and provisioning capabilities. For this requirement and to make available the RUBiS resources for future testing within the Panlab federation, the so called Resource Adapters (RA) where built.

For each resource there is a corresponding RA which exposes configuration parameters to the end user. As displayed in Figure 2, all the components are based on Virtual Machines managed by a XEN server. The implemented RAs instantiate all these Virtual Machines and configure the internal components according to end-user needs. The work load generator exposes parameters such as: used IP for the testbed, memory, hard disk size, number of clients, ramp up time for the requests and a parameter used during the execution of the experiment called Action which accepts the values start and stop. The Proxy Unit exposes parameters such as used IP for the testbed, memory, hard disk size, username, password and IP to connect to the RUBiS application resource. The RUBiS application and the RUBiS database have similar parameters to the above and additionally a MON_CPU_UTILIZATION parameter which is used to monitor the resource and a CPU_CAPACITY used to set the max cpu capacity of the resource.

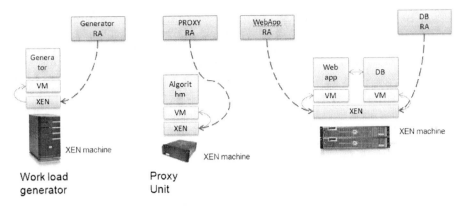

Fig. 2. The Resource adapters of the available testbed resources

```
rubis_app.radl ☒    rubis_cl.radl    rubis_proxy.radl    rubis_db.radl

Resource Adapter "rubis_app"
  Configuration Parameters { // Visible Parameters to VCT user
      APP_VLANID;
      APP_IP;
      APP_GW;
      APP_MEM;
      APP_DISKSPACE;
      APP_DBIP;
      CPU_CAPACITY description = "Set CPU Capacity " ;
      MON_CPU_UTILIZATION description = "Readonly.Returns resource utilization";
  }

  Binding Parameters { // Local Parameters used for resource configuration
      admin = "root";//root
      admin_pwd = "v██████";//█████
      admin_ip ="150.140.█████";//██████████ (where scripts are)
      admin_port ="22";//22
  }

  On Update {
      ProcessOnAllConfigurationParametersComplete = NO;
      RAProtocol SSH {
          Remote Machine = "admin_ip";
          RPort = "admin_port";
          RUsername = "admin";
          RPassword = "admin_pwd";
          RExecute{
              "/repo/scripts/shutdown_forget.sh appname;"
              "/repo/scripts/rubis/rubis_app.sh appname" << "APP_VLANID"
                      "APP_IP" "APP_GW" "APP_MEM" "APP_DISKSPACE" "APP_DBIP">>
              "/repo/scripts/rubis/changecpu.sh dbname" << "CPU_CAPACITY" >>
          }
      }
  }
```

Fig. 3. RADL definition for the RUBiS application resource

The resource adapters where defined using the Panlab's Resource Adapter Description Language (RADL)[4]. RADL is a concrete textual syntax for describing a Resource Adapter based on an abstract syntax defined in a meta-model. RADL is an attempt for describing a RA in a way that decouples it from the underlying implementation code. RADL's textual syntax aims to be easier to describe a RA than code in Java or other target language. RADL is useful in cases when there is a need to configure a resource that offers an API for configuration. The user can configure the resource through some Configuration Parameters. The RA "wraps" the parameters and together with the Binding Parameters, the RA can configure the resource. A Binding Parameter is a variable that is assigned locally by the resource provider, e.g. a local IP address. This approach was also adopted for developing the RUBiS RAs. Figure 3 displays the RADL definition for the RUBiS application server.

The Configuration Parameters section describes the exposed parameters to the end user. The Binding Parameters are used for internal purposes of the local testbed configuration. The On Update section describes what the rubis_app RA does when it receives a provisioning update command from the upper layers. The RA will use the SSH protocol to connect to the internal machine and execute scripts on it.

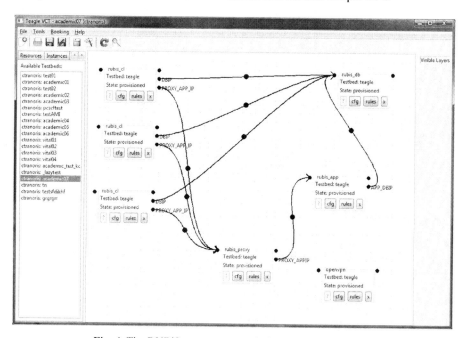

Fig. 4. The RUBiS use case setup designed in the VCT tool

The tools to deploy, monitor and run the experiments are those offered by the Panlab architecture [1]. The Resource Adapters where loaded in the testbed PTM and made available through Panlab's repository. Figure 4 displays the use case setup as can be done inside the VCT tool of Panlab. The resources are available in the left side of the tool. Three rubis client where selected one rubis proxy, one rubis application and one rubis database. Interconnections where made also between these components in order to assign reference values to all resources. For example the RUBiS clients need to know about the IP of the proxy which hosts the algorithm. The proxy needs to know the IP of the RUBiS application which also needs a reference to the RUBiS database.

4   Running and Operating the Experiment

The scenario during the experiment utilizes the Federation Computing Interface (FCI) API that Panlab provides [5]. Federation Computing Interface (FCI) is an API for accessing resources of the federation. It is an SDK for developing applications that access VCT requested resources through the Panlab office services during operation of testing. It is quite easy to embed it into your application/ SUT in order to gain control of the requested resources during testing. The FCI is delivered to the customer after the generation of the SLA and not only does it contain the necessary libraries but also the alias of the resources that are used in the VCT scenario. This allows the User-Application/SUT to access the testbed resources during execution of the experiment in order to manage and configure various environment parameters or to get status of the resources.

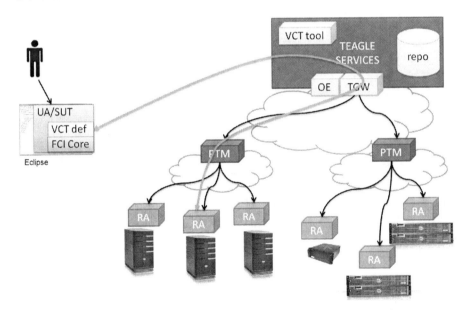

Fig. 5. Designing the algorithm to operate resources during execution

In our testing scenario there is a need to configure resources or even get monitoring status data properly after the VCT is provisioned and while the testing is in progress. Figure 5 displays this condition where the System Under Test (SUT) is our algorithm. FCI automatically creates all the necessary code that the end user can then inject inside the algorithm's code. The end-user needs just to ender his credentials in order

FCI to generate the necessary wrapper classes and functions that are capable of accessing the reserved, provisioned resources. An example is given in the following code listing in Java:

```java
//an example Java federation program
public class Main {
  public static void main(String[] args) {
    //An example for VCT: academic07
    academic07 myvct = new academic07();
    myvct.getuop_rubis_cl_91().setRAMP_UP_TIME("55000");
    myvct.getuop_rubis_cl_91().setACTION( "start" );
    myvct.getuop_rubis_cl_91().setNUM_CLIENTS( "300" );
    int moncpu =
myvct.getuop_rubis_db_33().getMON_CPU_UTILIZATION();
```

Assuming that we have given the name academic07 for our VCT definition, the java listing displays how we can access the resources of this VCT. FCI creates a java class, called academic07() that we can instantiate in order to get access to the resources. Additionally, for each resource that participates in the VCT java classes are able to provide access. For example the command `myvct.getuop_rubis_cl_91().set-ACTION( "start" );` starts the RUBiS client of the rubis_cl_91 resource. The command myvct.getuop_rubis_db_33().getMON_CPU_UTILIZATION(); is able to give back the CPU usage of the database resource.

5   Conclusions

The results of running an experiment in Panlab are encouraging in terms of moving the designed algorithms from simulating environments to near production environments. What is really attractive is that such algorithms can be tested in a best-effort environment with real connectivity issues that cannot be easily performed in simulation environments. The presented use case example demonstrated the usage of existing experimental facilities in this case by exploiting the Panlab framework. The interesting of this experiment is that it extends the framework to allow the system under test to directly request or release resources that it uses.

5.1   Results

First results of running such an experiment although not comparable currently with similar approaches are really encouraging in terms of moving the designed algorithms from simulating environments to near production environments. Using the existing deployed RUBiS facility makes the setup and scaling up of such a testbed much easier. What is really attractive is that such algorithms can be tested in a best-effort environment with real connectivity issues that cannot be easily performed in simulation environments. The scenario presented can be easily scaled up with many clients and web applications. Also, the proxy under test can be replaced by one or more load balancers.

5.2   Testbed Availability

The resources for creating similar scenarios are going to be available under the Panlab Office offerings. Currently there are a limited amount of resources that are capable of hosting the RUBiS environment. We expect to make more resources available as demand increases.

Acknowledgments. The work presented in this paper has been performed during PII a Seventh Framework Program (FP7) project funded by EU.

References

1. Website of Panlab and PII European projects, supported by the European Commission in its both framework programmes FP6 (2001-2006) and FP7 (2007-2013): http://www.panlab.net
2. European Commission, FIRE website: Last cited: November 21, 2010, http://cordis.europa.eu/fp7/ict/fire
3. RUBiS, http://rubis.ow2.org/
4. RADL, http://trac.panlab.net/trac/wiki/RADL
5. Federation Computing Interface(FCI), http://trac.panlab.net/trac/wiki/FCI

Multipath Routing Slice Experiments in Federated Testbeds

Tanja Zseby[1], Thomas Zinner[2], Kurt Tutschku[3], Yuval Shavitt[4],
Phuoc Tran-Gia[2], Christian Schwartz[2], Albert Rafetseder[3], Christian Henke[5],
and Carsten Schmoll[1]

[1] FOKUS - Fraunhofer Institute for Open Communication Systems, Berlin, Germany
[tanja.zseby|carsten.schmoll]@fokus.fraunhofer.de,
[2] University of Wuerzburg, Institute of Computer Science, Wuerzburg, Germany,
[thomas.zinner|christian.schwartz|phuoc.trangia]@informatik.uni-wuerzburg.de
[3] University of Vienna, Professur "Future Communication" (endowed by Telekom
Austria), Austria
[kurt.tutschku|albert.rafetseder]@univie.ac.at
[4] Tel Aviv University, School of Electrical Engineering, Tel Aviv, Israel
shavitt@eng.tau.ac.il
[5] Technical University Berlin, Chair for Next Generation Networks, Berlin, Germany
c.henke@tu-berlin.de

Abstract. The Internet today consist of many heterogeneous infrastructures, owned and maintained by separate and potentially competing administrative authorities. On top of this a wide variety of applications has different requirements with regard to quality, reliability and security from the underlying networks. The number of stakeholders who participate in provisioning of network and services is growing. More demanding applications (like eGovernment, eHealth, critical and emergency infrastructures) are on the rise. Therefore we assume that these two basic characteristics, a) multiple authorities and b) applications with very diverse demands, are likely to stay or even increase in the Internet of the future. In such an environment *federation* and *virtualization* of resources are key features that should be supported in a future Internet. The ability to form slices across domains that meet application specific requirements enables many of the desired features in future networks.

In this paper, we present a Multipath Routing Slice experiment that we performed over multiple federated testbeds. We combined capabilities from different experimental facilities, since one single testbed did not offer all the required capabilities. This paper summarizes the conducted experiment, our experience with the usability of federated testbeds and our experience with the use of advanced measurement technologies within experimental facilities. We believe that this experiment provides a good example use case for the future Internet itself because we assume that the Internet will consist of multiple different infrastructures that have to be combined in application specific overlays or routing slices, very much like the experimental facilities we used in this experiment. We also assume that the growing demands will push towards a much better measurement instrumentation of the future Internet. The tools used in our experiment can provide a starting point for this.

J. Domingue et al. (Eds.): Future Internet Assembly, LNCS 6656, pp. 247–258, 2011.

1 Introduction

Multipath Routing Slices constitute a new transport service in future generation networks. *Network Virtualization (NV)* techniques [5,17] allow the establishment of such separate slices on top of a joint physical infrastructure (substrate). NV enables the parallel and independent operation of application-specific virtual networks (e.g. for banking, gaming, web) with their own virtual topology, naming, routing and resource management on top of a shared physical infrastructure. Virtual networks are denoted in NV as *slices* [15]. Slices that are not dedicated to a single application and that implement a general data transport service are designated as *routing slices* [13]. Routing slices as an architectural concept is known as *Transport Virtualization (TV)* [23,24]. These concepts have roots in the work on active networks, where the control plane of a router enabled applications fine-grained control of their own routing [6,11] and sharing of the resources at the routers using either constant or ad-hoc slices [16].

Slices, and routing slices in particular, are made up of shared resources that can be contributed by different administrative authorities. Thus, routing slices can be thought of as a *federation* [15] of networking resources, i.e., a combination of fractions of (virtual) links and (virtual) routers.

Due to the fine grained granularity of networking resources, routing slices have appealing features. Multiple paths between a single source and destination pair may exist in a slice and can be pooled in a *Multipath Routing Slice*. Such a multipath routing slice allows the use of alternative paths if a failure occurs and therefore improves resilience. A further application field of multipath transmissions is to obtain higher capacity between a source and destination pair. Packets can be distributed on the paths so that paths are used concurrently . As a result Multipath Routing Slices may pose the feature of *location transparency*, i.e., they permit data transport resource to be accessed without knowledge of their physical or network location.

Besides the establishment of routing slices and the instrumentation of federated environments with measurement functions, federation has also further challenges. The control and verification of service level agreements (SLAs) between domains as well as inter-domain security have to be addressed in federated testbeds as well as in the real Internet. Measurement functions can help to support this. Inter-domain SLA validation would profit from common data formats and data exchange among providers (e.g. [8]). Intrusion detection systems can increase situation awareness (and with this overall security) by sharing information. Nevertheless, the operators of the testbeds we considered in our setup are willing to cooperate. This is not always the case for (potentially competing) network operators in general. For this it is helpful to calculate cost and gain of sharing information with neighbors. Making these values explicitly known to the stakeholders can help to provide incentives for cooperation.

Although the concepts of Routing Slices and multipath routing slices are apparently favorable for future networks, the development of a network architecture (together with its protocols and mechanisms) based on these concepts is rather complex. Some questions that arise in the development process are for instance: a) how can (virtual) resources be configured to collaborate in slices, b) how can the performance of paths be measured to select them for pooling, or c) how does the performance of the system scale with the number of available networking resources?

Answering such questions comprehensively by means of mathematical analysis, simulations or experiments in local laboratories is often not possible. Applying analytical methods often requires assumptions that reduce the applicability of solutions. Simulation requires not only the modeling of the problem space but also requires knowledge about and integration of potential parameters that can influence results. If results depend on many parameter, the applied level of abstraction might be too coarse. Working with models requires many simplification that can lead to unrealistic results. Tests in labs suffer from scalability limits since physical distances and the number of resources are limited. Also the acquisition of specific measurement equipment is often difficult in local labs due to the high costs of such hardware. In short, as network scientists, we need larger testbeds in order to supplement theoretical analysis and validate theoretical results by experiments in large-scale highly distributed environments and under real network conditions.

The experimental facilities as provide by the the European FIRE program [1], the US-American GENI [10] and VINI systems [4], or the German G-Lab [20], aim at fulfilling these requirements. A federation of them provides the required scalability features (e.g. large distances between entities) and allows the use of special equipment and features that are only available in specific testbeds. The federation of testbeds can give new insights for federating resources in general and therefore for the design of networking architectures that are made up of federated resources, like multipath routing slices. However, todays testbeds are typically customized to particular user groups and offer different capabilities and interfaces. The federation of them still requires research on how these facilities should interconnect with each other in order to unleash the true benefits of federation resulting from the broader set of available features and functions.

In this paper, we present a multipath routing slice experiment that we performed over multiple federated testbeds. Since there was not a single testbed that could offer all the capabilities we needed, we combined capabilities from different experimental facilities (G-Lab, PlanetLab Europe, and VINI). While a measurement instrumentation of testbeds is essential to path selection in multipath routing slices, we additionally require highly precise measurements in our experiment. Therefore, our contribution is threefold. We present in the paper a) our experiment (setup, results, findings), b) our experience with the usability of federated testbeds and c) the use of advanced measurement technologies in experimental facilities.

The paper is structured as follows: in Section 2, we introduce the objectives and requirements of the multipath routing slice experiments. Section 3 describes the federation and setup of testbed systems for the experiment. Section 4 outlines the results of the experiments which validate an analytical performance model for multipath transmissions. Section 5 describes the lessons learned during the

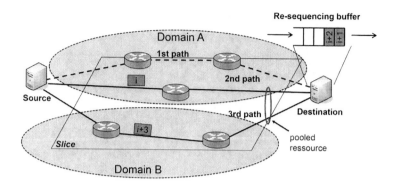

Fig. 1. Multipath transmission slice

experimental setup in the federated experimental facilities. Section 6 discusses the sharing and joined use of measurement equipment and tools. Finally, Section 7 provides a brief summary and outlook to the enhancements of federated facilities.

2   Experiment Objectives and Requirements for a Concurrent Multipath Transport

Alternative multipath transport services in future federated networks might employ concurrent or consecutive packet transmission. Concurrent transmission techniques have recently been receiving a lot of attention due to their advantages in resource pooling [22]. The strong interest has also been witnessed by recent research projects such as Trilogy [21] or state-of-the-art protocols like SCTP-CMT [7] and Multipath-TCP [14]. For the case of Concurrent Multipath Transmissions (CMT), different transport resources are pooled together and appear to be one single virtualized transport resource. However, different path characteristics such as one way delay, capacity or jitter of the pooled resources lead to a different behavior than in the case of a single resource. Different path delays inevitably lead to out-of-order arrivals at the destination. An effect that does not, or at least does not appear within this dimension, on a physical link. The right order within the packet stream can be restored by a re-sequencing buffer, as proposed and analyzed theoretically in [24]. The investigated architecture and the measurement setup is illustrated in Figure 1. The federation of resources is outlined by the use of concurrent transmission paths from different domains.

We build a model for such a re-sequencing buffer and try to predict the buffer occupancy based on network conditions. The main purpose of our experiment is the validation of the proposed model. For that, one way delay distributions of the different paths, i.e., the input parameter of the model, and the re-sequencing buffer occupation, i.e., the output parameter of the model, have to be measured. In order to conduct the desired measurements the experiment has the following requirements concerning testbeds: a) The possibility to set up a routing overlay to emulate the multipath transport. b) A large distributed set of nodes in order to get a high diversity of different path delay values. This enables a verification of the model with an adequate amount of different configurations. c) Advanced measurement methods for high precision and hop-by-hop one-way delay measurements.

3 Experiment Setup

We investigate the capabilities of different testbeds in order to find a suitable testbed that fulfills the needs of our experiment. Table 1 describes the differences of PlanetLab Central (PLC), PlanetLab Europe (PLE), German Lab (G-Lab), and the VINI testbed. It can be seen that a single testbed is not able to cope with the tight requirements for our multipath experiment. G-Lab allows exclusive reservation and installation of arbitrary software but is only distributed within Germany, has a limited access, and currently provides no federation method. PLE, PLC, and VINI can be federated by the Slice Federation Architecture (SFA), but only VINI provides a routing configuration service. We therefore used manually configured overlay routing on application layer to combine PLE and G-Lab with VINI. PLE is the only network that additionally provides the advanced measurement tools that we need for our experiment. In order to verify results we used both, passive and active measurement tools. Active measurements provide a statement about the network situation and require the injection of test traffic. Passive measurements measure the experiment's traffic itself and therefore provide a statement about the real treatment of the traffic in the network. We used the network of distributed ETOMIC nodes, which is federated with PLE under the OneLab federated experimental facilities [3] and multi-hop packet tracking [18], which is available in PLE. Figure 2 shows a screenshot of the visualization for the passive measurements with the packet tracking tool. The tool is available at [2].

Our setup, depicted in Figure 1 consists of a source, a destination and different paths between source and destination. The packet forwarding was realized either by application layer packet forwarding over different hosts or different paths configured in VINI. For our packet layer forwarding setup we used ETOMIC nodes which provide high precision GPS synchronized timestamps and

Table 1. Comparison of different experimental facilities

Feature	G-Lab	PLE	PLC	VINI
Scope	Germany	Europe	World	Mainly US
Exclusive Reservation	Yes	No	No	No
Routing	with own tools	with own tools	with own tools	Yes
Bandwidth and QoS	with own tools	Planned	Planned	Yes (service)
Openness/Federation	No (tests planned)	Yes (SFA)	Yes (SFA)	Yes (SFA)
Tools/Packet Tracking	individually	Yes (service)	individually	individually
Clock Sync	NTP	NTP, some GPS	NTP	NTP

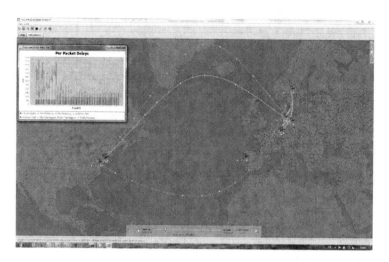

Fig. 2. Visualization for the passive measurements with the packet tracking tool

thus enabled us to measure one way delays of each packet. However, the experimental setup was very complicated since the different paths had to be set up manually. For the experiments with VINI the setup of a multipath transmission experiment was much easier. Since our end nodes did not provide the required measurement precision we utilized multi-hop packet tracking [18] for conducting one way delay measurements.

For the performed experiment we emulated a concurrent multipath transmission via two different paths. We transmitted 100.000 packets over each of the used paths. The packets were scheduled in a round robin manner with an inter-packet-time of 10 ms on each path. We measured the buffer occupancy at the receiver and the measured one way delay, once measured actively and once measured passively, as described above. The results of our experiments are discussed in the next section.

4   Experimental Results for Multipath Routing Slices

This section summarizes results of the experiments we conducted for a concurrent multipath transmission as outlined in [24]. We sent every ten milliseconds two packets from the source to the destination via two different paths. First, we discuss active measurements performed in PLE with support of the ETOMIC measurement system. After that we outline the passive measurements [18] conducted within GLAB and VINI.

4.1   Active Measurements with PLE and ETOMIC

For the measurements we used ETOMIC nodes with DAG cards located in Pamplona and Elte as source and destination. The packets were transmitted via two different paths, one via a PLE node located in Vrije, one via a node in Tromso.

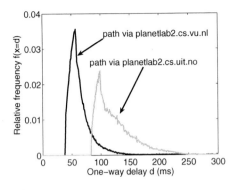
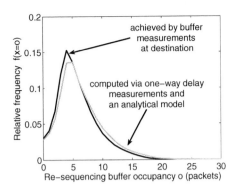

Fig. 3. Actively measured one-way delays as relative frequencies for two different paths within PLE

Fig. 4. Comparison between actively measured and estimated re-sequencing buffer occupancy

The one-way delay in milliseconds is depicted as relative frequencies in Figure 3. It can be seen that the path delay via Vrije is smaller than the path delay via Tromso. Further, the one way delay values range in an interval of more than 100 milliseconds, i.e. the delay values for the packets are highly variable during the measurements. Based on these measurements, the occupancy of the re-sequencing buffer can be approximated by the analytical model.

Figure 4 illustrates the observed re-sequencing buffer occupancies in packets. It can be seen, that the probability for an empty buffer is very low, and that most likely five packets are stored within the buffer. This is due to the fact that packets sent via Tromso experience higher one way delays than packets sent via Vrije. In addition, higher buffer occupancies may also occur.

Further, the estimated re-sequencing buffer occupancy is also depicted in Figure 4. It can be seen that the gap between the buffer occupancy computed with the analytical model and the measured buffer occupancy is very small. Thus, we can conclude, that for the given scenario the prediction of the model is very accurate.

4.2   Passive Measurements with VINI and GLAB

For these experiments we configured one path via a GLAB node in Darmstadt and a second one via VINI. The one-way delays of the packets were captured by passive multipoint measurements, cf. [18].

The measured one-way delay is depicted as relative frequencies in Figure 5. The figure shows that the path delay via GLAB is smaller than the path delay via VINI and rather constant. Further, the one way delay values on the VINI path range in an intervals of more than 100 milliseconds, i.e. the delay values for the packets are highly variable during the measurements. That is due to the fact that we injected additional random delay on this path.

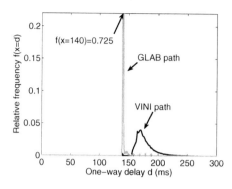

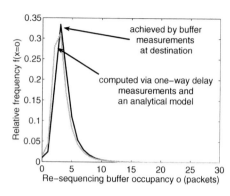

Fig. 5. Passively measured one-way delays for two different paths within GLAB/VINI

Fig. 6. Comparison between measured and estimated re-sequencing buffer occupancy

The measured buffer occupancy as well as the the analytical approximation is depicted in Figure 6. It can be seen tat most likely up to 10 packets have to be stored in the buffer. In addition, due to the varying one way delays higher buffer occupancies may also occur. Again the gap between measured and computed buffer occupancy is very small, i.e. the model is again very accurate.

The discussed measurements were conducted separately, i.e., under different conditions. However, the significance of the results could be improved by conducting passive and active measurements concurrently.

5   Lessons Learned for the Usage of Federated Experimental Facilities

In this section we summarize our experience in the federation of experimental facilities in the course of our experiments. First, we detail the lessons we learned while using different testbeds, then we describe the observations we made for each of the used testbeds.

5.1   Challenges while Preparing the Experiments

First we present the challenges which occurred during the experiments.

Booking of Resources With the SFA software it was possible to book nodes in PlanetLab, PlanetLab Europe and in the VINI Testbed. The Glab testbed is designed as an exclusive testbed for experimenters that participate in the G-Lab project. Thus, G-Lab does not provide a federation interface yet. Therefore, G-Lab nodes were booked separately and connected manually to the overlay. Although G-Lab was designed as an exclusive resource, some tests for the integration with SFA and the PII Framework are planned. By federating the different facilities the efforts needed for resource booking upon the different testbeds could be reduced.

Configurable Routing Slice A configurable routing topology was essential for our experiments. As we had to employ a routing infrastructure over multiple testbeds to match our requirements it would have been beneficial to use a federation framework like SFA or the PII framework to configure the topology, but both frameworks do net yet support overlay creation. In order to resolve this issue, we established tunnels between the nodes of the different testbeds. Especially PlanetLab and PlanetLab Europe lack a standardized way to configure the topology and the routing protocol. Further, only one virtual interface is configurable and the configuration is restricted, e.g. one cannot change flow or routing table entries. Thus, we utilized application layer overlay routing techniques for transmitting packets via different paths, i.e., we set up a routing topology manually. VINI on the other hand provides an easy-to-use topology creation, configuration of interfaces and the use of arbitrary routing protocols. Further, it also allows the emulation of link characteristics and the booking of guaranteed bandwidth. However, VINI provides only a few number of nodes, which are mainly located in the US.

Observation tools Our experiments are strongly dependent on precise observation tools that capture the experiment result and environmental conditions. By using different testbeds, we had the possibility to use adequate tools which allowed separated active and passive measurements. However, due to the different methods, the comparability of the results is reduced. Here, common observation methods or a common understanding how the provided tools differ would enhance the comparability and, thus, the value of the gained results. An initiative pushing activities in this field is the OneLab project which launched Free T-Rex [2], a website dedicated to information about *Free Tools for Future Internet Research and Experimentation*. The Advanced Network Monitoring Equipment (ANME) deployed by the Onelab project within Planetlab Europe includes precise network cards for active delay measurements using ETOMIC and the continuous monitoring platform (CoMo) for passive measurements. This enables high precision active and passive measurements in parallel and thus allows a comparison between active and passive measurement methodologies.

Clock Synchronization For our experiment we required precise active and passive one way delay measurements. Nevertheless, the achievable accuracy of such measurements depends on the synchronization status of the involved observation points. The ETOMIC boxes used in our experiment are GPS time synchronized and meet our precision requirements. A general clock synchronization service across testbeds, in the best case supported by GPS-based clocks, would help to provide more accurate measurements.

5.2   Observations on the Single Testbeds

Regarding the use of the particular testbeds we provide the following observations.

G-Lab provides an exclusive resource. We can book nodes exclusively and can generate a much better controllable environment. We can install and use arbitrary software on the G-Lab nodes. We assume that such features are of interest for many experimenters, but the closed nature of G-Lab makes it unavailable for experimenters that do not participate in a G-Lab project. A further disadvantage of G-Lab is that it has a comparatively small number of nodes and the nodes are only in Germany, therefore it is not suitable for experiments that require real Internet conditions with regard to scale, delay values, and geographical distribution of nodes.

PlanetLab Europe offers many additional functions, research tools and hardware to support active and passive high precision measurement. Such an infrastructure helps experimenters to perform measurements and retrieve accurate results. But neither PlanetLab nor PlanetLab Europe provide routing support. Due to this we had to invest a lot of effort to manually set up tunnels for a routing topology.

VINI is very well suited to provide routing support. VINI also supports SFA, so nodes could be booked via SFA with the same credentials we used for PlanetLab Europe. The disadvantages of VINI is that it only provides a few nodes, which are mainly in the US. Furthermore, not all nodes have public Internet access, which makes the configuration more complicated. Another issue that came up during our experiments is that the VINI infrastructure is too good, i.e., the delay on a path is very low. Since we had no GPS clock synchronization we could not capture the delay between two hops in the precision required for the transmission speed.

6   Sharing and Standardizing Measurement and Observation Tools

As part of our work we have seen the need for all the heterogeneous experimental facilities to standardize experiment measurements and observation tools, not only to capture the outcome of the experiment but also to log the experimental conditions. Free T-REX [2] provides a platform for testbed users, testbed operators and developers to offer their measurement results and software tools to the public and to share their experience. Further, free T-Rex seeks to employ standardized instruments to improve the comparability and openness of scientific results in the field of future Internet research. The platform gives an overview of available tools in future Internet experimental facilities and, based on user feedback, the tools' feasibility for experiment requirements can be assessed. Another objective is to create links to relevant groups and support standardization efforts in the field of research experiment observation.

Free T-Rex offers such valuable resources like access to the MoMe [12] trace and tool database and measurement services, the employed packet tracking service [18], TopHat [9], and the DIMES [19] infrastructure.

7   Conclusion

In this paper, we outlined how the federation of multiple experimental facilities can contribute to an improved design of future, federated Internet architectures. We described how federated transmission resources can be exploited for multipath routing slices and how this may form a new concept for network architectures. For that, we validated an analytical model of a multipath routing slice mechanism by measurements in federated testbeds. We showed why isolated testbeds are insufficient for some of the experiments and identified difficulties and requirements for federated testbeds. Our experiment would not have been possible in available non-federated testbeds. In this way, the federation of testbeds enabled a truly comprehensive evaluation of the proposed multipath routing slice mechanisms. The experiments and their results have demonstrated that the federation of experimental facilities is very powerful to accelerate the design of future networks. Although the advantages are striking, the usage of the testbeds and the federation of the facilities is still painful. Booking of resources was easy, but the configuration of a federation on the routing layer or below this layer and of the shared measurement equipment requires still huge efforts. If such federations were made easier, the full power of federated testbeds and of federation for future networks might be truly unleashed.

Acknowledgements The research leading to these results has received funding from the European Union's Seventh Framework Program (FP7/2007-2013) under grant agreement n° 216366 for the NoE project "Euro-NF". In particular, it was funded through the Euro-NF specific joint development and experimentation project "Multi-Next". Furthermore, we would like to express our appreciation for the support through the experimental facilities GLAB, PLE and ETOMIC and the projects VINI and ONELAB. Further, the authors deeply want to thank Andy Bavier for his support during the course of this work.

References

1. FIRE - Future Internet Research & Experimentation (2010), Information available at http://ict-fire.eu/
2. Free T-REX: Free Tools for Future Internet Tools and Experimentation (2010), Information available at http://www.free-t-rex.net/
3. Onelab - Future Internet Testbeds (2010), Information available at http://onelab.eu/
4. VINI - A Virtual Network Infrastructure (2010), Information available at http://vini-veritas.net/

5. Anderson, T., Peterson, L., Shenker, S., Turner, J.: Overcoming the internet impasse through virtualization. IEEE Computer, 34–41 (April 2005)
6. Anerousis, N., Hjlmtysson, G.: Service level routing on the Internet. In: IEEE GLOBECOM'99, vol. 1, pp. 553–559 (2002)
7. Becke, M., Dreibholz, T., Yyengar, J., Natarajan, P., Tuexen, M.: Load Sharing for the Stream Control Transmission Protocol (SCTP), Internet-Draft (2010), http://tools.ietf.org/html/draft-tuexen-tsvwg-sctp-multipath-00
8. Boschi, E., Denazis, S., Zseby, T.: A measurement framework for inter-domain sla validation. Comput. Commun. 29, 703–716 (2006), doi:10.1016/j.comcom.2005.07.026
9. Bourgeau, T., Augé, J., Friedman, T.: TopHat: supporting experiments through measurement infrastructure federation. In: Proceedings of the International Conference on Testbeds and Research Infrastructures for the Development of Networks and Communities, TridentCom (2010)
10. GENI Consortium. GENI - Global Environment for Network Innovations (2006), Information available at http://www.geni.net/
11. Kornblum, J., Raz, D., Shavitt, Y.: The active process interaction with its environment. Computer Networks 36(1), 21–34 (2001)
12. MoMe. Cluster of European Projects aimed at Monitoring and Measurement (2010), Information available at http://www.ist-mome.org/
13. Nakao, A., Peterson, L., Bavier, A.: A Routing Underlay for Overlay Networks. In: Proc. of the ACM Sigcomm 2003 Conference, Karlsruhe, Germany (Aug. 2003)
14. Yoshifumi Nishida and Philip Eardley. Charter of the Multipath TCP Work Group (MPTCP) (Mar 2010), Information available at http://tools.ietf.org/wg/mptcp/
15. Larry Peterson, Soner Sevinc, Jay Lepreau, Robert Ricci, John Wroclawski, Ted Gaber, and Stephen Schwab. Slice-Based Facility Architecture (2009), Information available at http://svn.planet-lab.org/attachment/wiki/GeniWrapper/sfa.pdf
16. Psounis, K.: Active Networks: Applications, Security, Safety, and Architectures. IEEE Communications Surveys 2(1) (1999), http://www.comsoc.org/pubs/surveys/1q99issue/psounis.html
17. Scott Rixner. Network virtualization: Breaking the performance barrier. ACM Queue, (Jan./Feb. 2008)
18. Santos, T., Henke, C., Schmoll, C., Zseby., T.: Multi-Hop Packet Tracking for Experimental Facilities. In: Demo Sigcomm 2010, New Delhi, India (Aug. 2010)
19. Shavitt, Y., Shir, E.: DIMES: Let the internet measure itself. ACM SIGCOMM Computer Communication Review 35(5), 71–74 (2005)
20. Phuoc Tran-Gia. G-Lab: A Future Generation Internet Research Platform (2008), Information available at http://www.future-internet.eu/
21. Trilogy. Trilogy: Architecting the Future Internet (2010), Information available at http://www.trilogy-project.org/
22. Wischik, D., Handley, M., Braun, M.B.: The Resource Pooling Principle. SIGCOMM Comput. Commun. Rev. 38, 47–52 (2008), doi:10.1145/1452335.1452342
23. Zinner, T., Tutschku, K., Nakao, A., Tran-Gia, P.: Re-sequencing Buffer Occupancy of a Concurrent Multipath Transmission Mechanism for Transport System Virtualization. In: Proc. of the 16. KiVS 2009, Kassel, Germany (Mar. 2009)
24. Zinner, T., Tutschku, K., Nakao, A., Tran-Gia, P.: Using Concurrent Multipath Transmission for Transport Virtualization: Analyzing Path Selection. In: Proceedings of the 22nd International Teletraffic Congress (ITC), Amsterdam, Netherlands (Sep. 2010)

Testing End-to-End Self-Management in a Wireless Future Internet Environment

Apostolos Kousaridas[1], George Katsikas[1], Nancy Alonistioti[1], Esa Piri[2], Marko Palola[2], and Jussi Makinen[3]

[1] University of Athens
Athens, Greece
scan.di.uoa.gr
{akousar, katsikas, nancy}@di.uoa.gr
[2] VTT Technical Research Centre of Finland
Oulu, Finland
{Esa.Piri, Marko.Palola}@vtt.fi
[3] Octopus Network
Oulu, Finland
www.octo.fi
jussi.makinen@octo.fi

Abstract. Federated testbeds aim at interconnecting experimental facilities to provide a larger-scale, more diverse and higher performance platform for accomplishing tests and experiments for future Internet new paradigms. In this work the Panlab experimental facilities and specifically the Octopus network testbed has been used in order to experiment on the improvement of QoS features by using the Self-NET software for self-management over a WiMAX network environment. The monitoring and configuration capabilities that different administrative domains provide has been exploited in order to test network and service layers cooperation for more efficient end-to-end self-management. The performance results from the experiments that have been performed prove that the proposed self-management solution and the mechanisms for the selection of the appropriate network or service level adaptation improve end-to-end behaviour and QoS features.

Keywords: Experimentation, Testing Facilities, self-Management, Future Internet, WiMAX, Quality of Service

1 Introduction

Several network management frameworks have been specified during the last two decades by various standardization bodies and forums, like IETF, 3GPP, DMTF, ITU, all trying to specify interfaces, protocols and information models by taking into consideration the respective network infrastructure i.e., telecom world, the Internet and cellular communications. The current challenge for the network management systems

J. Domingue et al. (Eds.): Future Internet Assembly, LNCS 6656, pp. 259–270, 2011.

is the reduction of human intervention in the fundamental management functions and the development of the mechanisms that will render the Future Internet network capable of autonomously configuring, optimizing, healing and protecting itself, handling in parallel the emerging complexity. In the autonomic network vision, each network device (e.g., router, access point), is potentially considered as an autonomic element, which is capable of monitoring its network-related state and modifying it based on policy rules that the network administrators have specified.

The scope of this work is to experiment on the improvement of QoS features (e.g., packet loss, delay, jitter) by using a self-management framework over a live network environment and exploiting monitoring and configuration capabilities that different administrative domains provide (i.e. access network and service layer). The effectiveness and the feasibility of various parameters optimization of existing network protocols avoiding manual effort are also tested. The implemented and tested self-management framework has been designed by the Self-NET project [1]. It is based on the so called closed control loop or Monitor-Decide-Execute Cycle (MDE) and consists of the Network Element Cognitive Manager (NECM) and the Network Domain Cognitive Manager (NDCM) [2].

The experimentation work has been carried out as cooperation with Self-NET and PII projects [3] by utilizing Octopus Network [4] testing resources, which are part of Panlab federation [5] of interconnected testing facilities.

The remainder of the paper is organized as follows: The Panlab experimental facilities that have been used as well as their configuration are described in section 2. Section 3 presents the mechanisms that have developed for service-aware network self-management framework. Finally, the experimentation results that have been collected from the tests and the improvement of the performance by using the self-management mechanisms are highlighted in section 4, while section 5 concludes this paper.

2   Experimental Facilities Decription

The testing facility connecting a fixed WiMAX network to the service-aware network is shown in Fig. 1. The WiMAX network environment consists of Airspan Micro-MAX base station (BS) [7] and Airspan ProST subscriber station (SS) located on the Octopus testbed at Oulu [4]. The BS and SS operate in a laboratory environment with short distance direct line-of-sight condition, which keeps the signal strength relatively stable and strong throughout the measurement cases. As regards the Self-NET provision side at Greece Distributed Internet Traffic Generator (D-ITG) [8] has been used, which is a software tool that generates traffic at both UoA end machines. This is a Java based platform that manipulates two independent entities, the first is ITGSend process that undertakes the traffic generation and the latter is ITGRecv process that captures the packets to the receiver. Traffic sender can concurrently generate multiple flows with user-defined parameters that can be analyzed from the receiver to extract traffic QoS features (e.g. packet loss, delay, jitter). There are also some contributory entities that assist in improving the traffic simulation by providing log information

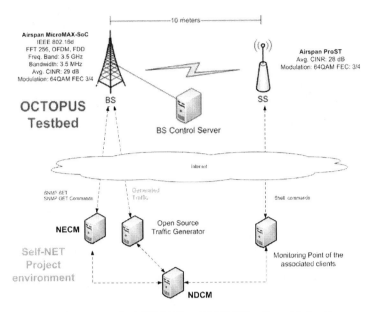

Fig. 1. Octopus testbed WiMAX and Self-NET software federation

(ITGLog), printing and plotting specific metrics (ITGDec, ITGPlot) and remotely controlling the traffic generation (ITGApi). The most well-known network, transport, and application layer protocols are supported by this platform such as TCP, UDP, ICMP, DNS, Telnet, and VoIP (G.711, G.723, G.729, Voice Activity Detection and Compressed RTP).

The Self-NET project carries out experiments over the WiMAX testbed, remotely via the Internet. The experiment required development of an additional BS control software and deployment of IP routing and tunneling between Octopus and Self-NET environments.

We implemented a BS control software (i.e. NECM) to allow dynamically collect WiMAX link information from the BS and to control Quality of Service (QoS) settings on the fly. The NECM changes QoS service classes by setting a new configuration to the BS using Simple Network Management Protocol (SNMP).

IEEE 802.16 standards specify various packet scheduling schemes to ensure required QoS of different traffic types. For example, transmission delay constraints of real-time multimedia streaming are much stricter than that of bulk data transfer. IEEE 802.16d [5], the employed WiMAX testbed is based on, specifies four different scheduling types, namely Unsolicited Grant Service (UGS), Real-time Polling Service (rtPS), Non-real-time Polling Service (nrtPS), and Best Effort (BE). UGS and rtPS are for real-time traffic where maximum latency and jitter can be set in addition to minimum reserved and maximum sustained traffic rates. BE and nrtPS are for delay-tolerant data transmission. However, nrtPS provides assured bandwidth for the traffic flow whereas BE does guarantee nothing for the traffic flow but packets are transmitted if bandwidth available.

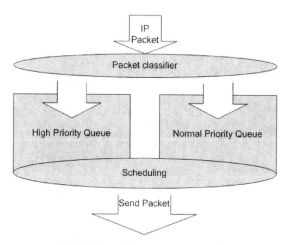

Fig. 2. Downlink packet scheduling

In the employed BS, the aforementioned scheduling types are supported only in the uplink through a request/grant scheduling. In the downlink, the BS supports a scheduling type where several traffic flows can simply be treated with different priorities only, not assuring delay or bandwidth requirements. This downlink scheduling type is capitalized on in our experiments. During default scheduling operation of the BS downlink, all traffic is treated equally by the packet classifier and put to the same normal priority transmission queue where BE scheduling is employed to. The BS controller can be commanded to configure the BS to handle particular traffic flows with higher priority. In this case the BS has two transmission queues of different priorities, as illustrated in Fig. 2. In the downlink scheduling, the packets can be classified to different transmission queues of various priorities based on the IP packet's source and/or destination MAC address, IP address, or port number. In our experiments, we used port numbers to classify the IP traffic flows. We found that during the reconfiguration of the BS service classes packet transmission between BS and SS was temporarily stagnated, however, resulting in break times constantly below a second.

The Self-NET project experiments also required setting up IP routing and tunneling from and to the WiMAX link. Two routers are dedicated on the Octopus testbed for tunneling and routing IP traffic. The user traffic from the Self-NET experimentation is tunneled by using two IP tunnels over the Internet and rerouted over the WiMAX air interface at the Octopus testbed. For the test environment provisioning, the IP tunneling (IPIP) and routing was setup at both ends, which requires two routers at the user premises – one for sending data to the uplink and receiving the downlink flows and one for sending to the downlink and receiving from the uplink.

As depicted in Fig. 3, there are two IPIP tunnels established at the overall topology in order to deploy the federation of these two testbeds. The first tunnel connects the WiMAX BS with the UoA BS Connector (10.1.3.3 – 10.1.3.1) while the second one connects the WiMAX SS with the UoA SS Connector (10.1.3.4 –10.1.3.2), creating an internal 10.1.3.0/24 network between these network entities. The traffic sent from the UoA BS Connector (10.1.1.1) is routed over the IPIP tunnel to the WiMAX BS

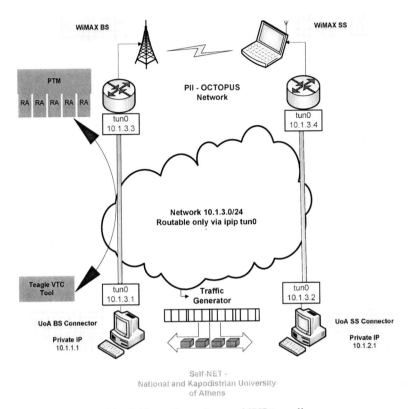

Fig. 3. Network topology and IPIP tunneling

and after the Wireless transmission (DL) to the WiMAX SS, the UoA SS Connector (10.1.2.1) receives the packets via the second tunnel. The respective procedure occurs for the UL, while the UoA SS Connector traffic is tunneled to the WiMAX SS, transmitted to the WiMAX BS and routed again through IPIP tunnel to the UoA BS Connector. During the traffic exchange, the public IPs' are opaque, as the routing procedure explicitly uses the private addresses.

Fig. 3 illustrates also the Panlab federation tools [5] such as Panlab Testbed Manager (PTM), which was installed on Octopus Network to allow Teagle Virtual Customer Testbed (VCT) tool to carry out the topology setup operation. Resource Adapter Description Language (RADL) [9] was used to generate source code for each Resource Adaptor (RA), where, for example, the WiMAX network elements can be considered as available and configurable resources. We decided to use a separate RA for each IP tunneling machine, BS and SS. The RAs managing tunneling send commands to respective machines via SSH to setup both tunneling and routing. The default values are stored in each RA and the user of the VCT tool needs to input only public IP addresses and user credentials for the two external tunneling machines in order to setup the IP tunnels and routes.

3   Mechanism for Service-Aware Network Self-Management

The allocation of Monitoring-Decision Making-Execution (Cognitive) Cycle phases at the NECM and NDCM agents is presented in this section, in order to enable network and service layers cooperation for more efficient end-to-end self-management (Fig. 1). The term cooperation is used to describe the collection of the service-level monitoring data and the usage of service-level adaptation actions for efficient network adaptation.

The NECM of the WiMAX BS constantly **monitors** network device statistics (e.g., UL/DL used capacity, TCP/UDP parameters, service flows), which are periodically transmitted to the corresponding NDCM. The latter one retrieves also associated clients perceived QoS (delay, packet loss, and jitter), the type of service (VoIP, FTP, Video) that each client consumes as well as service profile information from the service providers. The Service-level NECM undertakes to collect service-level data. The Service-level NECM could be placed at the service provider's side, even at premises of network operators. We should point that the Service-level NECM performs also service management tasks (e.g., service composition, discovery) by exploiting the Cognitive Cycle (Monitoring-Decision Making-Execution) paradigm. This type of functionality is not part of this work.

The **decision making** engine of the NDCM filters the collected monitoring data from the network and the service level in order to identify faults or optimization opportunities (e.g., high packet loss) according to the specified rules or QoS requirements. In the specific use case the goal of the NDCM Decision making engine is the identification of high average packet error rate (PER) values for the end clients that consume a VOIP service. The second step is the selection of the appropriate configuration action. The following actions are taken into consideration by the NDCM:

- Change the codec that $k_1 \in \Re$ flows use.
- Change the priority of $k_2 \in \Re$ flows at the WiMAX BS.
- Change the priority of $k_3 \in \Re$ flows at the WiMAX BS and the codec of $k_4 \in \Re$ flows.

Two schemes for the selection of the optimal action have been proposed and they are described below (Fig. 4 and Fig. 5).

According to the decision making output the configuration action is transferred either to the WiMAX BS NECM in order to **execute** the change priority action via SNMP set command or to the Service-level NECM in order to execute the codec update. Our scheme is based on the available monitoring and configuration capabilities that network elements provide.

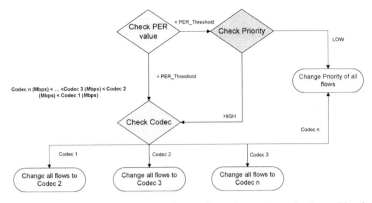

Fig. 4. Decision-making algorithm for configuration action selection – Simple

Fig. 4 presents the simple version of the decision taking scheme Firstly, the PER value is checked in order to select the 'Change Priority' or 'Change Codec' action. If the PER is lower than a pre-defined threshold (PER-threshold) the NDCM decides to change all flows from low priority to high priority service class at the WIMAX BS side. If the priority value is already set as high, then the NDCM proceeds to the 'Change Codec' action. In that case NDCM will check the specific codec that all flows use. According to the Codec type the NDCM decides the transition to a codec that achieves higher data compression, resulting in less data rate requirements; thus reducing packet error rate value. If the clients use the less demanding codec, then the change priority solution is checked. Finally, if none of the above actions are effective then the NDCM will search for an alternative configuration action.

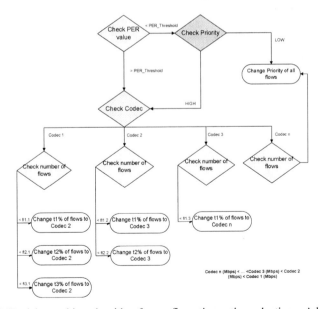

Fig. 5. Decision making algorithm for configuration action selection – Advanced

The above figure (Fig. 5) illustrates the advanced version of the scheme presented above. Specifically, each 'Change codec' action takes into consideration the number of flows $fla.b$, where b denotes the codec type and a the flow threshold of type b which traverse the network and adapts only a percentage of the underlying flows ($ta.b\%$).

4   Performance Results

In this section we provide the performance results that prove the QoS features improvement (e.g., average delay, average jitter, packets dropped) after the reconfiguration actions (e.g., due to an increase of the packet loss rate of VoIP traffic). The configuration actions that have been used are:

- The change of the prioritization scheme at the WiMAX BS side (e.g., from low priority to high priority service class). The following port-based priorities have been set:
  - High Priority: Port range [9850, 10100]
  - Low Priority: Port range [10101, 10250]
- The change of the VoIP codec between the service provider and the end user (service-level adaption). The data rates of each VoIP codec are:
  - G.711.1: 48 kbps
  - G.711.2: 40 kbps
  - G.729.3: 8 kbps
  - G.729.2: 7 kbps
  - G.723.1: 5 kbps

Table 1. Critical thresholds of Packet Loss sharp increment

Codec Type	Threshold of Flows Number
G.711.1 - (fl1.1)	29
G.711.2 - (fl1.2)	46
G.729.2 - (fl1.3)	63
G.729.3- (fl1.4)	97
G.723.1 - (fl1.5)	120

As it is described in Section 3, the decision making schemes that have been proposed for the selection of the appropriate action use a list of thresholds (i.e. *PER-threshold, fla.b, ta.b%*). In order to estimate these thresholds accurately and to avoid setting arbitrary values, a first phase of testing took place. Specifically, various number of VoIP flows have been injected into the Octopus Network and different combinations of codec types and priorities (high, low) have been set in order to measure the arising packet error rate, and consequently calculate the appropriate threshold values. The packet loss rate increases, while the number of VoIP flows does, too. However, the increase rate is not linear since there is a critical value for the number of flows that causes a sharp increase of the Packet Loss (over *PER-threshold* = 4%). This value varies among the different codec types, as it is depicted in Table 1, where the codec thresholds for different flow numbers are presented (*fla.b*).

The following tables depict the improvement on specific QoS features after the re-configuration actions due to an increase of the packet loss rate of VoIP traffic. Table 2 presents the reduction of the packet loss rate after the change of the prioritization (from low priority to high priority service class) at the WiMAX BS of the 28 VoIP flows that use G.711.1 codec.

Table 2. QoS features improvement using high priority service class – Simple scheme

	G.711.1 – Low Priority	G.711.1 – High Priority
Number of flows	28	28
Total packets	28607	26767
Average delay	1.028651 s	1.018491 s
Average jitter	0.012321 s	0.013235 s
Average bitrate	2546.360118 Kbit/s	2580.231640 Kbit/s
Average packet rate	2491.719455 pkt/s	2301.527932 pkt/s
Packets dropped	573 (2.004 %)	12 (0.045 %)

Table 3. QoS features improvement after total VoIP codec change from G.711.1 to G.711.2 (in the case that service class prioritization change is not effective) – Simple scheme

	G.711.1 – Low Priority	G.711.1 – High Priority	G.711.2 – Low Priority
Number of flows	32	32	32
Total packets	30565	30602	19558
Average delay	0.514 s	0.761 s	0.42 s
Average jitter	0.012 s	0.012 s	0.016 s
Average bitrate	2789.06Kbit/s	2717.74Kbit/s	3148.41Kbit/s
Average packet rate	2485.90pkt/s	2504.07 pkt/s	1582.36 pkt/s
Packets dropped	3442 (10.12 %)	7929 (20.58 %)	20 (0.10 %)

Table 3 depicts the QoS features improvement after a service level adaption of the 32 G.711.1 VoIP flows that traverse the WiMAX BS and face high packet error rate. The modification of the service class prioritization at the BS side (from low priority to high priority class) is not effective, thus an alternative configuration action has been deduced. Specifically, the change of all VoIP codecs between the service provider and the end user, selecting the G.711.2 codec, reduces the number of the dropped packets.

Since the total codec change may be a simple but greedy solution, an advanced adaptation scheme is also proposed and deployed in order to reduce Packet Loss ratio without sacrificing the provided QoS. This scheme is based on the partial codec adaptation according to the number of the VoIP flows (Fig. 5).

The three tables below showcase the QoS features improvement after the exploitation of the advanced scheme for the selection of the adaptation (Fig. 5). It should be mentioned that the adaptation ratios presented are indicative, as there is a wide range of such ratios according to the codec type and the number of VoIP flows (from 10% to 100%). More specifically, in Table 4, the 27 G.711.1 flows are adapted to 21 G.711.1 and six G.711.2 flows, so this rational adaptation (20%) results to a satisfactory Packet Loss ratio without changing all the codecs.

Table 4. QoS features improvement after partial (20%) VoIP codec change from G.711.1 to G.711.2 – Advanced scheme

	G.711.1 – Low Priority	80% G.711.1 – 20% G.711.2 – Low Priority
Number of flows	27	27
Total packets	29158	27668
Average delay	0.996881 s	1.019370 s
Average jitter	0.012377 s	0.013391 s
Average bitrate	2719.482 Kbit/s	2600.848 Kbit/s
Average packet rate	2558.313 pkt/s	2380.823 pkt/s
Packets dropped	1301 (4.461%)	79 (0.285%)

Table 5. QoS features improvement after partial (50%) VoIP codec change from G.711.1 to G.711.2 – Advanced Scheme

	G.711.1 – Low Priority	50% G.711.1 – 50% G.711.2 – Low Priority
Number of flows	29	29
Total packets	29494	25126
Average delay	1.075899 s	1.070250 s
Average jitter	0.013444 s	0.014543 s
Average bitrate	2502.232 Kbit/s	2596.203 Kbit/s
Average packet rate	2539.245 pkt/s	2152.005 pkt/s
Packets dropped	2621 (8.886%)	13 (0.05173%)

Table 5 presents the changes of the traffic measurements after a 50% codec adaptation. The 29 G.711.1 flows are replaced with 14 G.711.1 and 15 G.711.2 flows and this adaptation contributes to about 8.5% Packet Loss reduction.

The last partial adaptation example is depicted in Table 6, where the adaptation ratio reaches 70% of the flows. The 35 G.711.1 flows are altered to 11 G.711.1 and 25 G.711.2 flows while the resulted Packet Loss scores a 40% reduction.

Table 6. QoS features improvement after partial (70%) VoIP codec change from G.711.1 to G.711.2 – Advanced scheme

	G.711.1 – Low Priority	30% G.711.1 – 70% G.711.2 – Low Priority
Number of flows	35	35
Total packets	31308	25220
Average delay	1.085282 s	1.161476 s
Average jitter	0.013925 s	0.016809 s
Average bitrate	2758.579 Kbit/s	2534.948 Kbit/s
Average packet rate	2646.066 pkt/s	2092.565 pkt/s
Packets dropped	13338 (42.6%)	613 (2.43%)

5   Conclusion

In this paper, we have presented the cooperation between Self-NET and Panlab projects and specifically the usage of Panlab testing facilities (i.e. Octopus testbed) for the experimentation on networks self-management, by using the mechanisms that the Self-NET project has designed. The experiments that have been carried out by using the Octopus wireless network environment prove both the feasibility of the proposed architecture and the QoS improvement (e.g., packet error rate reduction) that could be achieved by applying the appropriate adaptation considering the network conditions.

Different wireless links and networks have different capabilities and often service implementers and providers do not have a possibility to test their service over various networks of different access technologies. Our empirical experiments show how a remote wireless link such as WiMAX can be remotely used. However, in order to provide a wireless link as a bookable resource for a large set of customers, the establishment of the tunnels between the wireless link and the remote user of the link and a correct configuration of the routes need to be automated. This can be achieved by using the tools developed by Panlab testbed federation. Scalability issues and interactions with other network management tasks is part of our future work.

References

1. Self-NET project, http://www.ict-selfnet.eu
2. Kousaridas, A., Nguengang, G., Boite, J., Conan, V., Gazis, V., Raptis, T., Alonistioti, N.: An experimental path towards Self-Management for Future Internet Environments. In: Tselentis, G., Galis, A., Gavras, A., Krco, S., Lotz, V., Simperl, E., Stiller, B. (eds.) Towards the Future Internet - Emerging Trends from European Research, pp. 95–104 (2010)

3. Website of Panlab and PII European projects, supported by the European Commission in its both framework programmes FP6 (2001-2006) and FP7 (2007-2013): http://www.panlab.net
4. Octopus Network test facility, http://www.octo.fi
5. IEEE 802.16 Working Group (ed.): IEEE Standard for Local and Metropolitan Area Networks. Part 16: Air Interface for Fixed Broadband Wireless Access Systems. IEEE Std. 802.16-2004 (October 2004)
6. Wahle, S., Magedanz, T., Gavras, A.: Conceptual Design and Use Cases for a FIRE Resource Federation Framework. In: Towards the Future Internet - Emerging Trends from European Research, pp. 51–62. IOS Press, Amsterdam (2010)
7. Airspan homepage, http://www.airspan.com
8. Distributed Internet Traffic Generator,
   http://www.grid.unina.it/software/ITG/index.php
9. Resource Adapter Description Language,
   http://trac.panlab.net/trac/wiki/RADL

Part V:

Future Internet Areas: Networks

Introduction

Although the current Internet has been extraordinarily successful as a ubiquitous and universal means for communication and computation, there are still many unsolved problems and challenges some of which have basic aspects. Many of these aspects could not have been foreseen when the first parts of the Internet were built, but they do need to be addressed now. The very success of the Internet is creating obstacles to the future innovation of both the networking technology that lies at the Internet's core and the services that use it. In addition, the ossification of the Internet makes the introduction and deployment of new network technologies and services very difficult and very costly.

The aspects, which are considered to be fundamentally missing, are:

- Mobility of networks, services, and devices.
- Guaranteeing availability of services according to Service Level Agreements (SLAs) and high-level objectives.
- Facilities to support Quality of Service (QoS) and Service Level Agreements (SLAs).
- Trust Management and Security, privacy and data-protection mechanisms of distributed data.
- An addressing scheme, where identity and location are not embedded in the same address.
- Inherent network management functionality, specifically self-management functionality.
- Cost considerations, whereby the overhead of management should be kept under control since this is a critical part of life-cycle costs.
- Facilities for the large scale provisioning and deployment of both services and management, with support for higher integration between services and networks.
- Facilities for the addition of new functionality, including the capability for activating a new service on-demand, network functionality, or protocol (i.e. addressing the ossification bottleneck).
- Support of security, reliability, robustness, mobility, context, service support, orchestration and management for both the communication resources and the services' resources.
- Support of socio-economic aspects including the need for appropriate incentives, diverse business models, legal, regulative and governance issues.
- Energy awareness.

The content of this book includes three chapters covering some of the above research challenges in Future Internet. It also includes a tie to a paper from the Socio-economics area.

The "Challenges for Enhanced Network Self-Manageability in the Scope of Future Internet Development" chapter examines perspectives from the inclusion of the autonomicity and self-manageability features in the scope of Future Internet's (FI) deployment. Apart from the strategic importance for further evolution, we also discuss some major future challenges among which is the option for an effective network management (NM), as FI should possess a considerably enhanced network manageability capability. It analyses a new network manageability paradigm that allows network elements (NEs) to: be autonomously inter-related/controlled; be dynamically adapted to changing environments, and; learn the desired behaviour over time. As self-organizing and self-managing systems have a considerable market impact, we identify benefits for all market actors involved. In addition, we incorporate some recent, but very promising experimental findings, mainly based on the context of a specific use-case for network coverage and capacity optimization, highlighting the way towards developing specific NM-related solutions, able to be adopted by the real market sector.

The "Efficient Opportunistic Network Creation in the Context of Future Internet" chapter is dedicated to the design of Opportunistic Networks. In the Future Internet era, mechanisms for extending the coverage of the wireless access infrastructure and service provisioning to locations that cannot be served otherwise or for engineering traffic whenever the infrastructure network is already congested will be required. Opportunistic Networks are a promising solution towards this direction. Opportunistic Networks are dynamically created, managed and terminated. During the creation phase, nodes that will constitute the Opportunistic Network needs, are selected and assigned with appropriate spectrum and routing patterns. Accordingly, this chapter focuses on the Opportunistic Network creation problem and particularly on the efficient selection of nodes to participate therein. A first step towards the formulation and solution of the Opportunistic Network creation problem is made, whereas indicative results are also presented in order to obtain some proof of concept for the proposed solution.

The "Bringing Optical Networks to the Cloud: an Architecture for a Sustainable Future Internet" chapter describes how to combine optical network technology with Cloud technology in order to achieve the challenges of Future Internet. The extent of Internet growth and usage raises critical issues associated with its design principles that need to be addressed before it reaches its limits. Many emerging applications have increasing requirements in terms of bandwidth, QoS and manageability. Moreover, applications such as Cloud Computing and 3D-video streaming require optimization and combined provisioning of different infrastructure resources and services that include both network and IT resources. Demands become more and more sporadic and variable, making dynamic provisioning highly needed.

As a huge energy consumer, the Internet also needs to have energy-saving functions. Applications critical for society and business or for real-time communication demand a highly reliable, robust, and secure Internet. Finally, the Future Internet needs to support sustainable business models, in order to drive innovation, competition, and research. Combining optical network technology with Cloud technology is key to addressing these challenges. In this context, we propose an integrated approach: realizing the convergence of the IT models and optical-network-provisioning models will help bring revenues to all the actors involved in the value chain. Premium advanced networks and IT managed services integrated with the vanilla Internet will ensure a sustainable Future Internet, which enables demanding and ubiquitous applications to co-exist.

The "Deployment and Adoption of Future Internet Protocols" chapter from the Socio-Economics Area addresses the deployability of network protocols. The main message of this chapter is that implementation, deployment, and adoption need to be thought about carefully during the design of the protocol, as even the best technically designed protocol can fail to get deployed. Initial, narrow, and subsequent widespread scenarios should be identified and mental experiments performed concerning these scenarios in order to improve the protocol's design. It presents a new framework, which, when used by a designer would improve the chances that their protocol will be deployed and adopted. This framework was applied to two emerging protocols: Multipath TCP and Conex. Multipath TCP is designed to be incrementally deployable by being compatible with existing applications and existing networks, whilst bringing benefits to end users. For Congestion Exposure (Conex), a reasonable initial deployment scenario is a combined CDN-ISP that offers a premium service using Conex, as it requires only one party to deploy Conex functionality.

Alex Galis

Challenges for Enhanced Network Self-Manageability in the Scope of Future Internet Development

Ioannis P. Chochliouros[1,*], Anastasia S. Spiliopoulou[2], and Nancy Alonistioti[3]

[1] Head of Research Programs Section, Network Strategy and Architecture Dept.,
Hellenic Telecommunications Organization S.A. (OTE),
99 Kifissias Avenue, 15124 Maroussi, Athens, Greece
ichochliouros@oteresearch.gr
[2] Lawyer, General Directorate for Regulatory Affairs,
Hellenic Telecommunications Organization S.A. (OTE),
99 Kifissias Avenue, 15124 Maroussi, Athens, Greece
aspiliopoul@ote.gr
[3] Lecturer, National and Kapodistrian University of Athens,
Dept. of Informatics and Communications, 15784, Panepistimiopolis, Ilissia, Athens, Greece
nancy@di.uoa.gr

Abstract. The work examines perspectives from the inclusion of the autonomicity and self-manageability features in the scope of Future Internet's (FI) deployment. Apart from the strategic importance for further evolution, we also discuss some major future challenges among which is the option for an effective network management (NM), as FI should possess a considerably enhanced network manageability capability. We examine a new network manageability paradigm that allows network elements (NEs) to: be autonomously interrelated/controlled; be dynamically adapted to changing environments, and; learn the desired behaviour over time, based on the original context of the Self-NET research project effort. As self-organizing and self-managing systems have a considerable market impact, we identify benefits for all market actors involved. In addition, we incorporate some recent, but very promising experimental findings, mainly based on the context of a specific use-case for network coverage and capacity optimization, highlighting the way towards developing specific NM-related solutions, able to be adopted by the real market sector. We conclude with some essential arising issues.

Keywords: Autonomicity, cognitive networks, Future Internet (FI), network manageability, Network Management (NM), self-configuration, self-manageability, self-management, situation awareness (SA).

1 Introduction – Moving Towards the Future Internet

There is an extensive consensus that the Internet, *as one of the most critical infrastructures of the 21ˢᵗ century,* can critically affect traditional regulatory theories as

---

* Corresponding Author.

J. Domingue et al. (Eds.): Future Internet Assembly, LNCS 6656, pp. 277–292, 2011.

well as existing governance practices [1]. But, as the future of the Internet comes into consideration, in parallel with the appearance and/or the development of modern infrastructures, even greater challenges appear, with many concerns relevant to privacy, security and governance and with a diversity of issues related to Internet's effectiveness and inclusive character. Future related facilities will "attract" more users to innovative services requiring greater mobility and bandwidth, higher speeds and improved interactivity through the launch of many interactive media- and content-based applications [2]. Nevertheless, such claims necessitate a more secure, reliable, scalable and easily manageable Internet architecture. If well deployed, the *Internet of the future* can bring novelty, productivity gains, new markets and growth.

In fact, innovative functionalities with more enhanced performance levels are necessary to sustain the real-time requirements of a multitude of novel applications. Furthermore, the Internet underpins the whole global economy. The diversity and sheer number of applications and business models supported by the Internet have also largely affected its nature and structure ([3], [4]).

The *Future Internet (FI)* will not be "more of the same", but rather "appropriate entities" incorporating new technologies on a large scale that can unleash novel classes of applications and related business models [5]. If today's Internet is a crucial element of our economy, FI will play an even more vital role in every conceivable business process. It will become the productivity tool "*par excellence*". At present, there are many so called "*Future Internet*" initiatives around the world working on defining and implementing a new architecture for the Internet intended to overcome existing limitations mostly in the area of networking ([6], [7]). The complexity of the FI, bringing together large communities of stakeholders and expertise, requires a structured mechanism to avoid fragmentation of efforts and to identify goals of common interest. Appropriate action is therefore invaluable to pull together the different initiatives, in order to provide more potential options and/or opportunities for the market players involved. Europe remains an international force in advanced information and communication technologies (ICT) and has massively adopted broadband and Internet services [8]. The European Union (EU) is actually a potential leader in the FI sector [9]. Leveraging FI technologies through their use in "smart infrastructures" offer the opportunity to boost European competitiveness in emerging technologies and systems, and will make it possible to measure, monitor and process huge volumes of information. This can also give the means to "overcome" fragmentation and to construct a related critical mass at European level, while fostering competition, openness and standardisation, involving consumer/citizen, ensuring trust, security and data protection with transparent and democratic governance and control of offered services as guiding principles ([10], [11]).

1.1   Autonomicity and Self-Management Features in Modern Network Design

The face of the Internet is continually changing, as new services appear and become globally noteworthy, while market actors are adapting to these challenges through suitable business models [12]. The current Internet has been founded on a basic architectural premise, that is: *a simple network service can be used as a "universal means" to interconnect intelligent end systems* [13]. Thus, it is centred on the network layer being capable of dynamically selecting a path from the originating source of a packet to its ultimate destination, with no guarantees of packet delivery or traffic characteristics. The continuation of simplicity in the network has pushed complexity into the end-points, thus allowing Internet to reach an impressive scale in terms of interconnected devices. However, while the scale has not yet reached its limits, the growth of functionality and the growth of size have both slowed down. It is now a common belief that current Internet is reaching both its architectural capability and its capacity limits (i.e.: addressing, reachability, new demands on quality of service (QoS), service/application provisioning, etc.). The next generation network architecture will be flexible enough to support a range of application visions in a dynamic way, ensuring convergence between technology, business and regulatory concerns. Enhanced communication services will open many possibilities for innovative applications that are not even envisioned today. Challenges for the *Network of the Future* may refer to a great variety of factors, including but not limited to: Dependability and security; scalability; services (i.e.: cost, service-driven configuration, simplified services composition over heterogeneous networks, large scale and dynamic multi-service coexistence, exposable service offerings/catalogues); monitoring; Service Level Agreements (SLAs) and protocol support for bandwidth (dynamic resource allocation), latency and QoS; automation (e.g. automated negotiation), and; the option for *autonomicity*. The resolution of these challenges would bring benefits to network and to service-application providers, in terms of: Simplified contracting of new business; establishing/identifying reference points for resource allocation and re-allocation; enabling flexibility in the provisioning and utilization of resources; offering the ability to scale horizontally, and; providing a natural complement to the virtualization of resources - *by setting up and tearing down composed services, based on negotiated SLAs*. This also involves benefits for service providers/consumers, in terms of: Ready identification-selection of offerings; potential to automate the negotiation of SLA Key Performance Indicators (KPIs) and pricing; reduced cost and time-to-market for services; scalability of composed services, and; flexibility and independence from the underlying network details.

In addition, a current trend for networks is that they are becoming service-aware. Service awareness itself has many aspects, including the delivery of content and service logic, fulfilment of business and other service characteristics such as QoS and SLAs and the optimization of the network resources during the service delivery. Thus, the design of networks and services is moving forward to include higher levels of automation, autonomicity, *including self-management*. Conversely, services themselves are becoming network-aware. Networking-awareness means that services are executed and managed within network execution environments and that both services and network resources can be managed uniformly in an integrated way. It is commonly acknowledged that the FI should have a considerably enhanced *network manageability capability*, and be an inseparable part of the network itself. Manageability of the current network typically resides in client stations and servers, which interact with network elements (NEs) via protocols such as SNMP (Simple Network Management Protocol). The limitations of this approach are reduced scaling properties to large networks and the need for extensive human supervision and intervention. A new network manageability paradigm is thus needed that allows NEs to be autonomously interrelated and controlled; adapts dynamically to changing environments, and; learns the desired behaviour over time. The effective design of monitoring protocols so as to support detection mechanisms critical for the elaboration of *self-organizing networks* has to be based on a clear understanding of engineering "trade-offs" with respect to local vs. non-local and aggregated information, *for instance*. In fact, several issues identified in current network infrastructures impose the need for the introduction of an innovative architectural design. Furthermore, the diversity of services as well as the underlying hardware and software resources comprise management issues highly challenging, meaning that currently, a diversity in terms of hardware resources leads to a diversity of management tools (distinguished per vendor). In addition, security risks currently present in network environments request for immediate attention. This could be achieved by building trustworthy network environments to assure security levels and manage threats in interoperable frameworks for autonomous monitoring.

1.2   The Vision of a Modern Self-Managing Network

The future vision is that of a *self-managing network* whose nodes/devices are designed in such a way that all the so-called traditional network management functions, defined by the *"FCAPS"* management framework (Fault, Configuration, Accounting, Performance and Security) [14], as well as the fundamental network functions such as routing, forwarding, monitoring, discovery, fault-detection and fault-removal, are made to automatically "feed" each other with information such as goals and events, to effect feedback processes among the different functions. Such processes allow reactions of various functions in the network (also including its individual nodes/devices), to achieve and maintain well-defined network goals [15].

Self-management capabilities may relate to a great variety of significant issues, such as: (i) Cross-domain management functions, for networks, services, content, together with the design of cooperative systems providing integrated management functionality of system lifecycle, self-functionality, SLA and QoS; (ii) Embedded management functionality in all FI systems (such as: in-infrastructure/in-network/in-service and in-content management); (iii) Mechanisms for dynamic deployment of new management functionality without interruption of actually running systems; (iv) Mechanisms for dynamic deployment of measuring and monitoring probes for services'/network's behaviour, *including traffic*; (v) Mechanisms for conflict and integrity-issues detection/resolution across multiple self-management functions; (vi) Mechanisms, tools and methodology construction for the verification and assurance of diverse self-capabilities that are *"guiding systems"* and their adaptations, correctly; these can also relate to mechanisms for allocation & negotiation of different available resources; (vii) Increased level of self –awareness/-knowledge/-assessment and self-management capabilities for FI resources; (viii) Increased level of self-adaptation and self-composition of resources to achieve autonomic and controllable behaviour; (ix) Increased level of resource management, *including discovery, deployment utilization, configuration, control and maintenance*; (x) Self-awareness capabilities to support objectives of minimizing system life-cycle costs and energy footprints; (xi) Orchestration and/or integration of management functions, and; (xii) Capabilities for the control relationships between self-management and self -governance of the FI.

In such an evolving environment, *it is required the network itself to help detect, diagnose and repair failures, as well as to constantly adapt its configuration and optimize its performance.* Looking at *Autonomicity and Self-Manageability*, the former (i.e. control-loops and feed-back mechanisms/processes, as well as the information/knowledge flow used to drive control-loops), becomes an enabler for network self-manageability [16]. Furthermore, new wireless sensor network technologies provide options for inclusion of additional intelligence and the capability, for the network elements and/or domains to *"sense, reason and actuate"*. Suitable systems with communication and computational capabilities can be integrated into the fabric of the Internet, providing an accurate reflection of the real world, delivering fine-grained information and enabling almost real-time interaction between the virtual world and real world. In particular, autonomous self-organizing systems are beginning to emerge and to be widely established [17]. Such systems *"can adapt autonomously"* to changing requirements and reduce the reliance on centrally planned services, *especially if they are effectively joined with new network management techniques.* Operators may use these tools to guarantee QoS service in a period of exploding demand and rising network congestion at peak times. The trend in building dependable real-life systems and smart infrastructures today is *"to move from monolithic, centralized and strictly hierarchical systems to highly distributed networked systems with local and global autonomy"*. When they are deployed in complex processes, these systems exhibit promising features and capabilities such as modularity and scalability, low cost, robustness and adaptability. Some of the challenges for operators/service providers include management (especially in self-organized wireless environments), resilience and robustness, automated re-allocation of resources, operations' abstractions in the underlying infrastructure, QoS guarantees for bundled services and optimization of operational expenditures (OPEX).

Ubiquitous and self-organizing systems are not only disruptive technologies that impact the way how market actors organize core processes as well as existing structures in value chains and industry, but have also considerable impact. The present Internet model is based on clear separation of concerns between protocol layers, with intelligence moved to the edges, and with the existent protocol pool targeting user and control plane operations with less emphasis on management tasks [18]. The area of FI is considered as a representative example of a *"complex adaptive organization"* (or *"entity"*), where the involved partners have diverse goals and tension to maximize their gains. There is a need for new ways to organize, control and structure communication systems, according to new management schemes and networking techniques without neglecting the advantages of current Internet. Among the core drivers for the FI are increased reliability, enhanced services, more flexibility, and simplified operation. The latter calls for including *Network Management (NM)* issues into the design process for FI principles. In general, NM is a service (or application) that employs a diversity of tools, applications, and devices to assist human network managers in monitoring and maintaining networks. Thus, NM should be an integral part of the future network infrastructure. Management is a key factor in manageability, usability, performance, etc., and is an important factor to the operational costs of any "network entity". FI requires a new management approach, promoted mainly by the necessity of support interoperability between heterogeneous, complex and distributed systems, while it should remain open for further and continuous improvement without the necessity of another disruptive modification in the future. Furthermore, as NM is important for the reliable and safe operation of networks, it is also crucial for the success of the FI. In the scope of these challenges, the *Self-NET Project* (*https://www.ict-selfnet.eu/*) aims to integrate the self-management and cognition features and the inevitable part of FI evolution.

2   Network Management Activities in the Self-NET Scope

The *Self-NET Project* designs, develops and validates an innovative paradigm for cognitive self-managed elements of the FI. Self-NET engineers the FI, based on cognitive behaviour with a high degree of autonomy [19] by proposing and examining the operation of self-managed FI elements around a novel "feedback-control cycle" (i.e. the "Monitoring/Decision-Making/Execution" or "MDE" cycle) as shown in Fig.1. Thus, dynamic distribution of resources according to network needs at specific time intervals can be pursued by introducing the "MDE" cycle to overcome bottlenecks and ensure seamless service provisioning – *even in case of services with high bandwidth requirements.* The completion of the aforementioned objective can make certain better QoS, *beyond the original best-effort status,* and simultaneously eases operational and network management functionalities.

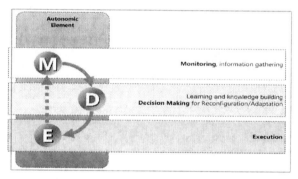

Fig. 1. The Distributed Cognitive Cycle for Systems and Network Management (DC-SNM)

Cognitive management in FI elements introduces innovative techniques regarding converged infrastructures with ultra-high capacity access networks and converged service capability across heterogeneous environments. Besides, the introduction of cognition in networks can contribute towards overcoming structural limitations of current infrastructures -*which render it difficult to cope with a wide variety of networked applications, business models, edge devices and infrastructures*- so as to guarantee higher levels of scalability, mobility, flexibility, security, reliability and robustness. Self-NET principle design is based on high autonomy of NEs in order to allow distributed management, fast decisions, and continuous local optimization of existing networks or of specific network parts [20]. The three distinct phases of the Generic Cognitive Cycle Model-GCCM (i.e.: the MDE cycle) are as shown in Fig. 2.

Cognitive capabilities can enable the perception of the NEs environment and the decision upon the necessary action (e.g. configuration, healing, protection measures, etc.). As current management tasks are becoming overwhelming, Self-NET embeds new management capabilities into NEs to take advantage of the increasing knowledge that characterizes the daily operation of FI users [21]. Among the main Self-NET's efforts is *"to tackle complexity"* by following the well-known "divide and conquer" approach, that is by: "Breaking down the overall network management task into smaller manageable tasks" and assigning them to individual NEs; showing NEs how to tackle the relevant issues; giving NEs the ability to "learn" in order to solve new, emerging (and occasionally "unforeseen") problems; facilitating NEs to cooperatively solve problems that require a sort of coordination, and; enhancing FI with inherent management capabilities *(i.e. "making FI self-manageable")*. NEs with cognitive capabilities aim at fast localised decision-making and (re-)configuration actions, as well as learning capabilities that improve elements behaviour. An essential target of the Project effort is to develop innovative cross-layer design optimization approaches that alleviate the shortcomings and duplication of functionalities in different protocol layers of the present IP stack. Furthermore, Self-NET also provides a peer-to-peer style distribution of responsibilities among self-governed FI elements, therefore over-coming the barrier of current client-server and proxy-based models in the operation of mobility management, broadcast-multicast, and QoS mechanisms. A "key-objective" is the provision of a holistic architectural & validation framework that unifies net-working operations and service facilities [22]. FI design is required to provide an-swers to a number of current Internet's deficits, especially when the danger of in-creased complexity is more than evident. *Self-management and autonomic capabili-ties* can so alleviate this "drawback" by: providing inherent management capabilities; increasing flexibility, and; allowing an ever-evolving Internet. Towards realizing this aim, Self-NET considers that a DC-SNM along with a hierarchical distribution over the network can "map" self-management capabilities over FI architectures [23]. DC-SNM further facilitates the promotion of distributed-decentralized management over a hierarchical distribution of management and (re-) configuration making levels: (i) to (autonomic) NEs; (ii) to network domain types, and; (iii) up to the service provider realm, *hence allowing high autonomy of NEs with cognitive capabilities aimed at fast localised (re-)configuration actions and decision-making.* This brings about the in-triguing issue of orchestrating the cognitive cycles (MDE) at higher levels of the self-management distribution.

The *"decomposition"* of NM into responsibility areas (as shown in Fig.2) can pro-vide the principle on which universal management architecture can be developed, having as a main goal the efficient handling of complexity towards FI environments. This, combined with the introduction of cognitive functionalities at all layers, can allow decisions/configurations at shorter time-scales [24], where each element has embedded cognitive cycle functionalities and also the ability to manage itself and make appropriate local decisions. For an efficient and scalable NM, where various actors may participate, a distributed approach is thus adopted. Dynamic network (re-) configuration in many cases is based on cooperative decision of various FI elements and distributed NM service components. Hints and requests/recommendations are exchanged among the related layers, in order to *"identify"* a new situation-action for a "targeted" execution. The automated (dynamic) incorporation of various layers requirements into the management aspects also provides novel features to NM [25].

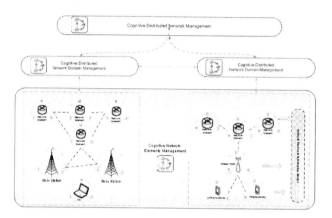

Fig. 2. The Distributed Cognitive Cycle for Systems and Network Management (DC-SNM).

In the context of the Self-NET Project, the introduction of a hierarchical cognitive cycle to enable multi-tier self-management in various NEs and dynamic network compartments provides a quite promising approach to alleviate management overhead, ensure dynamic adaptation to service requirements, situation aware NM and reconfiguration, while coping with the fragmentation of contemporary centralised NM, dedicated to specific types of networks ([26], [27]). NM is a wide area, including device monitoring, service levels and application management, security, ongoing maintenance, troubleshooting, planning, and other tasks – *ideally all coordinated and supervised by an experienced and reliable "entity"* (known as the *"network administrator"*). For businesses of all sizes, it is imperative to consider a NM solution *that is easy to use, quick to deploy, and offers low total cost of ownership*. Adoption of appropriate cognitive techniques on different platforms (or on parts of them) can be the "kick-off" that will encourage the creation of new networking infrastructures. Furthermore, it is essential to perform NM activities in a distributed way by incorporating self-organization and self-management principles [28]. Although there is a diversity of external and influencing available definition on self-management related work [29], the term *"self-management"* is applied here as *"the general term describing all autonomic and cognition-based operations in a system"*. Six distinct methods are identified with specific realizations and purposes; they all serve to demonstrate concepts inherent in the system properties ([19], [22]).

3   Challenges and Benefits for the Market Sector

The implementation-inclusion of suitable cognitive techniques/systems on diverse platforms can be the *"first step"* to support development of new networking infrastructures [30]. The introduced -*by the Self-NET-* functionalities can implicate major benefits for all relevant "actors" (i.e. for both operators and users), as follows:

Automatic network planning and reduction of management time of complex network parameters (and/or structures): Both present and future anticipated high proliferation of different services that a communications network should offer and support; this imposes a decisive challenge for any network operator involved, while implicating an appropriate *"adjustment"* of network performance together with *"optimization"* of the network resources usage. Daily (human) network manager activities include many tedious and time-consuming tasks, to make certain that the network delivers the desired services to its users. In many cases, the network operator is obliged to search through vast amounts of monitoring data to find any *"inconveniences"* to his network behaviour and to ensure a proper services' delivery. Embedding self-management functionalities in future NEs and establishing cognition at the diverse network levels (e.g., NEs, network compartments and domains) can automate the detection of any abnormal (or "adverse") behaviour, the remoteness of the relevant source(s), the diagnosis of the corresponding fault(s) and the expected repair of the conceived problematic situation. For a variety of reasons affecting the competitive presence of an operator in the market sector, it is a matter of high importance for the network to be able to "predict" irregular events (like faults or intrusions) and so to react, *accordingly*, in due time. Thus, applying self-aware techniques in a modern network environment can ease network composition and network planning procedures and can ensure the automatic adaptation of networks/services to capabilities of the network components.

Options for reduction of network operational cost: Any infrastructure that can perform automated operational tasks to optimize its network efficiency and the quality of service(s) offered, can contribute to the objective of reducing actual network operational expenditures (OPEX). The option for automating several procedures can be remarkably beneficial to network operators as it facilitates various complex (and resource-consuming) processes, currently deployed at a large time-scale and requiring significant human intervention. This also allows for a more inexpensive and simpler network deployment: That is, by applying self-management techniques intending to optimize the network in terms of coverage, capacity, performance etc., operators can decrease their operational expenditures by limiting the manual effort required for network operation and can actively utilize their NEs (or resources) more efficiently. Such techniques can also simplify network maintenance and fault management.

Options for easy "network adaptation" (e.g., in new traffic models and schemes): Traffic management of a communications network is mainly based on integrated and centrally coordinated deployment of specific measures and suitable rules, in response to the current network operating state and/or in anticipation of future needs and relevant conditions. Traffic management configuration of large wireless networks consisting of multiple, distributed NEs of varying technologies, is challenging, time-consuming, prone to possible errors and requires highly expensive control & management equipment from any market actor. Even when it is originally deployed, it involves continuous upgrading/modifications to provide a consistent and a transparent service environment, to sustain high QoS, to recover from faults and to maximize the overall network performance, especially when congestion phenomena appear.

Seamless users' experience in dynamic network selection: In competitive markets, end-users wish to have access to a network offering adequate coverage and services of high quality, on a real-time basis. Self-management can offer decentralized monitoring and proper decision-making techniques so that appropriate optimization hints can be extracted, in terms of determining the optimum course of actions to improve network performance and stability and to guarantee service continuity.

Enhanced service provision & adaptability: Dynamic detection of operational deficiencies and/or poor QoS delivered to the end-user, both imply for specific remediate actions to compensate for the related problematic situation(s). Improving the overall network quality also increases subscribers' satisfaction & trust. Thus, the optimization of all procedures in order to "minimize" (or occasionally to *"delete"*) service failures and to ensure continuity of service delivery is a critical matter for the user and the operator, in a competitive market. Besides, it is quite important for the entire network to incorporate options and other NM facilities to fulfil any requirement for novel service features, such as network (or service) reconfiguration capabilities, broadband management and support of an increased set of services/facilities offered.

Enabling effective networking under highly demanding conditions: A continuous and dynamically updated NM (proactively and reactively adapted to the network dynamics) is an appropriate tool for such purpose. That is, instead of using manual techniques, a fully automated, transparent and intelligent traffic management functionality can be much more beneficial. The Self-NET infrastructure can be used to provide an efficient real-time traffic management in a large network, thus maximizing network performance and radically decreasing human intervention. Several among the application areas can cover cases of traffic congestion, network attachments, link failures, performance degradation, mobility issues, multi-service delivery enhancements and involve intelligent autonomic congestion management and traffic routing, dynamic bandwidth allocation & dynamic spectrum re-allocation [22].

The continuity of service availability influences directly the technical approach of service realization and is an important parameter affecting network planning; indeed, the network should possess fitting techniques to *"adapt itself"* to an essential (occasionally prescribed) functional state. To this aim, the network should be able to gather information about various entities (elements, domains, sectors) and/or distinct modules, to detect their operational state(s) and to react to any deviations from the proposed "desired" state. Applying self-aware mechanisms can conduct to network performance optimization in terms of coverage and capacity, optimization of QoS delivered to the end-user and reduction of human intervention [31]. This option can contribute to guarantee some critical features including, but not limited to: (i) High availability & seamless services' continuity; (ii) Connectivity *anywhere and anytime*; (iii) Robustness and stability/steadiness of the underlying network; (iv) Scalability in terms of features-functions; (v) Balance between cost network-related benefits (OPEX reduction and optimized network functionalities), and; (vi) Heterogeneity support.

4   Experimental Results for Network Coverage and Optimization

In current practice, wireless network planning is a difficult and challenging task, involving expert knowledge and profound understanding of the factors affecting the performance of a wireless system. Several monitoring parameters should be taken into account for optimal coverage and capacity formation, while diverse configuration actions can be available that in many cases are interrelated, *as regards the consequences*. In established approaches, frequency planning is conducted as part of the deployment procedure for full network segments (or domains). The assignment of operating frequencies and/or channels to NEs is also a part of the broader frequency planning procedure. To eliminate conflicts in frequency assignment, the process is centrally coordinated in one that assumes and requires global knowledge and control over the concerned network segment(s)/domain(s). In this context, the latter option implies that the administrative entities are fully aware of the channel assigned to each individual NE and are fully capable of "adjusting" such assignments to their liking in a centrally coordinated manner. As a result, conflicts may be avoided or, *at least*, minimized, when a central entity coordinates and manages the entire procedure.

During the Self-NET Project effort, an extended experimental work has also been performed upon several specific use cases that have all been selected as appropriate "drivers-enablers" for testing and validation activities. A characteristic use case, particularly studied, was relevant to the challenge for achieving "*coverage and capacity optimization*", for the underlying network. In fact, management systems of modern FI networks incorporate autonomic capabilities to effectively deal with the increasing complexity of communication networks, to reduce human intervention, and to promote localized resource management. The Self-NET framework is based on the Generic Cognitive Cycle, which consists of the "M-D-E" phases. The Network Element Cognitive Manager (NECM) implements the MDE cycle at the NE level, whilst the Network Domain Cognitive Manager (NDCM) manages a set of NECMs, thus implementing sophisticated MDE cycle features. In order to test the key functionalities of the proposed solution, specific NM problems have been taken into account, *under the wider scope of wireless networks coverage and capacity optimization family* [32]. In the proposed test-bed, a heterogeneous wireless network environment has been deployed, consisting of several IEEE 802.11 Soekris access points (AP) [33] and an IEEE 802.16 Base Station (BS) [34], each embedding a NECM. Moreover, several single radio access terminals-RATs (i.e. Wi-Fi) and multi-RATs (i.e. WiFi, WiMAX) were located in the corresponding area, consuming a video service delivered by VLC (video LAN client) -based service provider [35]. For the management of the NECMs, a NDCM has been deployed. The cognitive network manager installed per NE has undertaken several distinct actions, that is: (i) The deductions about its operational status; (ii) the proactive preparation of solutions to face possible problems, and; (iii) the fast reaction to any problem by enforcing the anticipated reconfiguration actions. Interaction of NECMs and NDCM enabled the localized and distributed orchestration of various NEs. In the experimentation phase we focused on the (re-)assignment of operating frequencies to wireless NEs and the vertical assisted handover of multi-

RATs. The demonstration scenario has been divided into: (i) The optimal deployment of a new WiFi AP; (ii) the self-optimization of the network topology through the assisted vertical handover of terminals from loaded to neighbouring -less loaded- APs or BS(s), and; (iii) the self-optimization of the network topology due to high interference situation. The MDE cycle was instantiated in both the NECM and the NDCM. The NECM periodically monitored its internal state and local environment by measuring specific parameters, thus building its local view. All NECMs have periodically transmitted the collected information to the NDCM in order to enable, the latter, to "build" the second level of situation awareness (SA) and have the domain level view. The topology used and the allocation of network devices in a realistic office environment (at OTE's R&D premises), have both been considered as shown in Fig.3. The topology has been selected in order to be "characteristic" and to depict conditions that are common to corporate environments and, especially to those that can occasionally host numerous nomadic end-users.

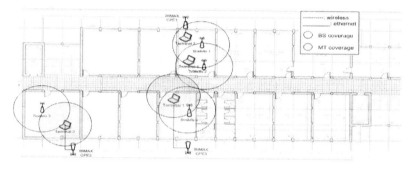

Fig. 3. Network Topology of the proposed Use Case for Coverage and Capacity Optimization.

Fig. 4 illustrates the total duration of the channel selection that takes place with the activation of an AP. It is shown that Soekris 1 and Soekris 4 need more time for channel selection. The most time consuming processes are *Execution* and *Communication*. The communication phase is responsible for Soekris 1 and Soekris 4 high delay. Specifically, both Soekris interact (i.e. communicate) with Soekris 2. The duration of *Execution phase* is high and the same for all devices due to technical and implementation reasons. Moreover, the *"decision-making" phase* (i.e. Channel Selection Objective Function) is too low for all NECMs, while the monitoring phase takes between 2.55-3.15 seconds. The communication between NECMs increases the duration of the communication phase, while the duration of the execution phase is increased, due to technical and implementation reasons.

Fig. 5 presents the duration of the mobile terminal re-allocation function (vertical assisted handover). This problem solving process is selected if a high load status has been identified. Similarly to the previous cases (i.e. the channel (re-)selection) the *communication phase* takes again the majority of time.

The proposed test-bed has demonstrated that the inclusion of the MDE cognitive cycle can provide several major operational benefits, such as: (i) Automated installation of wireless access devices with avoidance of overlapping among them; this is

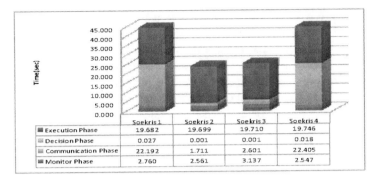

Fig. 4. Channel Selection Duration.

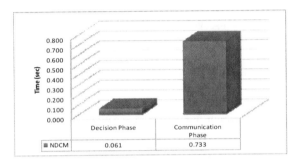

Fig. 4. Vertical Assisted Handover Duration.

done without any human intervention for channel selection. (ii) Automated channel (re-)selection which is made after consideration/evaluation of the existing interference in a specific location of the network. (iii) Automated optimization process for channel (re-)selection, according to specific parameters (i.e.: the number of end-users, the network traffic, the operational state of neighboring access devices, etc.); this can guarantee a more improved functionality for all network devices involved. (iv) Possibility for automated handover between end-users of heterogeneous wireless technologies (WiFi, WiMAX), especially in cases where there is "extreme" network traffic and/or overload, affecting the network functionality.

5   Conclusion

Evolution towards FI requests a more flexible architecture that will act as the "basis" for the disposal of a multiplicity of services-facilities with optimized quality levels, intending to attract/satisfy end-users. Such network infrastructures are characterized by the inclusion of embedded intelligence *"per element"* or *"per domain"*, targeting at a more distributed environment both in terms of management and operational activities. To this aim, cognitive networks with self-aware functionalities introduce a high level of autonomy, meaning that embedded and/or inherent management functionality in several components of FI systems composes management upon a *"per*

NE" and/or a "*per domain*" mechanism, rather than a centralized (traditional) network functionality. Compared to current network features, self-management techniques pave the way towards automated network processes such as the deployment of new NEs, the network reconfiguration (*in whole or in part*) and the selection/execution of the optimal corresponding solution (or "response") based on specific circumstances & remediation of identified malfunctions with the minimum potential service interruption. Consequently, new methods (related to embedded and/or autonomous management, virtualization of systems and network resources, advanced and cognitive networking of information objects), have to "re-define" the overall FI network architecture. To "encounter" such critical challenges, the main objective of Self-NET project effort is to describe and evaluate/analyze new paradigms for the management of complex and heterogeneous network infrastructures-systems (such as cellular, wireless, fixed and IP networks), taking into consideration the next generation Internet environment and the convergence perspective. This can efficiently integrate new operational capabilities in the "*underlying system*" by introducing innovative self-management attributes, resulting in noteworthy benefits for all actors involved. The Self-NET initiative develops self-management features that alleviate consequences of events for which the system would require various invocations of remedy actions and/or significant human intervention. This dynamic behavior and intelligence of handling various events (and/or situations) can potentially lead to an innovative and much promising beneficiary scope of the entire system's operations.

Acknowledgments. The present work has been composed n the context of the *Self-NET* ("*Self-Management of Cognitive Future Internet Elements*") European Research Project and has been supported by the Commission of the European Communities, in the scope of the 7th *Framework Programme ICT-2008, Grant Agreement No.224344.*

References

1. Commission of the European Communities: Communication on "A Public-Private Partnership on the Future Internet". European Commission, Brussels (2009)
2. Chochliouros, I.P., Spiliopoulou, A.S.: Broadband Access in the European Union: An Enabler for Technical progress, Business Renewal and Social Development. The International Journal of Infonomics (IJI) 1, 5–21 (2005)
3. Timmers, P.: Business Models for Electronic Markets. The International Journal on Electronic Markets and Business Media 8(2), 3–8 (1998)
4. Future Internet Assembly (FIA): Position Paper: Real World Internet (2009),
   http://rwi.future-internet.eu/index.php/Position_Paper
5. Afuah, A., Tucci, C.L.: Internet Business Models and Strategies: Text and Cases. McGraw-Hill, New York (2000)
6. European Future Internet Portal (2010), http://www.future-internet.eu/

7. Blumenthal, M.S., Clark, D.D.: Rethinking the Design of the Internet: The End-to-End Arguments vs. the Brave New World. ACM Trans. on Internet Techn. 1(1), 70–109 (2001)
8. Commission of the European Communities: Communication on "The Future EU 2020 Strategy". European Commission, Brussels (2009)
9. Tselentis, G., Domingue, L., Galis, A., Gavras, A., et al.: Towards the Future Internet-A European Research Perspective. IOS Press, Amsterdam (2009)
10. Organization for Economic Co-operation Development (OECD): The Seoul Declaration for the Future of the Internet Economy. OECD, Paris, France (2008)
11. Chochliouros, I.P., Spiliopoulou, A.S.: Innovative Horizons for Europe: The New European Telecom Framework for the Development of Modern Electronic Networks and Services. The Journal of the Communications Network (TCN) 2(4), 53–62 (2003)
12. Commission of the European Communities: Communication on "Future Networks and the Internet". European Commission, Brussels (2008)
13. Galis, A., Brunner, M., Abramowitz, H.: MANA Position Paper - Management and Service-Aware Networking Architecture (MANA) for Future Internet / Draft 5.0 (2008)
14. International Telecommunication Union-Telecommunication Standardization Sector: Rec. M. 3400: TMN Management Functions. ITU-T, Geneva, Switzerland (2000)
15. Pastor-Satorras, R., Vespignani, A.: Evolution and Structure of the Internet: A Statistical Physics Approach. Cambridge University Press, Cambridge (2004)
16. Dobson, S., Denazis, S., Fernandez, A., et al.: A survey of autonomic communications. ACM Trans. on Autonomous and Adaptive Systems (TAAS) 1(2), 223–259 (2006)
17. Boccalettia, S., Latora, V., Moreno, Y., Chavez, M., Hwang, D.-U.: Complex networks: Structure and Dynamics. Elsevier Physics Reports 424, 175–308 (2006)
18. Clark, D., Sollins, K., Wroclawski, J., Katabi, D., et al.: New Arch: Future Generation Internet Architecture (Final Technical Report). The US Air Force Research Laboratory (2003)
19. Chochliouros, I.P., Spiliopoulou, A.S., Georgiadou, E., Belesioti, M., et al.: A Model for Autonomic Network Management in the Scope of the Future Internet. In: Proceedings of the 48th FITCE International Congress, FITCE, Prague, Czech Republic, pp. 102–106 (2009)
20. Kousaridas, A., Polychronopoulos, C., Alonistioti, N., et al.: Future Internet Elements: Cognition and Self-Management Design Issues. In: Proceedings of the 2nd International Conference on Autonomic Computing and Communication Systems, pp. 1–6 (2008)
21. Raptis, T., Polychronopoulos, C., et al.: Technological Enablers of Cognition in Self-Manageable Future Internet Elements. In: Proceedings of The First International Conference on Advanced Cognitive Technologies and Applications COGNITIVE 2009, pp. 499–504. IARIA (2009)
22. Mihailovic, A., Chochliouros, I.P., Kousaridas, A., Nguengang, G., et al.: Architectural Principles for Synergy of Self-Management and Future internet Evolutions. In: Proceedings of the ICT Mobile Summit 2009, pp. 1–8. IMC Ltd, Dublin (2009)
23. Self-NET Project: Deliverable D1.1: System Deployment Scenarios and Use Cases for Cognitive Management of Future Internet Elements (2008), https://www.ict-selfnet.eu/
24. Agoulmine, N., Balasubramaniam, S., Botvitch, D., Strassner, J., et al.: Challenges for Autonomic Network Management. In: Proceedings of the 1st IEEE International Workshop on Modelling Autonomic Communications Environments (2006)
25. Strassner, J.: Policy-Based Network Management. Morgan Kaufmann Publishers, San Francisco (2003)
26. Elliott, C., Heile, B.: Self-organizing, self-healing wireless networks. In: Proceedings of IEEE International Conference on Personal Wireless Communications, pp. 355–362 (2000)

27. Chochliouros, I.P., Alonistioti, N., Spiliopoulou, A.S., et al.: Self-Management in Future Internet Wireless Networks: Dynamic Resource Allocation and Traffic Routing for Multi-Service Provisioning. In: Proceedings of MOBILIGHT-2009, pp. 1–12. ICST (2009)
28. Self-NET Project: Deliverable D5.1: First Report on Business Opportunities (2009)
29. Miller, B.: The autonomic computing edge: Can you CHOP up autonomic computing? IBM Corporation (2008)
30. Prehofer, C., Bettstetter, C.: Self-Organization in Communication Networks: Principles and Design Paradigms. IEEE Communications Magazine 43(7), 78–85 (2005)
31. Mihailovic, A., Chochliouros, I.P., Georgiadou, E., Spiliopoulou, A.S., et al.: Situation Aware Mechanisms for Cognitive Networks. In: Proceedings of the International Conference on Ultra Modern Telecommunications (ICUMT-2009), pp. 1–6. IEEE Computer Society Press, Los Alamitos (2009)
32. Pragad, A.D., Friderikos, V., Pangalos, P., Aghvami, A.H.: The Impact of Mobility Agent based Micro-Mobility on the Capacity of Wireless Access Networks. In: Proceedings of IEEE GLOBECOM-2007 (2007)
33. Soekris Engineering net5501, http://www.soekris.com/net5501.htm
34. RedMAX, Redline Communications: AN-100U/UX Single Sector Wireless Access Base Station User Manual (2008)
35. C.: open-source multimedia framework, player and server, http://www.videolan.org/vlc

Efficient Opportunistic Network Creation in the Context of Future Internet

Andreas Georgakopoulos, Kostas Tsagkaris, Vera Stavroulaki, and
Panagiotis Demestichas

University of Piraeus, Department of Digital Systems,
80, Karaoli and Dimitriou Street, 18534 Piraeus, Greece
{andgeorg,ktsagk,veras,pdemest}@unipi.gr

Abstract. In the future internet era, mechanisms for extending the coverage of
the wireless access infrastructure and service provisioning to locations that cannot be served otherwise or for engineering traffic whenever the infrastructure
network is already congested will be required. Opportunistic networks are a
promising solution towards this direction. Opportunistic networks are dynamically created, managed and terminated. During the creation phase, nodes that
will constitute the opportunistic network needs, are selected and assigned with
the proper spectrum and routing patterns. Accordingly, this paper focuses on the
opportunistic network creation problem and particularly on the efficient selection of nodes to participate therein. A first step towards the formulation and solution of the opportunistic network creation problem is made, whereas indicative results are also presented in order to obtain some proof of concept for the
proposed solution.

Keywords: Opportunistic Networks, Node Selection, Coverage Extension, Capacity Extension, Future Internet.

1   Introduction

The emerging wireless world will be part of the Future Internet (FI). All kinds of
devices and networks will have the interconnection potential. Thus, any object or
network element will have communication capabilities embedded and several objects
in a certain environment will be able to create a communication network. Challenges
such as the infrastructure coverage extension or the infrastructure capacity extension,
arise.

Opportunistic networking seems a promising solution to the problem of coverage
extension of the infrastructure in order to provide service to nodes which normally
would be out of the infrastructure coverage or to provide infrastructure decongestion
by extending its capacity. In general, opportunistic networks (ONs) involve nodes and
terminals which engage occasional mobility and dynamically configured routing patterns. They can comprise numerous network elements of the infrastructure, and termi-
nals/ devices potentially organized in an infrastructure-less manner. It is assumed that

J. Domingue et al. (Eds.): Future Internet Assembly, LNCS 6656, pp. 293–306, 2011.

ONs are operator-governed, coordinated extensions of the infrastructure in order to assist the infrastructure and not to be used as an alternative to the infrastructure. Operator governance of ONs is achieved through policies provided by the operator network management. The network management provides the overall operational framework of the ON but it is out of the scope of this work.

Further on, ONs will exist temporarily, i.e., for the time frame necessary to support particular network services and accommodate new FI-enabled applications (requested in a specific location and time). A region could be served by more than one ONs that could co-exist, under the coordination of the operator. At the lower layers, the operator designates the spectrum that will be used for the communication of the nodes of the ON (i.e., the spectrum derives through coordination with the infrastructure). On the other hand, the network layer capitalizes on context, policy, profile, and knowledge awareness to optimize routing and service/ content delivery. Additionally, mechanisms for the efficient, dynamic creation of ONs are needed to be developed which should comply node selection according to specific, predefined characteristics.

To that respect, ONs can be seen as part of the Cognitive Control Network (CCN) as illustrated in Fig. 1 and is proposed by ETSI in [1]. The CCN involves an emerging group of functionalities aiming at introducing cognition mechanisms to the evolving wireless world. The spontaneous creation of ONs can comprise the outcome of advanced decision making provided by such cognitive mechanisms. Framed within this statement, this work discusses on the ON creation as a means to provide extended coverage to the infrastructure and/or deplete congested parts of it, and in particular, it introduces a node selection algorithm and evaluates its effectiveness by means of simulation.

The rest of the article is structured as follows: the second section discusses the related work in the area of node selection and coexistence of ONs with infrastructure elements. Third section discusses the high-level solution of the opportunistic networking concept and defines the phases through the lifecycle of the ON (i.e. ON suitability determination, ON creation, ON maintenance and ON termination). The fourth section focuses on the ON creation phase and provides the algorithmic solution of the opportunistic node selection problem statement with respect to indicative scenarios such as the opportunistic coverage extension or the opportunistic capacity extension. The fifth section provides simulation result sets in order to evaluate the proposed algorithm and strengthen the proof of concept. Finally, the article concludes with key findings and future work.

2   Related Work

Various approaches concerning node selection for wireless sensor or mesh networks have been already discussed. For example, random node selection in unstructured P2P networks is discussed in [2] while authors in [3] address the relay selection problem in cooperative multicast over wireless mesh networks. Also, in [4], analytical and simulation approaches are used in order to investigate the relationship between the lifetime of sensor networks and the number of reporting nodes and to provide the trade-off between maximizing network lifetime and the fastest way to report an event in a wireless sensor node.

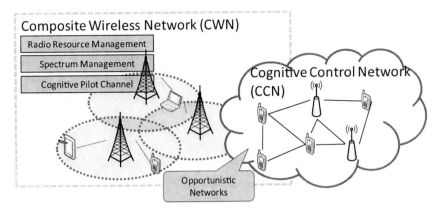

Fig. 1. The emerging cognitive wireless world

In [5], the selection and navigation of mobile sensor nodes is investigated by taking into consideration three metrics including coverage, power and distance of each node from a specified area. In [6], a grid-based approach for node selection in wireless sensor networks is analyzed, in order to select as few sensors as possible to cover all sample points. In [7], the issue of server selection is being investigated by proposing a node selection algorithm with respect to the worst-case link stress (WLS) criterion. These works are proposing specific sensor node selection algorithms by taking into consideration attributes such as the area of coverage, the navigation/ mobility issues of moving sensors, or the minimization of the number of relays.

Despite the fact that ONs may resemble mesh networking or ad hoc networking, certain differences do exist. For example, mesh networking is not used for the expansion of the coverage of the infrastructure, but for the wireless coverage of an area using various Radio Access Technologies (RATs) [8]. Hence, they are not operator-governed. Moreover, ad hoc networking uses peer nodes to form an infrastructure-less, self-organized network [9], but does not necessarily have the logic of specific node selection for the efficient network creation according to a pre-specified set of parameters. With the proposed ON approach, issues like these and limitations of ad hoc and mesh networking are trying to be addressed and resolved. ON comes with a bundle of benefits that could evolve in areas where infrastructure is difficult to exist or communication demand rises instantly and support is needed for successful handling.

Also, authors in [10] propose and implement a Virtual Network Service (VNS) as a value-added network service for the deployment of Virtual Private Networks (VPNs) in a managed wide area IP network while in [11] a survey regarding automatic configuration of VPNs takes place where auto-configuration mechanisms are discussed and compared. Virtual network provisioning across multiple substrate networks is studied in [12] where authors evaluate the resource matching, splitting, embedding and binding steps required for virtual network provisioning. However, Virtual Networks (VNs) such as VPNs intend to be used among public infrastructures to provide secure, remote access to users of e.g. an office network. Also, Virtual LANs (VLANs) which are another type of VNs are logical networks which are based on physical networks and can be used for grouping of hosts in the same domain regardless of their physical location. Compared to the proposed ON approach, it is clear that VNs may be temporarily established (e.g. a VPN which is established between home and office network for a certain amount of time) but they are not operator-governed nor are they coordinated extensions to places where infrastructure is not available.

To that respect, the contribution of this work is to propose a unified solution for ON creation which takes into consideration the dynamic nature of such networks and the application provisioning via the use of various kinds of nodes (e.g. cell phones, PDAs, laptops and other network-enabled devices). Thus, a fitness function is presented which is able to evaluate the eligibility of each candidate node and accept it or reject it from participating to the ON.

3   Solution Approach Based on ONs

The proposed ON approach is based on four discrete phases. These phases include the ON suitability determination, the ON creation, the ON maintenance and the ON termination. To that respect, each phase is previewed in the following subsections.

3.1   ON Suitability Determination

Specifically, the suitability determination phase is rather crucial, because based on the observed radio environment, the node capabilities, the network operator policies and the user profiles, the outcome of this phase will be to decide whether it is suitable to set-up an operator-governed ON or not, at a specific time and place. In order to evaluate the suitability, the detection of opportunities for ON establishment with respect to total nodes and potential radio paths should be taken into consideration as main inputs. As a result, the network operator needs to be aware (by discovery procedures) of the nodes' related information.

Each node is distinguished by a set of characteristics. Node characteristics will include the capabilities (including available interfaces, supported RATs, supported frequencies, support of multiple connections, relaying/bridging capabilities) and status of each candidate node in terms of resources for transmission (status of the active links), storage, processing and energy. Moreover, the operator needs to be aware of the location and the mobility level of each node. A prerequisite of each case (e.g. opportunistic coverage extension or opportunistic capacity extension) is that the nodes need to have some type of access to the infrastructure, or to have some type of access to a decongested Access Point (AP).

Furthermore, application requirements and the similarity level of the requested applications (i.e. common application interests) have to be taken into consideration by defining the involved applications, their resource requirements, and their appropriateness for being provided through ONs. Also, the potential gains from a possible ON creation should be considered in order to provide the main output of this phase which will be the request for ON creation.

3.2   ON Creation

The next phase of the ON lifecycle is the ON creation. The ON creation phase is responsible for providing mechanisms and decisions for effectively creating the ON based on the output received from the suitability determination phase. It focuses on selecting participant nodes according to the previously mentioned set of characteristics and on choosing optimal radio paths in order to ensure optimal QoS. Finally, it performs all the required procedures to effectively connect ON members with each other and to ensure continuity of service for the members with regard to the infrastructure. The main output of the ON creation phase will be the selected nodes and the selected links/interfaces/RATs/spectrum.

3.3   ON Maintenance

Once the ON is created, it will have to adapt dynamically during all its operational lifetime to changing environment conditions (e.g. context, operator's policies, user profiles). In order to achieve this, after the successful completion of the creation phase, the maintenance phase has to be initiated. Generally, the maintenance phase will have to monitor nodes, spectrum, operator's policies, QoS on a frequent basis and to decide whether it is suitable to proceed to a reconfiguration of the ON. Thus, the maintenance phase is responsible for the opportunistic network management and reconfiguration. Monitoring mechanisms during the operation of the ON would be introduced in order to accommodate any alterations that are made in the ON and act accordingly.

3.4   ON Termination

The termination phase will eventually take the decision to release the ON, thus triggering all the necessary procedures and associated signaling. It is distinguished according to the reason of termination. As a result, there may be termination of the ON due to cessation of application provision, termination due to inadequate gains from the usage of the ON and forced termination.

   Fig. 2 illustrates the previously mentioned phases and their possible interconnections. According to this, the suitability determination phase issues a creation request which triggers the ON creation. Once the ON is successfully created, frequent monitoring is needed in order to maintain the flawless and high-quality operation of the ON. If needed, reconfiguration procedures may take place by re-creating the ON. Moreover, the maintenance phase may issue a termination request in the cases where the ON experiences sudden and unsolved drop in QoS, or when the application provision has successfully ended, thus the ON is not needed anymore.

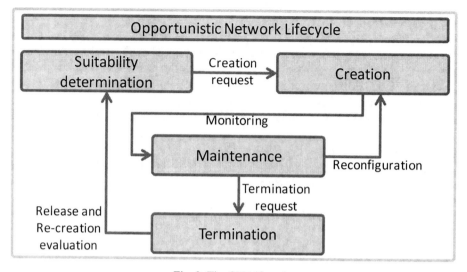

Fig. 2. The ON Lifecycle

4   Opportunistic Network Creation

Following the ON lifecycle overview solution, this work focuses on the part of the ON creation phase and more specifically to the selection of nodes which will form the ON. In order to be able to create the ON and enable service provisioning to end-users, it is needed to gain awareness of the status of candidate, relay nodes (i.e. nodes that can be used as routers, even when they do not need to use an application) and the application nodes (i.e. the nodes that use a specific application). Moreover, gateway nodes have to be defined in order to provide connectivity between the ON and the infrastructure (e.g. Macro Base Station -MBS), upon request.

To that respect, some indicative business scenarios have been identified in order to elaborate on the ON paradigm. Fig. 3 illustrates the opportunistic coverage extension scenario. According to this scenario, a node which acts as a traffic source like a laptop or a camera is out of the coverage of the infrastructure. As a result, a solution would comprise the creation of an ON in order to serve the out of infrastructure coverage node. Opportunism primarily lies in the selection for participation in the ON of the appropriate subset of nodes, among the candidate nodes that happen to be in the vicinity, based on profile and policy information of the operator and the use of spectrum that will be designated by the network operator, for the communication of the nodes of the ON. Through the opportunistic approach multiple benefits for key players derive such as, the end user gets access to the infrastructure in situations where it normally would not be possible, while the access provider may experience increased cashflow as more users are being supported.

Another indicative scenario would comprise the notion of the opportunistic capacity extension. According to this scenario, a specific area experiences traffic congestion issues and an ON is created in order to route the traffic to non-congested APs.

Access providers are benefited from the fact that more users can be supported since new incoming users that otherwise would be blocked can now be served, while end users experience improved QoS since congestion situations can be resolved as illustrated in Fig. 4.

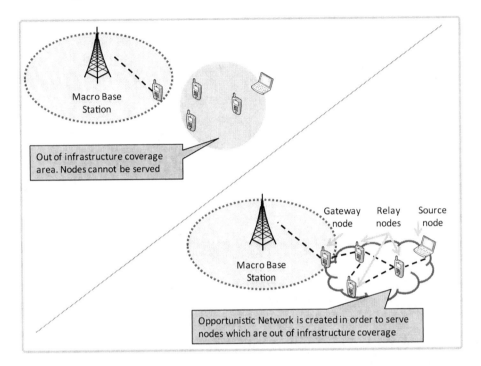

Fig. 3. Opportunistic coverage extension scenario. An ON is created in order to serve the out of infrastructure coverage nodes.

To that context, in order to gain awareness of the status of the candidate nodes in the vicinity, a monitoring mechanism is needed, that will be able of monitoring aspects that has to take into account, each node's related information in order to be able to calculate a fitness function. The monitored aspects are:

- Energy level of the node
- Availability level of the node –taking into account node's capabilities (including available interfaces, supported RATs, supported frequencies, support of multiple connections, relaying/ bridging capabilities, gateway capabilities-wherever applicable), status of each node in terms of resources for transmission (status of the active links), storage, processing, mobility levels –location, supported applications (according to node capabilities and application requirements).
- Delivery probability of the node

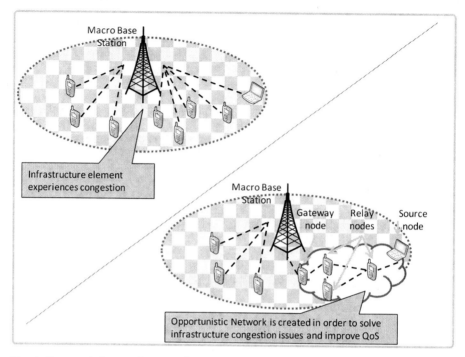

Fig. 4. Opportunistic capacity extension scenario. An ON is created in order to alleviate congestion in the infrastructure and improve QoS.

The aforementioned node characteristics provide the input to a fitness function which according to a specified threshold, decides on whether to accept or reject a candidate node for being part of the ON. The accepted nodes are eligible of forming the ON. Relation (1) below shows the fitness function considered in this paper for the node selection:

$$Fitness\ Function = x_i * [\ (\ e_i * w_e\ ) + (\ a_i * w_a\ ) + (\ d_i * w_d\ )\ ]\ . \tag{1}$$

where e_i denotes the energy level of node i, a_i denotes the availability level of node i at a specific moment and d_i denotes the delivery probability of packets of node i. Also x_i acts as multiplier according to Relation (2):

$$x_i = \begin{cases} 1, e_i > 0 \cap a_i > 0 \\ 0, e_i = 0 \cup a_i = 0 \end{cases} . \tag{2}$$

On the other hand, for the definition of weights of the fitness function, the Analytic Hierarchy Process (AHP) has been used. As the AHP theory suggests [13], initially a decision maker decides the importance of each metric as AHP takes into account the decision maker's preferences. For our example it is assumed that energy is a bit more important than availability and the delivery probability is less important from both energy and availability. According to these assumptions a matrix is completed. The matrix contains the three factors (i.e. energy, availability and delivery probability) along with their levels of importance. As a result, the weight for energy is w_e=0.53, the weight for availability is w_a=0.33 and the weight for delivery probability is w_d=0.14. Furthermore, the theory explains that in order to have consistent results an λ_{max} attribute must be greater than the absolute number of the proposed factors (i.e. 3 for our assumption). The derivation of the λ_{max} attribute is also explained in the AHP technique [13] and it is used for the checking of sanity of the weights. In our example the λ_{max} turns to be 3.05, hence the derived results are on safe ground. Finally, a Consistency Ratio is calculated. Saaty in [13] argues that a Ratio > 0.10 indicates that the judgments are at the limit of consistency though Ratios > 0.10 could be accepted sometimes. In our case, the Consistency Ratio is around 0.04, well below the 0.10 threshold, so our estimations tend to be trustworthy according to the AHP theory.

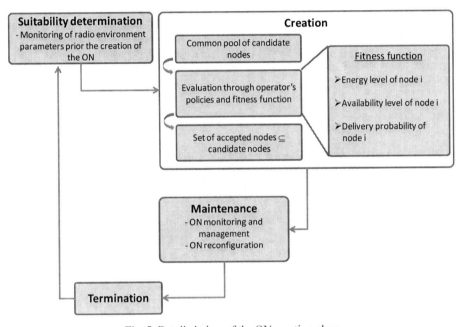

Fig. 5. Detailed view of the ON creation phase

Additionally, the detailed view of the ON creation phase is depicted in Fig. 5 where the responsibilities of the creation phase are visually explained. These responsibilities include the evaluation and selection of nodes and spectrum conditions which will provide the set of the accepted nodes to the ON. Also, the interconnections of the creation phase with the suitability determination and the maintenance are visible.

5   Results

In order to obtain some proof of concept for our network creation solution, a Java-based prototype has been developed which calculates the fitness function and informs the system on the accepted and rejected nodes. By using the derived number of accepted nodes, indicative results on the potential performance of these nodes when used in an ON have been also collected and analyzed using the Opportunistic Network Environment (ONE) [14], [15].

An indicative network topology of 60 total participant non-moving nodes is illustrated in Fig. 6. Each node features 2 interfaces, a Bluetooth (IEEE 802.15.1) [16] and a high-speed interface (e.g. IEEE 802.11 family [17]). According to the high-speed interface, each node has a transmission data rate of 15 Mbps. On the other hand, the Bluetooth interface has a transmission data rate of 1 Mbps but it is used for a rather short-range coverage (e.g. 10 meters). Also, every new message is created at a 30-second interval and has a variable size ranging from 500 to 1500 kilobytes, depending on the scenario. Messages are created only from specific 3 hosts which are acting as source nodes, and are transmitted to specific 3 nodes which are acting as destination nodes via all the other relay nodes. Moreover, the initial energy level of each node is not fixed for all nodes, but it may range from 20 to 90 percent of available battery.

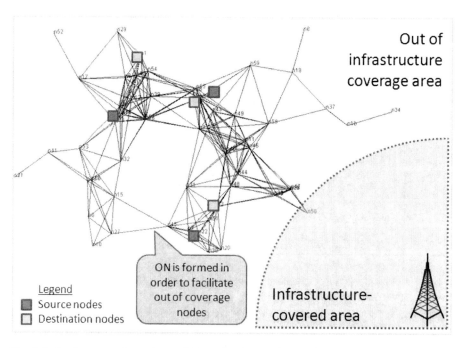

Fig. 6. Indicative network topology of 60 total nodes consisting of 3 source nodes, 3 destination nodes and 54 relay nodes

Fig. 7 shows results from the ONE simulator of the delivery probability for a variable value of nodes, ranging from 7 to 120 nodes. The routing protocol that was used for the specific simulation was the Spray & Wait protocol [18]. According to the observations, when the 500 kilobytes fixed message was used, higher delivery rates were observed for every number of accepted nodes. On the other hand, there is a tendency of significantly lower delivery rates as the message size increases to 1000 and 1500 kilobytes. Moreover, the results are compared with a simulation that ran for a variable message size ranging from 500 to 1000 kilobytes as illustrated in Fig. 8. So, it seems that as the message size is increased the delivery rates tend to decrease.

Fig. 9 compares the maximum lifetime of the ON with the expected delivery probability of the variable message size scenario (from 500 to 1000 kilobytes). The maximum lifetime of the ON corresponds to the last link that is expected to drop during the simulation. It is shown that as the ON expands (i.e. the number of accepted nodes according to the fitness function increases), the lifetime of the ON tends to reach a maximum level and then keep stable. On the other hand, as the ON expands, the observed delivery probability of messages tends to decrease.

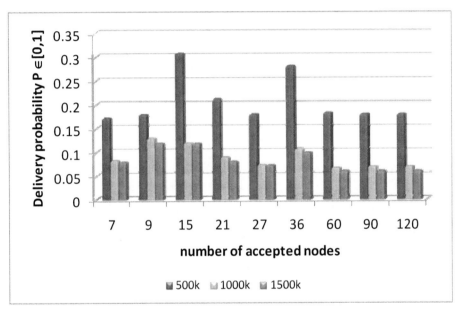

Fig. 7. Delivery probability rates for fixed message sizes ranging from 500 kilobytes to 1500 kilobytes

Also, Fig. 10 illustrates the average number of hops in the vertical axis and the number of nodes in the horizontal axis for 500, 1000, 1500 kilobytes of fixed message sizes and the variable message size ranging from 500 to 1000 kilobytes. In this chart, the tendency of increment of the average number of hops as the number of nodes increases is clear. The average number of hops corresponds to the summation of each number of hops observed by each accepted node divided by the actual number of accepted nodes.

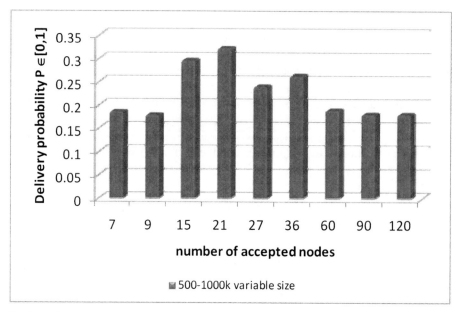

Fig. 8. Delivery probability rates for variable message sizes ranging from 500 kilobytes to 1000 kilobytes

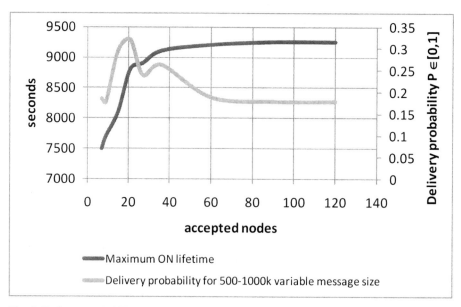

Fig. 9. Delivery probability rates compared to maximum network lifetime for a variable number of nodes ranging from 7 to 120 and a variable message size ranging from 500 kilobytes to 1000 kilobytes

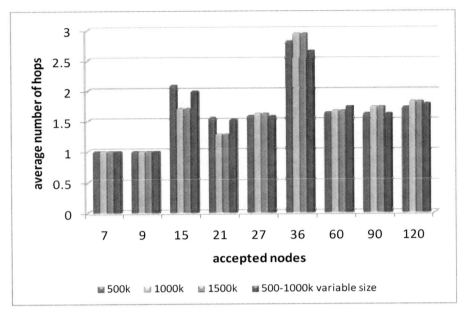

Fig. 10. Average number of hops for a variable number of nodes ranging from 7 to 120

6 Conclusion and Future Work

This work presents the efficient ON creation in the context of Future Internet. Operator-governed ONs are a promising solution for the coverage or capacity extension of the infrastructure by providing extra coverage or capacity wherever and whenever needed without the operator having to invest to expensive infrastructure equipment in order to serve temporary load surge in an area.

For the efficient creation of the ON, specific node attributes need to be taken into consideration in order to ensure flawless application streams. As a result, participant nodes are not chosen randomly but according to a set of evaluation criteria as proposed to this work. Nevertheless, according to the provided simulation result sets, it is shown that as the message size increases from 500 to 1500 kilobytes the delivery rates tend to decrease. This is also compared to the fact that the network lifetime according to the ONE simulator, is not increased from a specific number of nodes and above as it tends to reach the maximum level and stabilize after around 60 nodes. Finally, the average number of hops tends to increase as the number of nodes increases.

Future plans include the development of algorithmic approaches for the post-creation phases of the ON (i.e. the maintenance and the termination) in order to provide cognitive ON monitoring and reconfiguration mechanisms and handle successfully the normal or forced ON termination situations. ON management procedures are part of the post-creation phase (i.e. the maintenance) and as a result are a subject of future study.

Acknowledgment. This work is performed in the framework of the European-Union funded project OneFIT (www.ict-onefit.eu). The project is supported by the European Community's Seventh Framework Program (FP7). The views expressed in this document do not necessarily represent the views of the complete consortium. The Community is not liable for any use that may be made of the information contained herein.

References

1. European Telecommunications Standards Institute (ETSI), Reconfigurable Radio Systems (RRS), "Summary of feasibility studies and potential standardization topics", TR 102.838, V.1.1.1 (October 2009)
2. Vishnumurthy, V., Francis, P.: On Heterogeneous Overlay Construction and Random Node Selection in Unstructured P2P Networks. In: INFOCOM 2006, 25th IEEE International Conference on Computer Communications (2006)
3. Rong, B., Hafid, A.: A Distributed Relay Selection Algorithm for Cooperative Multicast in Wireless Mesh Networks. In: 5th International Conference on Mobile Ad-hoc and Sensor Networks, Fujian (2009)
4. Bouabdallah, F., Bouabdallah, N.: The tradeoff between maximizing the sensor network lifetime and the fastest way to report reliably an event using reporting nodes' selection. Computer Communications 31, 1763–1776 (2008)
5. Verma, A., Sawant, H., Tan, J.: Selection and navigation of mobile sensor nodes using a sensor network. Pervasive and Mobile Computing 2, 65–84 (2006)
6. Chen, H., Wu, H., Tzeng, N.: Grid-based Approach for Working Node Selection in Wireless Sensor Networks. In: IEEE International Conference on Communication (2004)
7. Han, S., Xia, Y.: Optimal node-selection algorithm for parallel download in overlay content-distribution networks. Computer Networks 53, 1480–1496 (2009)
8. Akyildiz, I., Wang, X., Wang, W.: Wireless mesh networks: a survey. Computer Networks 47, 445–487 (2005)
9. Akyildiz, I., Lee, W., Chowdhury, K.: CRAHNs: Cognitive radio ad hoc networks. Ad Hoc Networks 7, 810–836 (2009)
10. Lim, L.K., Gao, J., Eugene, T.S., Chandra, P., Steenkiste, P., Zhang, H.: Customizable virtual private network service with QoS. Computer Networks 36, 137–151 (2009)
11. Rossberg, M., Schaefer, G.: A survey on automatic configuration of virtual private networks. Computer Networks (2011)
12. Houidi, I., Louati, W., Ameur, W., Zeghlache, D.: Virtual network provisioning across multiple substrate networks. Computer Networks 55, 1011–1023 (2011)
13. Saaty, T.L.: The Analytic Hierarchy Process. McGraw-Hill, New York (1980)
14. Keranen, A., Ott, J., Karkkainen, T.: The ONE Simulator for DTN Protocol Evaluation. In: SIMUTools '09 2nd International Conference on Simulation Tools and Techniques, Rome (2009)
15. Keranen, A., Ott, J.: Increasing reality for dtn protocol simulations. Tech. Rep., Helsinki University of Technology, Networking Laboratory (2007), http://www.netlab.tkk.fi/tutkimus/dtn/theone/
16. IEEE 802.15 WPAN Task Group 1 (TG1), http://www.ieee802.org/15/pub/TG1.html
17. IEEE 802.11 Wireless Local Area Networks, http://ieee802.org/11/
18. Spyropoulos, T., Psounis, K., Raghavendra, C.: Spray and Wait: An Efficient Routing Scheme for Intermittently Connected Mobile Networks. In: ACM SIGCOMM Workshop on Delay-Tolerant Networking, WDTN (2005)

Bringing Optical Networks to the Cloud: An Architecture for a Sustainable Future Internet

Pascale Vicat-Blanc[1], Sergi Figuerola[2], Xiaomin Chen[4], Giada Landi[5],
Eduard Escalona[10], Chris Develder[3], Anna Tzanakaki[6], Yuri Demchenko[11],
Joan A. García Espín[2], Jordi Ferrer[2], Ester López[2], Sébastien Soudan[1],
Jens Buysse[3], Admela Jukan[4], Nicola Ciulli[5], Marc Brogle[7],
Luuk van Laarhoven[7], Bartosz Belter[8], Fabienne Anhalt[9], Reza Nejabati[10],
Dimitra Simeonidou[10], Canh Ngo[11], Cees de Laat[11], Matteo Biancani[12],
Michael Roth[13], Pasquale Donadio[14], Javier Jiménez[15],
Monika Antoniak-Lewandowska[16], and Ashwin Gumaste[17]

[1] LYaTiss
[2] Fundació i2CAT
[3] Interdisciplinary Institute for BroadBand Technology
[4] Technische Univeristät Carolo-Wilhelmina zu Braunschweig
[5] Nextworks
[6] Athens Information Technology
[7] SAP Research
[8] Poznan Supercomputing and Networking Center
[9] INRIA
[10] University of Essex
[11] Universiteit van Amsterdam
[12] Interoute
[13] ADVA
[14] Alcatel-Lucent
[15] Telefónica I+D
[16] Telekomunikacja Polska
[17] Indian Institute of Technology, Bombay

Abstract. Over the years, the Internet has become a central tool for society. The extent of its growth and usage raises critical issues associated with its design principles that need to be addressed before it reaches its limits. Many emerging applications have increasing requirements in terms of bandwidth, QoS and manageability. Moreover, applications such as Cloud computing and 3D-video streaming require optimization and combined provisioning of different infrastructure resources and services that include both network and IT resources. Demands become more and more sporadic and variable, making dynamic provisioning highly needed. As a huge energy consumer, the Internet also needs to be energy-conscious. Applications critical for society and business (e.g., health, finance) or for real-time communication demand a highly reliable, robust and secure Internet. Finally, the future Internet needs to support sustainable business models, in order to drive innovation, competition, and research. Combining optical network technology with Cloud technology is key to addressing the future Internet/Cloud challenges. In this con-

J. Domingue et al. (Eds.): Future Internet Assembly, LNCS 6656, pp. 307–320, 2011.

text, we propose an integrated approach: realizing the convergence of the IT- and optical-network-provisioning models will help bring revenues to all the actors involved in the value chain. Premium advanced network and IT managed services integrated with the vanilla Internet will ensure a sustainable future Internet/Cloud enabling demanding and ubiquitous applications to coexist.

Keywords: Future Internet, Virtualization, Dynamic Provisioning, Virtual Infrastructures, Convergence, IaaS, Optical Network, Cloud

1 Introduction

Over the years, the Internet has become a central tool for society. Its large adoption and strength originates from its architectural, technological and operational foundation: a layered architecture and an agreed-upon set of protocols for the sharing and transmission of data over practically any medium. The Internet's infrastructure is essentially an interconnection of several heterogeneous networks called Autonomous Systems that are interconnected with network equipment called gateways or routers. Routers are interconnected together through links, which in the core-network segment are mostly based on optical transmission technology, but also in the access segments gradual migration to optical technologies occurs. The current Internet has become an ubiquitous commodity to provide communication services to the ultimate consumers: enterprises or home/residential users. The Internet's architecture assumes that routers are stateless and the entire network is neutral. There is no control over the content and the network resources consumed by each user. It is assumed that users are well-behaving and have homogeneous requirements and consumption.

After having dramatically enhanced our interpersonal and business communications as well as general information exchange—thanks to emails, the web, VoIP, triple play service, etc.—the Internet is currently providing a rich environment for social networking and collaboration and for emerging Cloud-based applications such as Amazon's EC2, Azure, Google apps and others. The Cloud technologies are emerging as a new provisioning model [2]. Cloud stands for on-demand access to IT hardware or software resources over the Internet. Clouds are revolutionizing the IT world [11], but treat the Internet as always available, without constraints and absolutely reliable, which is yet to be achieved. Analysts predict that in 2020, more than 80 % of the IT will be outsourced within the Cloud [9]! With the increase in bandwidth-hungry applications, it is just a matter of time before the Internet's architecture reaches its limits.

The new Internet's architecture should propose solutions for QoS provisioning, management and control, enabling a highly flexible usage of the Internet resources to meet bursty demands. If the Internet's architecture is not redesigned, not only mission-critical or business applications in the Cloud will suffer, but even conventional Internet's users will be affected by the uncontrolled traffic or business activity over it.

Today, it is impossible to throw away what has made the enormous success of the Internet: the robustness brought by the datagram building block and the end-to-end principle which are of critical importance for all applications. In this context, we argue that the performance, control, security and manageability issues, considered as non-priority features in the 70s [3] should be addressed now [6]. In this chapter, we propose to improve the current Internet's architecture with the advanced control and management plane that should improve the integration of both new optical transport network technologies and new emerging services and application that require better control over the networking infrastructure and its QoS properties.

In this chapter we explore the combination of Cloud-based resource provisioning and the virtualization paradigm with dynamic network provisioning as a way towards such a sustainable future Internet. The proposed architecture for the future Internet will provide a basis for the convergence of networks—optical networks in particular—with the Clouds while respecting the basic operational principles of today's Internet. It is important to note that for several years, to serve the new generation of applications in the commercial and scientific sectors, telecom operators have considered methods for dynamic provisioning of high-capacity network-connectivity services tightly bundled with IT resources. The requirements for resource availability, QoS guarantee and energy efficiency mixed with the need for an ubiquitous, fair and highly available access to these capacities are the driving force of our new architecture. The rest of this chapter exposes the main challenges to be addressed by the future Internet's architecture, describing the solution proposed by the GEYSERS[18] project and how it is a step towards the deployment and exploitation of this architecture.

2   Challenges

There are various challenges that are driving today's Internet to the limit, which in turn have to be addressed by the future Internet's architecture. We consider the following six challenges as priorities:

1. **Enable ubiquitous access to huge bandwidth:** As of today, the users/ applications that require bandwidth beyond 1 Gbps are rather common, with a growing tendency towards applications requiring a 10 Gbps or even 100 Gbps connectivity. Examples include networked data storage, high-definition (HD) and ultra-HD multimedia-content distribution, large remote instrumentation applications, to name a few. But today, these applications cannot use the Internet because of the fair-sharing principle and the basic routing approach. As TCP, referred to as the one-size-fits-all protocol, has reached its limits in controlling—alone—the bandwidth, other mechanisms must be introduced to enable a flexible access to the huge available bandwidth.

---

[18] http://www.geysers.eu

2. **Coordinate IT and network service provisioning:** In order to dynamically provision external IT resources and gain full benefit of these thanks to Cloud technologies, it is important to have control over the quality of the network connections used, which is a challenge in today's best-effort Internet. Indeed, IT resources are processing data that should be transferred from the user's premises or from the data repository to the computing resources. When the Cloud will be largely adopted and the data deluge will fall in it, the communication model offered by the Internet may break the hope for fully-transparent remote access and outsourcing. The interconnection of IT resources over networks requires well-managed, dynamically invoked, consistent services. IT and network should be provisioned in a coordinated way in the future Internet.

3. **Deal with the unpredictability and burstiness of traffic:** The increasing popularity of video applications over the Internet causes the traffic to be unpredictable in the networks. The traffic's bursty nature requires mechanisms to support the dynamic behavior of the services and applications. Moreover, another important issue is that the popularity of content and applications on the Internet will be more and more sporadic: the network effect amplifies reactions. Therefore, the future Internet needs to provide mechanisms that facilitate elasticity of resources provisioning with the aim to face sporadic, seasonal or unpredictable demands.

4. **Make the network energy-aware:** It is reported in the literature [10], that ICT is responsible for about 4 % of the worldwide energy consumption today, and this percentage is expected to rapidly grow over the next few years following the growth of the Internet. Therefore, as a significant contributor to the overall energy consumption of the planet, the Internet needs to be energy-conscious. In the context of the proposed approach, this should involve energy awareness both in the provisioning of network and IT resources in an integrated globally optimized manner.

5. **Enable secured and reliable services:** The network's service outages and hostile hacks have received significant attention lately due to society's high dependency on information systems. The current Internet's service paradigm allows service providers to authenticate resources in provider domains but does not allow them to authenticate end-users requiring the resources. As a consequence, the provisioning of network resources and the secure access of end users to resources is a challenge. This issue is even more significant in the emerging systems with the provisioning of integrated resources provided by both network and IT providers to network operators.

6. **Develop a sustainable and strategic business model:** Currently, the business models deployed by telecom operators are focused on selling services on top of their infrastructures. In addition, operators cannot offer dynamic and smooth integration of diversified resources and services (both IT and network) at the provisioning phase. Network-infrastructure resources are not understood as a service within the value chain of IT service providers. We believe that a novel business model is necessary, which can fully integrate the network substrate with the IT resources into a single infrastructure. In

addition, such business model will let operators offer their infrastructures as a service to third-party entities.

3   Model

In order to address the aforementioned challenges and opportunities, the proposed architecture introduces the three basic concepts featured by the future Internet:

- The **Virtual Infrastructure concept** and its operational model as a fundamental approach to enable the on-demand infrastructure services provisioning with guaranteed performance and QoS, including manageable security services.
- A **new layered architecture** for the Control and Management Plane that allows dynamic services composition and orchestration in the virtual infrastructures that can consistently address the manageability, energy-efficiency and traffic-unpredictability issues.
- The **definition of the new Role-Based operational model** that includes the definition of the main actors and roles together with their ownership and business relations and the operational workflow.

3.1   Virtual Infrastructures

A Virtual Infrastructure (VI) is defined as an interconnection of virtualized resources with coordinated management and control processes. The virtual infrastructure concept consists in the decoupling of the physical infrastructure from its virtual representation, which can either be aggregating or partitioning a physical resource. This concept has little to do with the way data is processed or transmitted internally, while enabling the creation of containers with associated nonfunctional properties (isolation, performance, protection, etc.). The definition of a VI as an interconnection of a set of virtual resources provides the possibility of sharing the underlying physical infrastructure among different operators, and granting them isolation. At the same time, dynamic VI-provisioning mechanisms can be introduced into the infrastructure definition, creating the possibility to modify the VI capabilities in order to align them with the VI usage needs at any given instant.

As stated above, optical network technologies are among the key components for the future Internet. They inherently provide plenty of bandwidth and, in particular, the emerging flexible technology supported by the required control mechanisms enable a more efficient utilization of the optical spectrum and on-demand flexible bandwidth allocation, hence addressing challenge #1. IT resources comprise another important category of future Internet shared resources aggregated in large-scale data centers and providing high computational and storage capacities. In order to build VIs, these resources are abstracted, partitioned or grouped into Virtual Resources (VRs), which are attached to the VI and exposed for configuration. Each VR contains a well-defined, isolated subset of capabilities of a given physical entity. Taking the example of an optical cross-connect, the VRs are built on its total amount of ports or its wavelengths, and control switching exclusively between these allocated ports and wavelengths.

3.2   A Novel Layered Architecture

The proposed architecture features an innovative multi-layered structure to re-qualify the interworking of legacy planes and enable advanced services including the concepts of Infrastructure-as-a-Service (IaaS) and service-oriented network-ing [4]. We aim to enable a flexible infrastructure provisioning paradigm in terms of configuration, accessibility and availability for the end users, as well as a sep-aration of the functional aspects of the entities involved in the converged service provisioning, from the service consumer to the physical ICT infrastructure.

Our architecture is composed of three layers residing above the physical in-frastructure as illustrated in Fig. 1. They are referred to as the **infrastructure-virtualization layer**, the **enhanced control plane**, that corresponds to the network management layer, and the **service middleware** layer. Each layer is responsible for implementing different functionalities covering the full end-to-end service delivery from the service layer to the physical substrate.

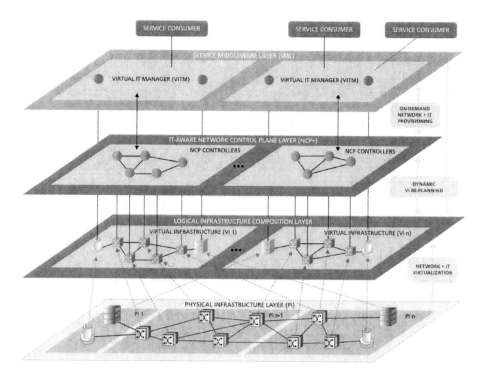

Fig. 1. Reference model as proposed in the GEYSERS project.

1. At the lowest level, the physical infrastructure layer comprises optical-network, packet network and IT resources from physical-infrastructure providers. Central to this novel architecture is the **infrastructure virtualization layer** which abstracts, partitions and interconnects infrastructure resources contained in the physical infrastructure layer. It may be composed of physical resources from different providers.
2. An **enhanced control plane** responsible for the operational actions to be performed over the virtual infrastructure (i.e., controlling and managing the network resources constituting the Virtual Infrastructure) is closely interacting with the virtualization layer.
3. Finally, a **service middleware layer** is introduced to fully decouple the physical infrastructure from the service level. It is an intermediate layer between applications running at the service consumer's premises and the control plane, able to translate the applications' resource requirements and SLAs into service requests. These requests specify the attributes and constraints expected from the network and IT resources.

The proposed architecture also addresses two major security aspects: secure operation of the VI provisioning process, and provisioning dynamic security services, to address challenge #5. Fig. 1 shows the reference model of our architecture as it has been modeled in the context of the GEYSERS project. The infrastructure-virtualization layer is implemented as the Logical Infrastructure Composition Layer (LICL) and the enhanced control plane as the NCP+. As an example, the figure shows two virtual infrastructures, each of them controlled by a single control-plane instance. The different notations used in the rest of this chapter are summarized in Tab. 1.

Table 1. Abbreviations

LICL	Logical Infrastructure Composition Layer
NCP	Network Control Plane
NCP+	Enhanced Network Control Plane
NIPS	Network + IT Provisioning Services
PIP	Physical Infrastructure Provider
SML	Service Middleware Layer
VI	Virtual Infrastructure
VIO	Virtual Infrastructure Operator
VIO-IT	Virtual IT Infrastructure Operator
VIO-N	Virtual Network Infrastructure Operator
VIP	Virtual Infrastructure Provider
VR	Virtual Resource

3.3   New Roles and Strategic Business Model

Given this virtualized network and IT architecture, new actors are emerging: the traditional carriers role is split among Physical Infrastructure Providers

(PIP), Virtual Infrastructure Providers (VIP) and Virtual Infrastructure Operators (VIO). PIPs own the physical devices and rent partitions of them to VIPs. These, in turn, compose VIs of the virtual resources rented at one or several PIPs, and lease these VIs to VIOs. VIOs can efficiently operate the rented virtual infrastructure through the enhanced Control Plane, capable of provisioning on-demand network services bundled with IT resources to meet challenge #2. New business relationships can be developed between Virtual IT Infrastructure Operators (VIO-IT) and Virtual Network Infrastructure Operators (VIO-N), but they can also converge to a single actor, with traditional operators moving their business towards higher-value application layers. This creates new market opportunities for all the different actors addressing challenge #6: infrastructure providers, infrastructure operators and application providers cooperate in a business model where on-demand services are efficiently offered through the seamless provisioning of network and IT virtual resources.

To illustrate this, we now describe a sample use case. A company hosts an Enterprise Information System externally on a Cloud rented from a Software-as-a-Service (SaaS) provider. It relies on the resources provided by one or more IT and network infrastructure providers. It also connects heterogeneous data resources in an isolated virtual infrastructure. Furthermore, it supports scaling (up and down) of services and load. It provides means to continuously monitor what the effect of scaling will be on response time, performance, quality of data, security, cost aspect, feasibility, etc. Our architecture will result in a new role for telecom operators that own their infrastructure to offer their optical network integrated with IT infrastructures (either owned by them or by third-party providers) as a service to network operators. This, on the other hand, will enable application developers, service providers and infrastructure providers to contribute in a business model where complex services (e.g., Cloud computing) with complex attributes (e.g., optimized energy consumption and optimized capacity consumption) and strict bandwidth requirements (e.g., real time and resilience) can be offered economically and efficiently to users and applications.

4   Virtual Infrastructures in Action

4.1   Virtual Infrastructure Life Cycle

The definition of the VI's life cycle is an important component of the proposed architecture that provides a common basis for the definition of the VI-provisioning workflow and all involved actors and services integration.

The VI life cycle starts with a VI request from the VIO, as represented on Fig. 2. A VI request may be loosely constrained, which considers the capabilities that the VI should provide but does not require a specific topology or strict resource attributes; or constrained to a topology, specifying the nodes that must be included and their capabilities and attributes. If a VI request is loosely constrained, the Virtual Infrastructure Provider (VIP) is responsible for deciding the final infrastructure topology. The VI request is not limited to a static definition of a VI, but temporal constrains can also be included. To specify VI requests,

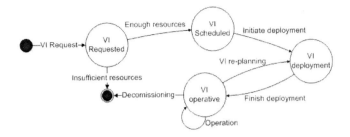

Fig. 2. VI Life Cycle

the *Virtual Infrastructure Description Language (VXDL)*[19] [7] can be used. It allows to formulate the topology and provisioning of a VI for its whole lifecycle [1]. After receiving the VI request, it is VIP's responsibility to split it into VRs request to different Physical Infrastructure Providers (PIPs), detailing the VRs that will be needed and their characteristics. Finally, each corresponding PIP will partition the physical resources into isolated virtual resources to compose and deploy the VI. Once the VI is in operation, the VIO can request for a VI re-planning to update its infrastructure so that it is optimized for a new usage pattern, involving for example traffic changes as predicted in challenge #3. A VI re-planning considers modifying the attributes of a node from the infrastructure (VR resizing), as well as including or excluding nodes into the VI. Depending on the initial SLAs agreed between the infrastructure operator and provider, a VI re-planning may involve a re-negotiation of the corresponding SLA or the current one may include it. The final stage in the VI life cycle is decommissioning, which releases the physical resources that were taken by the VIO.

4.2 Controlling the Virtual Infrastructures

Cloud applications are characterized by high dynamic and strict requirements in terms of provisioning of IT resources, where distributed computing and storage resources are automatically scaled up and down, with guaranteed high-capacity network connectivity. The enhanced Network Control Plane (NCP+) proposed in our architecture (Fig.1) offers integrated mechanisms for Network + IT Provisioning Services (NIPS) through the on-demand and seamless provisioning of optical and IT resources. These procedures are based on a strong inter-cooperation between the NCP+ and the **service middleware layer (SML)** via a service-to-network interface, named NIPS UNI during the entire VI service life cycle. The NIPS UNI offers functionalities for setup, modification and tear-down of enhanced transport network services (optionally combined with advance reservations), monitoring and cross-layer recovery. In particular, the NIPS UNI is based on a powerful information model that allows the SML to specify multiple aspects of the application requirements in the service requests. These requirements describe not only the characteristics of the required connectivity in terms

---
[19] http://www.ens-lyon.fr/LIP/RESO/Software/vxdl/home.html

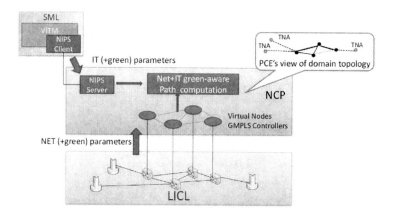

Fig. 3. Enhanced NCP—Network + IT energy-efficient service provisioning

of bandwidth, but could also specify further parameters, like the monitoring information required at the service layer or the mechanisms for cross-layer recovery. Following the service specification included in the requests, the network connectivity services are automatically tailored to the cloud dynamics, allowing for an efficient utilization of the underlying infrastructure. The integrated control of network and IT resources is achieved through different levels of cooperation between service plane and Network Control Plane, where part of the typical service plane's functionalities for the selection of the IT end points associated to an aggregate service can be progressively delegated to the NCP+. In anycast services the SML provides just a description of the required IT resources (e.g. in terms of amount of CPU), while the NCP+ computes the most efficient combination of end-points and network path to be used for the specific service. This feature requires a NCP+ with a global knowledge of the capabilities and availabilities of the IT resources attached to the network; this knowledge is obtained pushing a subset of relevant information from the SML to the NCP+, using the NIPS UNI. The NCP+ in the proposed architecture is based on ASON/GMPLS [8] and PCE [5] architectures and is enhanced with routing and signaling protocols extensions and constraints designed to support the NIPS. In particular, it is enhanced with the capability of advertising the energy consumption of network and IT elements, as well as availability, capabilities and costs of IT resources. The network-related parameters are collected from the LICL and flooded through the OSPF-TE protocol among the GMPLS controllers of the associated domain; on the other hand IT parameters retrieved through the NIPS UNI are communicated to a set of PCEs and not flooded among the GMPLS controllers. The path computation is performed by dedicated PCEs that implements enhanced computation algorithms able to combine both network and IT parameters with energy-consumption information in order to select the most suitable resources and find an end-to-end path consuming the minimum total energy (see Sec. 5). Figure 3 shows a high-level representation of the control plane: the routing algo-

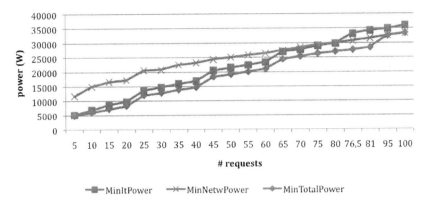

Fig. 4. Total Power consumption.

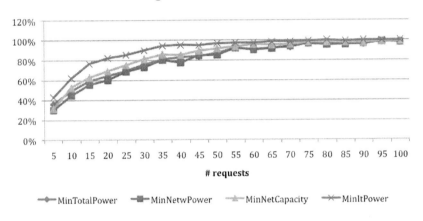

Fig. 5. Number of activated OXCs.

rithms at the PCE operate over a topological graph created combining network and IT parameters with "green" parameters, retrieved from the SML (IT side) and the LICL (network side).

Finally, another key element for the control plane is the interaction with the infrastructure-virtualization layer, in order to trigger the procedures for the Virtual Infrastructure's dynamic re-planning on the network side, besides the IT re-planning. In case of inefficiency of the underlying infrastructure, the control plane is able to request the upgrade or downgrade of the virtual resources, in order to automatically optimize the VI's size to the current traffic load.

5   Energy-Consumption Simulation Results

In this section we illustrate the potential of the proposed architecture in terms of energy-awareness, thereby addressing challenge #4. We evaluated an energy efficient routing algorithm(due to space limitations, the detailed algorithm is not

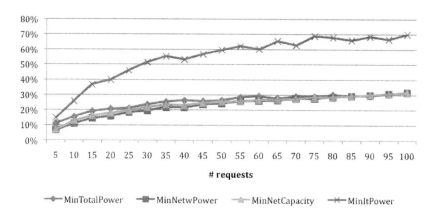

Fig. 6. Number of activated fibers.

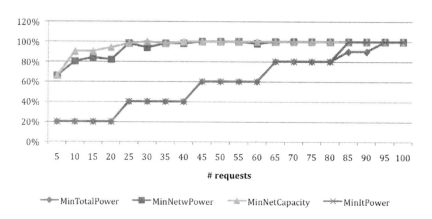

Fig. 7. Number of activated data centers.

shown here) from a networked IT use case: each source site has certain processing demands which need to be satisfied by suitable IT resources (in a data center). Note that we assume anycast routing, implying that the destination IT resource can be freely chosen among the available ones, since the requirement for the particular service is the accurate delivery of results, while the exact location of the processing is of no interest. Candidate IT resources reside at different geographical locations and connectivity of the source site to these IT resources is provided through the underlying optical network. Simulation results (see Fig. 4-6) indicate that our proposed algorithm can decrease the energy consumption by 10% compared to schemes where only IT infrastructure is considered and up to 50% when taking only the network into account, in an initial sample case study on an European network where we have run the calculations for 10 random demand vectors for each demand size going from 5- up to 100 connections for which, we have averaged the results[20]. (Please note that the savings are illustrative only, and depend on the case study assumptions.)

6 Conclusion

In this paper we have proposed a new way of configuring optical and IT resources within a new, layered architecture enabling the configuration of a virtual network infrastructure as a whole. Our goal is to address the six most critical challenges the Internet has to face urgently to support emerging disruptive applications and continue to grow safely. This chapter has presented the building blocks of this new architecture: the virtual infrastructure concept, a layered control and management architecture to complement the existing TCP/IP protocol stack and role-based operational model. Each of these building blocks has been further detailed and illustrated by test cases. The overall architectural blueprint complemented by the detailed design of particular components feeds the development activities of the GEYSERS project to achieve the complete software stack and provide the proof of concept of these architectural considerations. One approach of the project is to validate some concepts in a simulation environment. To make this proposal a viable solution for future production networks, a realistic and powerful test-bed will carry promising experiments. One of the test-bed's goals is to serve as a platform to implement and evaluate prototypes of the different software components creating and managing optical virtual infrastructures. The other goal is to evaluate the performance and functionality of such a virtualized infrastructure in a realistic production context.

Acknowledgement. This work has been founded by the EC ICT-2009.1.1 Network of the Future Project #248657.

References

1. Anhalt, F., Koslovski, G., Vicat-Blanc Primet, P.: Specifying and provisioning virtual infrastructures with HIPerNET. Int. J. Netw. Manag. 20(3), 129–148 (2010)
2. Buyya, R.: Market-Oriented Cloud Computing: Vision, Hype, and Reality of Delivering Computing as the 5th Utility. In: Proceedings of the 2009 9th IEEE/ACM International Symposium on Cluster Computing and the Grid, Washington, DC, USA. CCGRID '09, p. 1. IEEE Computer Society Press, Los Alamitos (2009), doi:10.1109/CCGRID.2009.97

---

[20] **MinTotalPower**: minimizing both network- and IT power; **MinNetworkPower**: only minimizing the network power; **MinNetCapacity**: minimizing the number of wavelengths needed to establish all requested lightpaths; **MinItPower**: only minimizing the energy consumed by the data centers.

3. Clark, D.: The design philosophy of the DARPA internet protocols. SIGCOMM Comput. Commun. Rev. 18, 106–114 (1988), doi:10.1145/52325.52336
4. Escalona, E., Peng, S., Nejabati, R., Simeonidou, D., Garcia-Espin, J.A., Ferrer, J., Figuerola, S., Landi, G., Ciulli, N., Jimenez, J., Belter, B., Demchenko, Y., de Laat, C., Chen, X., Yukan, A., Soudan, S., Vicat-Blanc, P., Buysse, J., Leenheer, M.D., Develder, C., Tzanakaki, A., Robinson, P., Brogle, M., Bohnert, T.M.: GEYSERS: A Novel Architecture for Virtualization and Co-Provisioning of Dynamic Optical Networks and IT Services. In: ICT Future Network and Mobile Summit 2011, Santander, Spain (June 2011)
5. Farrel, A., Vasseur, J.P., Ash, J.: A Path Computation Element (PCE)-Based Architecture. RFC 4655 (Informational) (Aug 2006), http://www.ietf.org/rfc/rfc4655.txt
6. Handley, M.: Why the Internet only just works. BT Technology Journal 24, 119–129 (2006), doi:10.1007/s10550-006-0084-z
7. Koslovski, G., Vicat-Blanc Primet, P., Charão, A.S.: VXDL: Virtual Resources and Interconnection Networks Description Language. In: Vicat-Blanc Primet, P., Kudoh, T., Mambretti, J. (eds.) GridNets 2008. LNICST, vol. 2, Springer, Heidelberg (2009)
8. Mannie, E.: Generalized Multi-Protocol Label Switching (GMPLS) Architecture. RFC 3945 (Proposed Standard) (Oct 2004), http://www.ietf.org/rfc/rfc3945.txt
9. Nelson, M.R.: The Next Generation Internet, E-Business, and E-everything, http://www.aaas.org/spp/rd/ch20.pdf
10. Pickavet, M., Vereecken, W., Demeyer, S., Audenaert, P., Vermeulen, B., Develder, C., Colle, D., Dhoedt, B., Demeester, P.: Worldwide energy needs for ICT: The rise of power-aware networking. In: Advanced Networks and Telecommunication Systems, 2008. ANTS '08. 2nd International Symposium on. pp. 1–3 (2008)
11. Rosenberg, J., Mateos, A.: The Cloud at Your Service, 1st edn. Manning Publications Co., Greenwich (2010)

Part VI:

Future Internet Areas: Services

Introduction

The global economy can be characterised under three main sectors. The primary sector involves transforming natural resources into primary products which then form the raw materials for other industries[1]. Examples of business in this area includes agriculture, fishing and mining. The secondary or industrial sector takes the output of the primary sector and produces finished goods which are then sold to either other businesses or to consumers[2]. The final sector is the tertiary or services sector where "intangible goods" or services are produced, bought and consumed[3]. Service provision is seen as an economic activity where generally no transfer of ownership is associated with the service itself and the benefits are associated with the buyers' willingness to pay. Public services are those where society pays through taxes and other means.

Over the last thirty years there has been a considerable shift, called tertiarisation[4], in industrialised countries from the primary and secondary sectors the service sector. Globally, we now find that the tertiary sector accounts for 63% of the world's 42 trillion Euro economy[5]. The economic importance of the service sector is a major motivation for services research both in the software industry and academia.

The Internet of Services is concerned with the creation of a layer within the Future Internet which can support the service economy. Two overarching requirements influence the scope and technical solutions created under the Internet of Services umbrella. Firstly, there is a need to support the needs of businesses in the area. Service oriented solutions can enable new delivery channels and new business models for the services industrial sector.

The Future Internet will be comprised of a large number of heterogeneous components and systems which need to be linked and integrated. For example, sensor networks will be composed on adhoc collections of devices with low-level interfaces for accessing their status and data online. Mobile platforms will need to access to external data and functionality in order to meet consumer expectations for rich interactive seamless experiences. Thus, a second driving requirement for the Internet of Services is to provide a uniform conduit between the Future Internet architectural elements through service-based interfaces.

Under the above broad requirements a number of research themes arise:

- **Architectural** – within a new global communications infrastructure there is a need to determine how a service layer would fit into an overall Future Internet architecture. For example, the boundary between the network and service layers and also how services would operate over connected objects which may form adhoc networks.
- **Management** – very quickly heterogeneity, dynamic contexts and scale lead to highly complex service scenarios where new approaches to managing the complexity are required. Here research focuses on describing services enabling automated

---

[1] http://en.wikipedia.org/wiki/Primary_sector_of_the_economy
[2] http://en.wikipedia.org/wiki/Secondary_sector_of_the_economy
[3] http://en.wikipedia.org/wiki/Tertiary_sector_of_the_economy
[4] http://www.eurofound.europa.eu/emire/GREECE/TERTIARIZATION-GR.htm
[5] http://en.wikipedia.org/wiki/List_of_countries_by_GDP_sector_composition

and semi-automated approaches to service discovery, composition, mediation and invocation.

- **Cloud Computing** – definitions vary but cloud computing is generally acknowledged to be the provision of IT capabilities, such as computation, data storage and software on-demand, from a shared pool, with minimal interaction or knowledge by users. Cloud services can be divided into three target audiences: service providers, software developers and users as follows[6]:

  - **Infrastructure as a Service** – offering resources such as a virtual machine or storage services.

  - **Platform as a Service** – providing services for software vendors such as a software development platform or a hosting service.

  - **Software as a Service** – offering applications, such as document processing or email to end-users.

Within this section we have three chapters which cover several of the issues outlined above. The ability to trade IT-services as an economic good is seen as a core feature of the Internet of Services. In the chapter Butler et al. "SLAs Empowering Services in the Future Internet" the authors discuss this in relation to Service Level Agreements (SLAs). In particular they claim a requirement for a holistic view of SLAs enabling their management through the whole service lifecycle: from engineering to decommissioning. An SLA management framework is outlined as a proposal for handling SLAs in the Future Internet. Evidence supporting the claims is provided through experiences in four industrial case studies in the areas of: Enterprise IT; ERP Hosting; Telco Service Aggregation; and eGovernment.

Ontologies are shared formal descriptions of a shared viewpoint over a domain which have attracted attention in recent years within the context of the Web. This work has led to the Semantic Web, and extension of the Web which is machine readable. Ontologies and semantics form a part of the next two chapters in this section.

As mentioned above there is an open question on how best to connect the network and service layers in a new communications infrastructure. Within the chapter Santos et al., "Meeting Services and Networks in the Future Internet" an ontology based approach is taken combined with a simplification of the network layer structure in order to facilitate network-service integration. More specifically, the approach, called FINLAN (Fast Integration of Network Layers), is a model which replaces several network layers with ontologies providing the foundations for an "Autonomic Internet".

Linked Data is the Semantic Web in its simplest form and is based on four principles:

- Use URIs (Uniform Resource Identifiers) as names for things.
- Use HTTP URIs so that people and machines can look up those names.
- When someone looks up a URI, provide useful machine-readable information, using Semantic Web standards.
- Include links to other URIs, so that other resources can be discovered.

---

[6]   See http://www.internet-of-services.com/index.php?id=274&L=0

Given the growing take-up of Linked Data for sharing information on the Web at large scale there has begun a discussion on the relationship between this technology and the Future Internet. In particular, the Future Internet Assemblies in Ghent and Budapest both contained sessions on Linked Data. The final chapter in this section Domingue et al., "Fostering a Relationship Between Linked Data and the Internet of Services" discusses the relationship between Linked Data and the Internet of Services. Specifically, the chapter outlines an approach which includes a lightweight ontology and a set of supporting tools.

John Domingue

SLAs Empowering Services in the Future Internet[1]

Joe Butler[1], Juan Lambea[2], Michael Nolan[1], Wolfgang Theilmann[3],
Francesco Torelli[4], Ramin Yahyapour[5], Annamaria Chiasera[6], and Marco Pistore[7]

[1] Intel, Ireland, {joe.m.butler, michael.nolan}@intel.com
[2] Telefónica Investigación y Desarrollo, Spain, juanlr@tid.com
[3] SAP AG, Germany, wolfgang.theilmann@sap.com
[4] ENG, Italy, francesco.torelli@eng.com
[5] Technische Universität Dortmund, Germany, ramin-yahyapour@udo.edu
[6] GPI, Italy, achiasera@gpi.it
[7] FBK, Italy, pistore@fbk.eu

Abstract. IT-supported service provisioning has become of major relevance in all industries and domains. However, the goal of reaching a truly service-oriented economy would require that IT-based services can be flexibly traded as economic good, i.e. under well defined and dependable conditions and with clearly associated costs. With this paper we claim for the need of creating a holistic view for the management of service level agreements (SLAs) which addresses the management of services and their related SLAs through the complete service lifecycle, from engineering to decommissioning. Furthermore, we propose an SLA management framework that can become a core element for managing SLAs in the Future Internet. Last, we present early results and experiences gained in four different industrial use cases, covering the areas of Enterprise IT, ERP Hosting, Telco Service Aggregation, and eGovernment.

Keywords: Service Level Agreement, Cloud, Service Lifecycle

1   Introduction

Europe has set high goals in becoming the most active and productive service economy in the world. Especially IT supported services have become of major relevance in all industries and domains. The service paradigm is a core principle for the Future Internet which supports integration, interrelation and inter-working of its architectural elements. Besides being the constituting building block of the so-called Internet of Services, the paradigm equally applies to the Internet of Things and the underlying technology cloud platform below. Cloud Computing gained significant attention and commercial uptake in many business scenarios. This rapidly growing service-oriented economy has highlighted key challenges and opportunities in IT-supported service provisioning. With more companies incorporating cloud based IT services as part of

---

[1]   The research leading to these results is partially supported by the European Community's Seventh Framework Programme ([FP7/2001-2013]) under grant agreement n° 216556.

J. Domingue et al. (Eds.): Future Internet Assembly, LNCS 6656, pp. 327–338, 2011.

their own value chain, reliability and dependability become a crucial factor in managing business. Service-level agreements are the common means to provide the necessary transparency between service consumers and providers.

With this paper, we discuss the issues surrounding the implementation of automated SLA management solutions on Service Oriented Infrastructures (SOI) and evaluate their effectiveness. In most cases today, SLAs are either not yet formally defined, or they are only defined by a single party, mostly the provider, without further interaction with the consumer. Moreover, the SLAs are negotiated in a lengthy process with bilateral human interaction. For a vivid IT service economy, better tools are necessary to support end-to-end SLA management for the complete service lifecycle, including service engineering, service description, service discovery, service composition, service negotiation, service provisioning, service operation, and service decommissioning.

We provide an approach that allows services to be described by service providers through formal template SLAs. Once these template SLAs are machine readable, service composition can be established using automatic negotiation of SLAs. Moreover, the management of the service landscape can focus on the existence and state of all necessary SLAs. A major aspect herein is the multi-layered service stack. Typically, a service is dependent on many other services, e.g. the offering of a software service requires infrastructure resources, software licenses or other software services.

We propose an SLA management framework that offers a core element for managing SLAs in the Future Internet. The framework supports the configuration of complex service hierarchies with arbitrary layers. This allows end to end management of resources and services for the business value chain. The scientific challenges include the understanding and modelling of the relationships between SLA properties. For instance assessing the performance of a service and its corresponding SLAs includes the whole stack and hierarchy. The deep understanding of this relationship needs to be modelled in a holistic way for reliability, performance etc.

The technical foundation to our approach is a highly configurable plug-in-based architecture, supporting flexible deployment options including into existing service landscapes. The framework's architecture mainly focuses on separation of concerns, related to SLAs and services on the one hand, and to the specific domain (e.g., business, software, and infrastructure) on the other.

With a set of four complementary use case studies, we are able to evaluate our approach in a variety of domains, namely ERP hosting, Enterprise IT, Service Aggregation and eGovernment. ERP Hosting is investigating the practicalities and benefits of holistic SLA planning and management when offering hosted ERP solutions for SMEs. Enterprise IT focuses on SLA-aware provisioning of compute platforms, managing decisions at provisioning time and runtime, as well as informing business planning. Service Aggregation demonstrates the aggregation of SLA-aware telecommunication and third party web services: how multi-party, multi-domain SLAs for aggregated services can best be offered to customers. eGovernment validates the integration of human-based services with those that are technology based, showcasing the automated, dynamic SLA-driven selection, monitoring and adjustment of third-party provisioned services.

The remainder of this paper is organized as follows. Chapter 2 introduces our reference architecture for an SLA management framework. Chapter 3 discusses the adoption of the framework, within the Future Internet but also in general System Management environments. Chapters 4-7 cover the respective use cases and evaluation results and Chapter 8 concludes the overall discussion.

2   Reference Architecture for SLA Management

The primary functional goal of our SLA management framework is to provide a generic solution for SLA management that (1) supports SLA management across multiple layers with SLA (de-)composition across functional and organizational Domains, (2) supports arbitrary service types (business, software, infrastructure) and SLA terms, (3) covers the complete SLA and service lifecycle with consistent interlinking of design-time, planning and run-time -management aspects; and (4) can be applied to a large variety of industrial domains.

In order to achieve these goals, the reference architecture follows three main design principles:

- a clear separation of concerns,
- a solid foundation in common meta models, and
- design for extensibility.

The primary building blocks of the architecture are the SLA Manager, responsible to manage a set of SLAs and corresponding SLA templates, and the Service Manager, responsible for service realizations. Both of these are realized as generic components which can be instantiated and customized for different layers and domains.

Figure 1 illustrates an example setup of the main components of the SLA@SOI framework and their relationships for three layers: business, software and infrastructure. The framework communicates to external parties, namely customers who (want to) consume services and 3rd party providers which the actual service provider might rely upon. Relationships are defined by stereotyped dependencies that translate to specific sets of provided and required interfaces.

On the highest level, we distinguish the Framework Core, Service Managers (infrastructure and software), deployed Service Instances with their Manageability Agents and Monitoring Event Channels. The Framework Core encapsulates all functionality related to SLA management, business management, and the evaluation of service setups. Infrastructure- and Software Service Managers contain all service-specific functionality. The deployed Service Instance is the actual service delivered to the customer and managed by the framework via Manageability Agents. Monitoring Event Channels serve as a flexible communication infrastructure that allows the framework to collect information about the service instance status.

Furthermore, the framework comes with a set of well linked meta-models, namely an SLA model, a service construction model (capturing provider-internal service aspects), the service prediction model, and an infrastructure model.

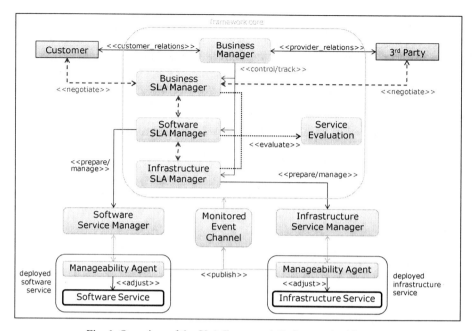

Fig. 1. Overview of the SLA Framework Reference Architecture

While all framework components come with default implementations they can also easily be extended or enhanced for more specific domain needs. Similarly, the provided meta-models come with clear extension mechanisms, e.g. to specify additional service level terms.

A typical negotiation/planning sequence realized via the framework starts with a customer querying for available products and corresponding SLA Templates. The framework then, initiates a hierarchical planning process that results in a specific offer. After agreement to an SLA offer, a provisioning sequence will be automatically triggered according to the SLA specification and will be executed in a bottom-up manner through the framework. Last, monitoring is conducted simultaneously at all framework layers; possible SLA warnings or violations either lead to local adjustment activities or are escalated to the next higher level. Further details on this architecture can be found at [2, 5].

3   Adoption Aspects

After introducing the reference architecture we now want to discuss various adoption issues. First we provide a sketch on how the architecture can be applied to the Future Internet. Second, we give an overview how SLA management relates to other management functions. Last, we provide a brief discussion on nun-functional aspects of the framework itself.

3.1   Adoption Considerations for the Future Internet

The SLA management framework architecture can easily be applied to different Future Internet scenarios. The SLA model is rich and extensible enough to be applied to e.g. infrastructure and networking resources, to sensor-like resources in the Internet of Things, to services in the Internet of Services, but also to describe people, knowledge, and other resources. Similarly, the service construction model can be adopted, which allows specification of arbitrary internal resource/service aspects.

Based on this model foundation, the framework components can be flexibly instantiated. Assuming to have Manageability Agents for the relevant artefacts in the Future Internet, a management environment consisting of SLA and Service Managers can be set up in different flavours. The setup can support autonomic scenarios, where specific SLA/Service managers are responsible for single artefacts, but also highly coordinated scenarios, where SLA/Service managers govern a larger set of entities collectively. Business aspects can be addressed at all layers by introducing a dedicated business manager. Service Evaluations can be also introduced at all layers: However there will most likely be different evaluation/optimization mechanisms for different actual domains.

Last, different framework instances can be flexibly created and connected as needed according to the requirements of the involved value chain stakeholders in the respective Future Internet scenario. In the following use-case chapters we also provide additional configuration examples of the framework.

3.2   Adoption Considerations for Cloud Computing

The SLA@SOI framework should become an intrinsic part of each cloud environment, whether it is about software-as-a-service, platform-as-a-service, or infrastructure-as-a-service. The Enterprise IT use-case (Section 3) is basically an infrastructure cloud use case that features SLA enabling. The ERP hosting use case (Section 4) contains many aspects of a software cloud.

3.3   Interlinkage with System Management

SLA-driven system management is the primary approach discussed in this paper. It actually tries to derive all kinds of management decisions from the requested or agreed service level agreements. However, there are also other management functions which are partially related to SLA management. As reference structure for these functions we use the 4 main categories of self-managing systems [3], namely self-configuring, self-healing, self-optimizing, and self-protecting.

Configuration management is closely related to SLA management. Possible configuration options are captured in the service construction model and are taken into consideration during the planning phase. Once, an SLA has been agreed and is to be provided, the configuration parameters derived during planning phase are used to set up the system. The same holds for replanning/adaptation cycles.

Self-healing is in the first place independent from SLA management as the detection and recovery from low-level unhealthy situations can be done completely independent from agreed SLAs. However, the detection of SLA violations and the automated re-planning could be also understood as self-healing process. Furthermore, low-level unhealthy situations might be used for predicting possible future SLA violations.

Self-optimizing as very closely related to SLA management and simply cannot be done without taking into account the respective constraints of the contracted SLAs.

Self-protecting is in the first place independant from SLA management. However, certain self-protecting mechanisms can be made part of an SLA.

3.4   Non-functional Properties of the SLA Framework Itself

The non-functional properties of the SLA@SOI framework play an important role for any eventual usage scenario. However, they heavily depend on the actual adoption style and cannot be simply described at the generic framework level.

The overhead introduced depends significantly on the granularity of the SLA management (how fine-grained the decomposition of an IT stack into services and SLAs is done) and the requested accuracy of the monitoring (significantly impacts on the number of events to be processed).

While the framework itself is fully scalable (in terms of its ability to run components in multiple instances), the actual scalability in a target environment depends on the number of interrelations between different artefacts. For example if many artefacts interrelate, their management cannot be easily parallelized but must be done from a central instance (e.g. a central SLA manager overseeing all SLAs in his domain).

Reliability aspects are mainly orthogonal and can be realized by state of the art redundancy mechanisms.

Last, flexibility has been a clear design goal of the framework, allowing for different setups of the framework instances which due to the underlying OSGI-based integration approach can be even changed during run-time.

4   Use Case – Enterprise IT

The Enterprise IT Use Case focuses on compute infrastructure provisioning in support of Enterprise services. We assume a virtualisation-enabled data centre style configuration of server capacity, and a broad range of services in terms of relative priority, resource requirement and longevity. As a support service in most enterprises, IT is expected to deliver application and data service support to other enterprise services and lines of business. This brings varied expectations of availability, mean-time-to-recover, Quality of Service, transaction throughput capacity, etc. A challenge for IT organisations in response to this, is how to deliver a range of service levels at the most optimal cost level. This challenge includes quick turnaround planning decisions such as provisioning and placement, run time adjustment decisions on workload migration for efficiency, and longer term strategic issues such as infrastructure refresh (in the case of internally managed clouds) and hosting service provider (external clouds).

This use case is therefore based around three distinct scenarios. The first scenario, titled "Provisioning", responds to the issue of efficient allocation of new services on IT infrastructure, SLA negotiation and provisioning of new services in the environment. The second scenario, "Run Time", deals with day-to-day, point in time operational efficiency decisions within the environment. These decisions maximise the value from the infrastructure investment. The final scenario, "Investment Governance" builds on the first two to demonstrate how they feed back into future business decisions. Taking a holistic cost view, it provides fine grained SLA based data to influence future investment decisions based on capital, security, compute power and energy efficiency.

In order to enable realistic and effective reasoning at provisioning and run time, a reference is included differentiates each of the supported Enterprise services in terms of their priority and criticality. This is the Enterprise Capability Framework or ECF.

From an implementation perspective, user interaction is via a web based UI, used by both IT customers and administrators. The Enterprise IT SLAT defines use case specific agreement terms which are loaded by the Business SLA manager to provide the inputs to provisioning requests in the form of PaaS services. Software services could potentially be selected by choosing a virtual machine template which contains pre-loaded applications, but software layer considerations are not considered core to this Use Case and are more comprehensively dealt with in the ERP Hosting Use Case. The Business SLA Manager passes service provisioning requests to the Infrastructure SLA Manager whose role is to carry out the creation of the new virtual machines which constitute the service along with monitoring and reporting for that service.

Evaluation of the framework is carried out with reference to parameters which align with IT and business priorities. The three scenarios on which the Use Case is based, are complementary and allow the framework to be assessed based on realistic objectives of an Enterprise IT function. Using Key Performance Indicators (KPIs) we evaluate the performance of the lab demonstrator in the areas of:

- IT enabling the Enterprise
- IT Efficiency
- IT Investment/Technology adoption

The Use Case identifies a hierarchy of KPIs which are measurable against established baseline and therefore result in a credible assessment of the impact of the SLA Management Framework in an Enterprise IT context.

Further details on this use case are available at [6].

5   Use Case – ERP Hosting

The ERP Hosting use case is about the dynamic provisioning of hosted Enterprise Resource Planning solutions. SLA management in this context promises great benefits to providers and customers: Providers are enabled to offer hosted solutions in a very cost-efficient and transparent way, which in particular offers new sales channels to-wards small and medium sized customers. Customers are enabled to steer their business in a more service-oriented and flexible manner that meets their business needs without spending too much consideration on IT matters. Furthermore, customers can flexibly negotiate the exact service details, in particular its service levels, so that they can eventually get the best fitting service for their needs.

The actual use case realizes a scenario with 4 layers of services. The top-level service considered is the so-called business solution. Such a solution typically consists of a software package (an application) but also some business-level activities, such as a support contract. At the next level, there are the actual software applications, such as for example a hosted ERP software package. At the next level, there are the required middleware components which are equally used for different applications. At the lowest layer, there are the infrastructure resources, delivered through an internal or external cloud. Each service layer is associated with a dedicated SLA, containing service level objectives which are specific to this layer. The business SLA is mainly about specifying support conditions (standard or enterprise support), quality characteristics (usage profile and system responsiveness), and the final price for the end customer. The Application SLA is mainly about the throughput capacity of the software solution, its response time, and the provider internal costs required for the offering. The Middleware SLA specifies the capacity of the middleware components, the response time guarantee of the middleware components and the costs required for the offering. The Infrastructure SLA specifies the characteristics of the virtual or physical resources (CPU speed, memory, and storage) and again the costs required for the offering.

The use case successfully applies the SLA framework by realizing distinct SLA Managers for the 4 layers and also 4 distinct Service Managers that bridge to the actual support department, the application, the middleware, and the infrastructure artefacts.

From a technical perspective, the most difficult piece in the realization of the whole use case was the knowledge discovery about the non-functional behaviour of the different components, e.g. the performance characteristics of the middleware. We collected a set of model-driven architecture artefacts, measurements, best practise rules and managed to consistently interlink them and to realize an overall hierarchical planning and optimization process. However, this process is still labour intensive and requires further automation tools in order to be applicable on a large scale.

From a business perspective the use case clearly proved tangible business benefits in different aspects. Time to market for quotation on service requests and provisioning of requested services can be significantly reduced. The dependability of provided services is increased proportional to the number of formally managed service level terms. The efficiency of service provisioning can be improved in the dimensions of environmental efficiency, resource efficiency, and process efficiency. The transparency for the end-to-end service delivery process involving different departments is largely improved due to the consistent interlinking of different views and artefacts. Last this transparency also supports an improved agility and allows for realizing changes much faster than in the past.

Further details on this use case can be found in [3]. Background information and a demo video are available at [7].

6   Use Case – Service Aggregation

The main aim of the Service Aggregation use case is the service-enabling of core Telco services and their addition with services from third parties (as Internet, infrastructure, media or content services). From the provider's point of view, they will be able to publish their services in the Service Aggregator and will be benefited in terms of reach new markets in which their services can be consumed and to be sold to the customers joined with reliable communication services offered by Telco providers. Customers can find the services and negotiate flexibly the terms of the consumption of the services included in the product. It is necessary to point out that negotiation takes place in three faces: Bank customer, Service Aggregator and Infrastructure provider. This implies the negotiation of the SLAs with quality of service aspects and the final price.

For the proof of concept we have chosen a combination of Telco capability (SMS) with a third party service (infrastructure) to build a product that is a service bundle to be offered to a Bank. There are two SLA@SOI framework instances implemented; the Service Aggregator and Infrastructure as a Service. Bank customer prototype uses several framework components mainly interfaces with Business Manager and Business SLA Manager. Service Aggregator and Infrastructure prototype have been implemented using business and infrastructure layers; additionally Service Aggregator integrates software layer (from SLA@SOI framework architecture). And finally Bank prototype is implemented using the top layer, business. Both providers utilize SLAT registries in their SLA Managers to publish the SLA templates of his services hierarchy. Business SLA template for SMS service includes some business terms like support, termination or price and other guarantee terms like availability, throughput, and response time. Communication of SLA templates between third party and Service Aggregator use advertising bus to share infrastructure templates. Then Service Aggregator can utilize local business SLA templates and third party business SLA templates to create the bundle of a product. Customer prototype is connected with Business Manager of the Service Aggregator to find and discover products; in this case it found the 'Communications and Infrastructure bundle' product. Customer retrieves the different SLA templates available for the product and the negotiation starts. Customer has to be previously registered and granted in the Service Aggregator. The negotiation is driven by the Business Manager of the Service Aggregator. It adopted provider requirements in shape of policy rules and promotions applicable to the final product. Infrastructure provider can also define the business negotiation of his services in the same way. When negotiation finished between the three actors, SLA has been signed between them in pairs and all the provision process is finished. The corresponding SLAs will be stored in SLA registries of each SLA manager used. In this way it is necessary to outline also is executed the provision of Telco web service wrappers by Software SLA Manager in an application server and also the provision of the infrastructure driven by Infrastructure SLA Manager (using the appropriate service manager). SMS wrappers deployed in the application server of the corresponding virtual machine has to connect and execute different tasks with core mobile network systems that are behind Telefónica Software Delivery Platform (SDP). The components that can be also connected in the use case are the monitors of the services (SMS and Infrastructure services). To take care about the violations, track interfaces are used to connect the adjustment components in each SLA Manager. Finally, Service Aggregator converts violations in penalties, and takes actions to adjust these violations and reports the situation to the customer.

Technical evaluation about SLA@SOI framework can be seen in a very positive way in terms of the functionality of the components and the outcome obtained by the use case. In the new ecosystems of Future internet of services the key will be the exporting and interconnection of services between different parties. It is necessary to care the service level agreements and the quality of the services guaranteed on those SLAs. SLA-aware aggregation of telecommunications services introduces a business opportunity for the agile and efficient co-creation of new service offerings and significant competitive advantages to all.

Further details on this use case including a demo video are available at [8].

7   Use Case – eGovernment

Public administrations often outsource human based services to 3rd party organizations. Such relationships are currently regulated with legal documents and human readable SLAs. The eGovernment use case aims at showing that the adoption of machine readable SLAs improves the agility and reduces costs, as it allows to automate several management activities also if the services are performed by humans.

In our proof of concept we considered a composed service allowing citizens to book medical treatments and to use and reserve at the same time the transport means for reaching the treatment place. Such a Health & Mobile Service is provided by a so called "Citizen Service Center" (CSC) and is composed of: a "Medical Treatment Service" provided by external Health Care Structures, a "Mobility Service" provided by specific Transport Providers and a "Contact Service" provided by phone call operators of the SCS and, when needed, also by a third party Call Center.. In this context, the SLA between the Government and the CSC regulates the provision of the health, mobile and contact services, as well as the expected overall satisfaction of the citizen. The SLA@SOI framework automates activities of the CSC that are usually performed manually or not performed at all, such as: the monitoring of SLAs, the allocation of the phone call operators, the dynamic selection of the mobility providers based on SLAs, the automatic re-negotiating of the SLA with the external Call Center.

The CSC fully adopts the SLA@SOI architecture. A SLA-aware BPEL Engine is used for the dynamic binding and execution of the composed service. A custom Human Service Manager allocates the human resources. A customised SLA Manager manages the negotiation with the Government and with the external Call Center. A Service Evaluation Component predicts, at negotiation time, the QoS obtainable with the available operators, in order to determine the SLA to negotiate with the external provider. At runtime the prediction feature warns about possible violations of the guarantee terms, triggering the automatic adjustment of internal human operators.

From the technical point of view, one of the main challenges of this use case has been the modelling of human-provided services, and the formalization of the strategies for handling human resources during negotiation and adjustment. This is still an ongoing task that has required several interviews with the operators working at the service providers. Also the SLAs defined in this use case have several peculiarities. For example, while typical software/hardware guarantee terms constraint the quality of each single execution of a service, in this use case the guarantee terms constraint the average value of KPIs computed for hundreds of executions measured on time periods of the order of months. Moreover, such measurements are relative to mobile time windows and periodic behaviours (consider, e.g., the difference between summer and winter in the delivery of the services considered in this use case). Extensions of the prediction model are under evaluation in order to cover new kinds of KPIs and guaranteed terms.

From the evaluation perspective, the application scenario is particularly critical due to sensitive data on the health status of the citizens and quite challenging for the key role of humans both in the provisioning and in the evaluation of the effectiveness of the platform. From the close interaction with the experts in the field, we derived an approach for the evaluation based on the feedback of the citizens and also of the operators (with focus groups and periodic interviews) in terms of: effort for the operators and citizens to interact with the system, usability and degree of acceptance of the system from the users, effectiveness of monitoring and adjustment functionalities. The results obtained from this evaluation is further integrated and compared with the trends in the real data extracted from the past behaviours of the systems at the service providers.

Overall, this scenario has been very valuable for SLA@SOI as a way to stress the project approach and core concepts to the case of human resources and human provided services.

Further details on this use case are available at [9].

8   Conclusions

Service level agreements are a crucial element to support the emerging Future Internet so that eventual services become a tradable, dependable good. The interdependencies of service level characteristics across layers and artefacts require a holistic view for their management along the complete service lifecycle. We explained a general-purpose SLA management framework that can become a core element for managing SLAs in the Future Internet. The framework allows the systematic grounding of SLA requirements and capabilities on arbitrary service artefacts, including infrastructure, network, software, and business artefacts. Four complementary industrial use cases demonstrated the applicability and relevance of the approach. Furthermore, the diverse and complementary nature of the Use Cases along with the consistent and structured evaluation approach ensures that the impact assessment is credible.

Future work concentrates on three aspects. Technology research will be deepened on the areas of SLA model extensibility, quality model discovery, business negotiation, and elastic infrastructure scaling. Use Case research will tackle additional scenarios, especially relevant for the Future Internet. Last, we plan to open up our development activities via an Open Source Project. The first framework version fully published as open source can be found at [5].

Open Access. This article is distributed under the terms of the Creative Commons Attribution Noncommercial License which permits any noncommercial use, distribution, and reproduction in any medium, provided the original author(s) and source are credited.

References

1. SLA@SOI Project Web Site. URL: http://www.sla-at-soi.eu
2. Theilmann, W., Happe, J., Kotsokalis, C., Edmonds, A., Kearney, K., Lambea, J.: A Reference Architecture for Multi-Level SLA Management. Journal of Internet Engineering 4(1) (2010), http://www.jie-online.org/ojs/index.php/jie/issue/view/8
3. Miller, B.: The autonomic computing edge: Can you CHOP up autonomic computing? Whitepaper IBM developerworks (March 2008), http://www.ibm.com/developer works/autonomic/library/ac-edge4/
4. Theilmann, W., Winkler, U., Happe, J., Magrans de Abril, I.: Managing on-demand business applications with hierarchical service level agreements. In: Berre, A.J., Gómez-Pérez, A., Tutschku, K. (eds.) FIS 2010. LNCS, vol. 6369, Springer, Heidelberg (2010)
5. SLA@SOI Open Source Framework. First full release by December 2010, http://sourceforge.net/projects/sla-at-soi
6. SLA@SOI project: Enterprise IT Use Case. http://sla-at-soi.eu/research/focus-areas/use-case-enterprise-it/
7. SLA@SOI project: ERP Hosting Use Case. http://sla-at-soi.eu/research/focus-areas/use-case-erp-hosting/
8. SLA@SOI project: Service Aggregator Use Case, http://sla-at-soi.eu/research/focus-areas/use-case-service-aggregator/
9. SLA@SOI project: eGovernment Use Case, http://sla-at-soi.eu/research/focus-areas/use-case-e-government/

Meeting Services and Networks in the Future Internet

Eduardo Santos[1], Fabiola Pereira[1], João Henrique Pereira[2],
Luiz Cláudio Theodoro[1], Pedro Rosa[1], and Sergio Takeo Kofuji[2]

[1] Federal University of Uberlândia, Brazil
eduardo@mestrado.ufu.br, fabfernandes@comp.ufu.br, lclaudio@feelt.ufu.br,
pedro@facom.ufu.br
[2] University of São Paulo, Brazil
joaohs@usp.br, kofuji@pad.lsi.usp.br

Abstract. This paper presents the researches for better integration between services and networks by simplifying the network layers structure and extending the ontology use. Through an ontological viewpoint, FINLAN (Fast Integration of Network Layers) was modeled to be able to deal with semantics in the network communication, cross-layers, as alternative to the TCP/IP protocol architecture. In this research area, this work shows how to integrate and collaborate with Future Internet researches, like the Autonomic Internet.

Keywords: Future Internet, Network Ontology, Post TCP/IP, Services

Introduction

In recent years it has been remarkable the Internet advancement in throughput and the development of different services and application features. Many of these are supported by the TCP/IP protocols architecture, however, the intermediate layers based on the protocols IP, TCP, UDP and SCTP were developed more than 30 years ago, when the Internet was used just for a limited number of hosts and with a few services support. Despite the development of the Internet and its wonderful flexibility and adaptability, there were no significant improvements in its Network and Transport layers, resulting in a communication gap between layers [7,8].

Integration of services and networks is an emerging key feature in the Future Internet and there are a lot of studies, proposals and discussions over questions related to a network able of supporting the current and Future Internet communication challenges. Some of these studies are related to: EFII, FIA, FIRE, FIND, GENI and other groups. Some of these groups are very expressive, for example FIRE, that includes discussions and projects like: ANA, AUTOI, BIONETS, CASCADAS, CHIANTI, ECODE, EFIPSANS, Euro-NF, Federica, HAGGLE, MOMENT, NADA, N4C, N-CRAVE, OneLab2, OPNEX, PERIMETER, PII, PSIRP, ResumeNet, Self-NET, SMART-Net, SmoothIT, TRILOGY, Vital++, WISEBED and 4WARD [2].

J. Domingue et al. (Eds.): Future Internet Assembly, LNCS 6656, pp. 339–350, 2011.

Considering the possibilities for improvements in the current TCP/IP architecture with collaboration for the Future Internet, this work is focused in one alternative to the TCP/IP protocols, at layers 3 and 4, in one perspective to meet the service requirements in a simplified and optimized way, taking into account the real world service needs. This research also shows one proposal to improve the communication between services and networks with semantics, disseminating the power of the meaning across the network layers.

1 Ontological Approach in FINLAN

The FINLAN (Fast Integration of Network Layers), presented in [16], is a post-IP research which eliminates the Network and Transport layers, meeting services directly to the network lower layers. Thereby, the networks are prepared to meet the requirements of services in a flexible and optimized way. For example, the work in [6] shows how FINLAN can deal with the requirement of delivery guarantee, presenting its ability to service adaptability related to the applications needs and, in [11], it is presented the FINLAN implementation and experimental results compared with TCP/IP.

In this area of possibilities, in [5, 7–10] our group discuss some fundamental aspects for the Post-IP research area and proposals to extend the use of ontology in computer networks to support the communication needs in a better way. Another aspect that can be placed in the context of the Future Internet is the use of ontology in networks. In current networks, the semantic communication generally is limited to the Application layer and this layer is restricted to sending meaning to the Network and Transport layers. Therefore, working with an ontological view over the Network and Transport layers has proved a promising object of research.

1.1 Ontological Layers Representation

The use of ontology at the intermediate layers permits the Internet Application layer better inform its needs to the intermediate layers. This increases the comprehension cross layers and contributes to get the upper and lower layers semantically closer. This helps the networks to improve the support to the communication needs, as the Application layer can inform, or request, requirements that the current TCP/IP layers 3 and 4 can not comprehend completely. Some of the communication needs that the lower layers can better support, by the ontology use in this work, are: Management, Mobility, QoE, QoS and Security.

This ontology at the intermediate layers is represented in FINLAN by the Net-Ontology and the DL-Ontology (Data Link) layers. The Net-Ontology layer has semantic communication, in OWL (Web Ontology Language), with its superior layer and the DL-Ontology layer. It is responsible to support the services needs of the superior layer. The DL-Ontology layer has semantic communication, also using OWL, with the superior and the Net-Ontology layers. It is responsible to support the Data Link communication to guarantee the correct delivery of data transfer between links. The main difference between these two layers is that the Net-Ontology layer is responsible to support service needs beyond simple data transfers. These layers, compared with the TCP/IP layers, are represented in Fig. 1, with examples of some Future Internet works that can be integrated with this approach at the intermediate layers.

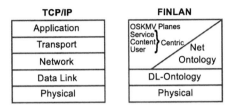

Fig. 1. TCP/IP and FINLAN Layers Comparison

From this layers comparison, some responsibilities of the actual TCP/IP Application, Transport and Network layers, are handled by the Net-Ontology layer, and some others by the Application layer. The applications, in FINLAN, inform their needs to the Net-Ontology layer using concepts. One application example is the encryption for security at the intermediate layer. In this example, in the actual TCP/IP protocols architecture, the layers 3 and 4 are not able to understand the security need in a context and its complexities usually must be controlled by the Application layer. However, in FINLAN, the Application layer can inform semantically this security need to the Net-Ontology layer. By this, the related complexities can be handled at the Net-Ontology layer level, instead of the Application layer level.

One layer level below, the difference between the TCP/IP Data Link and DL-Ontology layers is the semantic support, as the Data Link, in the TCP/IP, also does not support concepts. One application example is the services integration in heterogeneous environment to the devices mobility in 4G networks handovers, using the DOHand (Domain Ontology for Handover). In the experimental use of the DOHand, by Vanni in [17], the semantic possibilities for handover are reduced by the TCP/IP limitations. In this scenario, FINLAN approach contributes to the semantic communication between the DOHand and the DL-Ontology layer, for the handover in 4G networks.

In another application example, the FINLAN ontology can be used by the Management plane in OSKMV (Orchestration, Service Enablers, Knowledge, Management and Virtualisation planes), presented in [3], to better monitor the networks, as the semantic information can directly be handled by the Net-Ontology and DL-Ontology layers. Using the TCP/IP protocols architecture there are some limitations for the software-driven control network infrastructure, formed by the OSKMV, as the IP, TCP, UDP and SCTP protocols can not support some of the OSKMV needs directly in their protocols stack. These protocols generally just can send information at the data field and do not support semantic in their stacks.

This possibility, to use ontology at the intermediate networks layers, can also contribute to the translations of the MBT (Model-based Translator) software package, by the use of the FINLAN formal representation in OWL. Similar considerations about contributions can be done at the Service, Content and User Centric approaches, when using the current TCP/IP layers 3 and 4, and not one Clean-Slate approach. Through the use of FINLAN ontology layers, explicit represented in OWL, the OSKMV planes and the Service, Content and User Centric works can have the benefit to inform their needs to the Net-Ontology and DL-Ontology layers and better approximate the upper and lower layers semantically. For this, the next section gives one example of the FINLAN ontology approach use and the section 2 expands the possibilities.

1.2   FINLAN Ontology Example

Discussions about the use of ontology languages in the interface between network layers, instead of protocols, are pertinent. However, they do sense by the service concept and service provider definitions defended by Vissers, long time ago, in his defense about important issues in the design of data communications protocols [18]. Vissers' position also does sense for some current and Future Internet proposals by the separation of the internal complexities of each layer and exposing only the interfaces between them. In these concepts, the use of one ontology language also contributes to give a formal semantic power at the layers interface. This work uses OWL as formal language for this communication, as the OWL was adopted by a considerable number of initiatives and is a W3C standard. However, as the foundation is the ontology use at the intermediate layers, others formal ontology languages can be used.

One example of the FINLAN ontology use in the Future Internet research area is the possibility to support the AUTOI Functional Components communication with (and between) the network elements. So, the interactions between these components and their communication with the network intermediate layers can use OWL, instead of IP, UDP and TCP protocols. In a real integration, the migration of the end-user content service traffic presented in [14], section 4, for the Autonomic Management System (AMS) at the interaction *migrateService(contentService)*, is understood by the FINLAN Net-Ontology layer with:

```
<owl:Individual rdf:about="&Entity;Service-1">
    <rdf:type rdf:resource="&Entity;Service"/>
    <Name>Service-1</Name>
        <Migrate_service rdf:resource="&Entity;VirtualRouter-2"/>
</owl:Individual>
```

The Entity concept in the FINLAN ontology is one subclass of Thing (the OWL superclass) and the Service is one subclass of Entity.

```
<SubClassOf>
        <Class IRI="#Entity"/>
        <Class abbreviatedIRI=":Thing"/>
</SubClassOf>
<SubClassOf>
        <Class IRI="#Service"/>
        <Class IRI="#Entity"/>
</SubClassOf>
```

This work shows how FINLAN can contribute with Future Internet researches (using AutoI as integration example) and it is not scope to describe the ontology foundation concepts and the implementations to enable the network communication without using the IP, TCP, UDP and SCTP protocols, as these studies and results are presented in some of our previous works [4–10, 16].

2 Contributions to the Future Internet Works

The FINLAN project has adherence with some current efforts in the Future Internet research area, and the representation example above shows that the ontology cross layers can, naturally, be integrated with other initiatives. Its use is possible in a wide range of scenarios and is possible to visualize some characteristics and advantages. Attempting to the alignment with some Future Internet groups proposals, the next section extends possible collaborations that may be implemented in an integrated way with some works.

For better understanding, Fig. 2 illustrates an overview of the basic concepts of FINLAN ontology. The main classes are defined as follows:

- **DIS (Domain ID System):** is responsible for IDs resolution in FINLAN. Therefore, it manages the network's addressing;
- **Entity:** the class that represents all that can establish communication. For example: a service, a content, a network element and even a cloud computing;
- **ID:** the unique identifier of each entity;
- **Layer:** class representing a layer in FINLAN. It contains four sub-classes (Application, Net-Ontology, DL-Ontology and Physical) relating to the FINLAN layer structure, as illustrated in Fig. 1;
- **Necessity:** represents the requirements that an entity has during communication. For instance: delivery guarantee, QoS, security and others.

2.1 Collaboration to the AutoI Planes

One of the Autonomic Internet project expectations is to support the needs of virtual infrastructure management to obtain self-management of virtual resources which can cover heterogeneous networks and services like mobility, reliability, security and QoS. The FINLAN project can contribute in its challenges, some described in [1, 3, 12], to have the intermediate layers support for self-awareness, self-network and self-service knowledge.

Virtual Resources Management: The FINLAN ontology supports the network communication used by the AutoI vCPI (Virtual Component Programming Interface) [13], allowing a localized monitoring and management of the virtual resources. By this, the FINLAN ontology layers can comprehend communication needs as the instantiation, remotion and modification of virtual resources.

To support the information monitoring which is essential to the functions of self-performance and fault management, the FINLAN, through its ontology

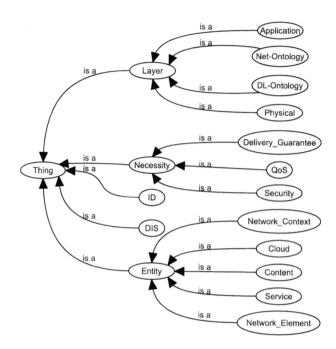

Fig. 2. Overview of FINLAN Ontology

design, can comprehend and do actions in information like: use of the CPU, memory assignment, packets lost and others. The invocation of the methods can be done by the AMSs, DOCs (Distributed Orchestration Component) or ANPI (Autonomic Network Programming Interface) and one OWL sample code for this communication is showed bellow:

```
<owl:Individual rdf:about="&Service;Component-1">
    <rdf:type rdf:resource="&Service;Interface"/>
    <pktsLost rdf:resource="&Entity;VirtualRouter-2"/>
    <stateCurrent rdf:resource="&Entity;VirtualRouter-2"/>
</owl:Individual>
<owl:Individual rdf:about="&Entity;VirtualRouter-2">
    <rdf:type rdf:resource="&Entity;NetworkElement"/>
    <hasVirtualLink rdf:resource="&Entity;VirtualRouter-1"/>
</owl:Individual>
```

In the FINLAN ontology, an Interface is a subclass of the Service Entity and two network elements, like virtual routers, can interact between them through the property hasVirtualLink.

Collect, Dissemination and Context Information Processing: FINLAN allows to create the Net-Ontology interface with AutoI to support the context-aware control functions for the self-management and adaptation in the CISP (Context Information Services Platform) needs. The context information in the FINLAN layers can act as an intermediate unity with its own semantic to reduce the number of interactions between the context sources and the context clients, diminishing the network effort in some cases.

The network context, for example, can interact with network elements and services according to the ontology concepts in the following code, where the context *NetContext-1* is in the network element *VirtualRouter-2* and the service *Service-1* has information to the network context *NetContext-1*.

```
<owl:Individual rdf:about="&Entity;VirtualRouter-2">
        <rdf:type rdf:resource="&Entity;NetworkElement"/>
        <isInsertedIn rdf:resource="&Entity;NetContext-1"/>
</owl:Individual>
<owl:Individual rdf:about="&Entity;NetContext-1">
        <rdf:type rdf:resource="&Entity;NetworkContext"/>
        <hasInformationTo rdf:resource="&Entity;Service-1"/>
</owl:Individual>
```

Management of Active Sessions: The AutoI open source implements a scalable and modular architecture to the deployment, control and management of active sessions used by virtual entities. It consists of one active element and the forwarding engine like a router [15].

Its integration with FINLAN can act in some components, like the Diverter, the Session Broker and the Virtualisation Broker. There are many others but these are essentials. As the communication between the AutoI modules is done through UDP transactions or TCP connections, FINLAN can collaborate in this scenario, facilitating the effort for the AutoI in the implementation of the network semantic support for them. The representation of the Session Broker communicating with a service is showed in the example below:

```
<owl:Individual rdf:about="&Service;SessionBroker">
        <rdf:type rdf:resource="&Service;Interface"/>
        <communicatesWith rdf:resource="&Entity;Service-1"/>
</owl:Individual>
```

The representation of the FINLAN project collaboration with AutoI is showed in Fig. 3, extended from the AutoI planes figure presented in [3].

2.2   Collaboration to the RESERVOIR Service Provider

The FINLAN project does not intend to conflict with the RESERVOIR proposals, as RESERVOIR also has the objective to supply an architecture and the reference implementation to a service oriented infrastructure. On the other hand, the FINLAN project aims to contribute with researches to build new technologies to replace the traditional TCP/IP layers 2, 3 and 4, giving semantic features to the network intermediate layers. In this proposal, is possible to use the RESERVOIR manifest formally defined in OWL to the contract and the SLA between the service provider and the infrastructure provider. As the manifest has specifications and rules well defined it facilitates the ontology creation to support them in FINLAN, and to create its individuals.

Specific information about components, as maximum and minimum, requisites for an instance (memory size, storage pool size, number of virtual CPUs,

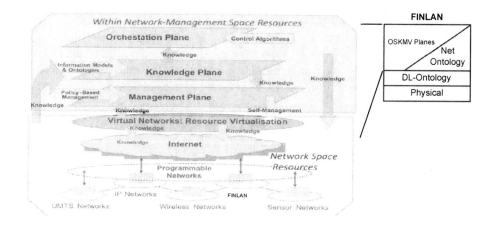

Fig. 3. FINLAN Collaboration with AutoI Planes

number of network interfaces and its bandwidths) are aspects to be worked in an effective collaboration. For this, the FINLAN ontology supports, for example, services that communicates with the ServiceCloud Entity, which has the need of the information stored in the manifest requirement. The ontology concept for this example is:

```
<owl:Individual rdf:about="&Entity;Service-2">
        <rdf:type rdf:resource="&Entity;Service"/>
        <communicatesWith rdf:resource="Entity;SCloud-1"/>
</owl:Individual>
<owl:Individual rdf:about="&Entity;SCloud-1">
        <rdf:type rdf:resource="&Entity;ServiceCloud"/>
        <hasNeedOf rdf:resource="&Requirement;Manifest-1"/>
</owl:Individual>
```

2.3   Collaboration to the Complexity Reduction for {User, Service, Content}-Centric Approaches

This work can collaborate to reduce the complexity of the network use by the user, service and content centric projects, as the ontology can offer better comprehension for the networks. About the proposals for a Clean-Slate solution this work also gives collaboration, by the OWL experiments at the intermediate network layers and the cross layers communication.

In the researches of service-centric, the FINLAN is placed to the level of the network elements and generates semantic support and answers to the higher layer needs. For example, to handle requests for services related to bandwidth, storage, encryption, location, indexing and others.

Related to the content-centric it is presented in [19] the difficulties of the current networks to support the objects concept. In order to reduce the complexity in this implementation it was brought to FINLAN the concept of content object providing the formalization to one Clean-Slate approach.

In this proposal, the objects Media, Rules, Behaviour, Relations and Characteristics, components of ALLOA (Autonomic Layer-Less Object Architecture), can use the FINLAN to easily interact with the network elements, simplifying the communication process with the lower layers.

Through the FINLAN Net-Ontology layer, requirements such as QoS and Security, can be requested to the network, making the {user, service, content}-centric approaches simpler, as shown in the sample code below:

```
<owl:Individual rdf:about="&Entity;PrivateContent">
        <rdf:type rdf:resource="&Entity;Content"/>
        <hasNeedOf rdf:resource="&Requirement;Security"/>
</owl:Individual>
<owl:Individual rdf:about="&Entity;MultimediaConference">
        <rdf:type rdf:resource="&Entity;Content"/>
        <hasNeedOf rdf:resource="&Requirement;QoS"/>
</owl:Individual>
```

3   Integration between Services and Networks

This section describes how to integrate this project in collaboration with others Future Internet works, continuing the examples with the AutoI integration. As the AutoI project has been fulfilling its purposes, it is observed that the evolution of TCP/IP layers to increase the networks communication possibilities, is a growing need and can not be disregarded to the future of the Internet infrastructure.

AutoI modules connections are performed in well defined form using connection handlers or similar classes that uses TCP/IP sockets. The AutoI integration with networks using semantics, instead of TCP/IP protocols, can be done using the FINLAN library. The benefit of this integration for the AutoI Orchestration plane is the use of the ontology at the intermediate layers to support the semantic needs to orchestrate the negotiation between the AMSs and the communications with the network elements. Other benefits are the possibility for the DOC to request different needs to the network layer.

3.1   Integration Using FINLAN Library

The FINLAN library uses raw sockets for sending and receiving packets directly between the data link and the Application layer. The implementation of the FINLAN proposal, supporting ontology, incorporates the concepts discussed in [16] and [6] and extends this functionality as it allows the understanding of the needs for establishing communication through the ontological model adopted.

For the example described in 1.2, to the service migration, the ANPI receives the *migrateService(contentService)* from the AMS. Based on the AutoI Java open source, in the ANPI demo, the *ANPISDD* class is prepared to use the IP and TCP (port 43702) protocols. So the SDD (Service Deployment Daemon) runs using the traditional TCP/IP stack, as in the following sample code extracted from the ANPISDD.java code.

```
public class ANPISDD extends Thread {
    private ServerSocket server;
    private int port = 43702;
    private Socket s = null;
    public static KnowledgePlane KP = null;
    private ANPIConnectionHandler HD = null;
...
    System.out.println("Listening " + this.port + "...");
    s1 = server.accept();  ...
```

With the use of the FINLAN library this communication can be done replacing
the IP and TCP protocols with the FINLAN representation using OWL over raw
sockets for the Net-Ontology and DL-ontology layers presented in Fig. 1. This
expands the semantic possibilities for the AutoI planes, through the intermediate
layers of the networks in the Future Internet, for the communication between
the Service Enabler plane and the Management/Knowledge plane implemented
by the AMS.

From a practical way, the FINLAN library implementation occurs through
the development of Net-Ontology and DL-Ontology layers, which depend on the
design of OWL concepts of the FINLAN proposal, illustrated in Fig. 2.

Fig. 4 shows how the Net-Ontology and DL-Ontology layers relate with the
developed ontology, Services and Physical layers of the network. These layers
are based on the formalization of the FINLAN concepts. This formalization was
modeled in Protégé, also used to generate the OWL, by the OWL API (version
3.0.0.1451). In this figure, the arrows show that the concepts in OWL are loaded
into Net-Ontology and DL-Ontology layers allowing the semantic communication
and network behavior control.

From the OWL concepts, the Net-Ontology layer is skillful to receive requests
for applications, which are transmitted via messages containing OWL tags, and
the needs of the data flow that will start. With the understanding of application
needs, the Net-Ontology layer, sends to the DL-Ontology layer another OWL
object with the requirements of data communication as a way of addressing, for
example. After the definitions of application requirements, the communication

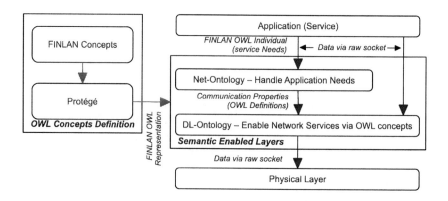

Fig. 4. Overview of FINLAN Library Implementation

is ready to be established, and the data is sent through the layers also using raw sockets.

At the current stage of development the implementation of FINLAN library is made in application level. Nevertheless, the future intentions are to implement the FINLAN ontology in Linux operating system kernel level, allowing the facilities in its use in different programming languages, since the methods proposed would be available at the operating system level.

4   Conclusions

This paper has presented the FINLAN ontology works in a collaboration perspective with some Future Internet projects. We have proposed to better meeting of services and networks by approaching services semantically to the network structure. It was showed how to integrate FINLAN with Future Internet projects, taking AutoI as example, and how the ontological approach can be applied to Future Internet works like monitoring and content-centric Internet.

Future work will implement the FINLAN ontology at the Linux kernel level and run performance and scalability experiments with different Future Internet projects open implementations. Further work also will do the extension of the scope of the ontological representation, by modeling the behavior of FINLAN to support requirements in contribution with different Future Internet projects.

We strongly believe that meeting services and networks through the reduction of network layers and, consequently, through the decreasing of users, services and content complexity is a possible way to achieve flexibility in future networks. Moreover, we expect that ontological approaches can help to build a Future Internet with its real challenges, requirements and new paradigms.

Acknowledgment. This work is a result of conceptual discussions and researches of all members of the FINLAN group. The authors would like to acknowledge the implementations and philosophical talks with this group. Also to thank the efforts to gather on the state-of-the-art of the Future Internet.

References

[1] Clayman, S., Galis, A., Chapman, C., Toffetti, G., Rodero-Merino, L., Vaquero, L.M., et al.: Monitoring Service Clouds in the Future Internet. In: Towards the Future Internet - Emerging Trends from European Research, p. 115 (2010)

[2] FIRE: FIRE White Paper (Aug. 2009), http://www.ict-fireworks.eu/fileadmin/documents/FIRE_White_Paper_2009_v3.1.pdf

[3] Galis, A., Denazis, S., Bassi, A., Giacomin, P., Berl, A., A., Fischer, o.: Management Architecture and Systems for Future Internet. In: Towards the Future Internet - A European Research Perspective, p. 112 (2009)

[4] Malva, G.R., Dias, E.C., Oliveira, B.C., Pereira, J.H.S., Kofuji, S.T., Rosa, P.F.: Implementação do Protocolo FINLAN. In: 8th International Information and Telecommunication Technologies Symposium (2009)

[5] Pereira, F.S.F., Santos, E.S., Pereira, J.H.S., Rosa, P.F., Kofuji, S.T.: Proposal for Hybrid Communication in Local Networks. In: 8th International Information and Telecommunication Technologies Symposium (2009)

[6] Pereira, F.S.F., Santos, E.S., Pereira, J.H.S., Rosa, P.F., Kofuji, S.T.: FINLAN Packet Delivery Proposal in a Next Generation Internet. In: IEEE International Conference on Networking and Services, p. 32 (2010)

[7] Pereira, J.H.S., Kofuji, S.T., Rosa, P.F.: Distributed Systems Ontology. In: IEEE/IFIP New Technologies, Mobility and Security Conference (2009)

[8] Pereira, J.H.S., Kofuji, S.T., Rosa, P.F.: Horizontal Address Ontology in Internet Architecture. In: IEEE/IFIP New Technologies, Mobility and Security Conference (2009)

[9] Pereira, J.H.S., Kofuji, S.T., Rosa, P.F.: Horizontal Addressing by Title in a Next Generation Internet. In: IEEE International Conference on Networking and Services, p. 7 (2010)

[10] Pereira, J.H.S., Pereira, F.S.F., Santos, E.S., Rosa, P.F., Kofuji, S.T.: Horizontal Address by Title in the Internet Architecture. In: 8th International Information and Telecommunication Technologies Symposium (2009)

[11] Pereira, J.H.S., Santos, E.S., Pereira, F.S.F., Rosa, P.F., Kofuji, S.T.: Layers Optimization Proposal in a Post-IP Network. International Journal On Advances in Networks and Services, in Press (2011)

[12] Rochwerger, B., Galis, A., Breitgand, D., Levy, E., Cáceres, J., Llorente, I., Wolfsthal, Y., et al.: Design for Future Internet Service Infrastructures. In: Towards the Future Internet - A European Research Perspective, p. 227 (2009)

[13] Rubio-Loyola, J., Astorga, A., Serrat, J., Chai, W.K., Mamatas, L., Galis, A., Clayman, S., Cheniour, A., Lefevre, L., et al.: Platforms and Software Systems for an Autonomic Internet. In: IEEE Global Communications Conference (2010)

[14] Rubio-Loyola, J., Astorga, A., Serrat, J., Lefevre, L., Cheniour, A., Muldowney, D., Davy, S., Galis, A., Mamatas, L., Clayman, S., Macedo, D., et al.: Manageability of Future Internet Virtual Networks from a Practical Viewpoint. In: Towards the Future Internet - Emerging Trends from European Research, p. 105 (2010)

[15] Rubio-Loyola, J., Serrat, J., Astorga, A., Chai, W.K., Galis, A., Clayman, S., Mamatas, L., Abid, M., Koumoutsos, G.: et al.: Autonomic Internet Framework Deliverable D6.3. Final Results of the AutonomicI Approach. AutoI Project (2010)

[16] Santos, E.S., Pereira, F.S.F., Pereira, J.H.S., Rosa, P.F., Kofuji, S.T.: Optimization Proposal for Communication Structure in Local Networks. In: IEEE International Conference on Networking and Services, p. 18 (2010)

[17] Vanni, R.M.P.: Integração de Serviços em Ambientes Heterogêneos: uso de Semântica para Comunicação Entre Entidades em Mudanças de Contexto. Ph.D. thesis, University of São Paulo - USP (2009)

[18] Vissers, C.A., Logrippo, L.: The Importance of the Service Concept in the Design of Data Communications Protocols. Proceedings of the IFIP WG6 1, 3 (1986)

[19] Zahariadis, T., Daras, P., Bouwen, J., Niebert, N., Griffin, D., Alvarez, F., Camarillo, G.: Towards a Content-Centric Internet. In: Towards the Future Internet - Emerging Trends from European Research, p. 227 (2010)

Fostering a Relationship between Linked Data and the Internet of Services

John Domingue[1], Carlos Pedrinaci[1], Maria Maleshkova[1], Barry Norton[2], and
Reto Krummenacher[3]

[1] Knowledge Media Institute, The Open University, Walton Hall, Milton Keynes,
MK6 7AA UK
{j.b.domingue, c.pedrinaci, m.maleshkova}@open.ac.uk
[2] Karlsruhe Institute of Technology, Karlsruhe, Germany
barry.norton@aifb.uni-karlsruhe.de
[3] Semantic Technology Institute, University of Innsbruck, 6020 Innsbruck, Austria
reto.krummenacher@sti2.at

Abstract. We outline a relationship between Linked Data and the Internet of
Services which we have been exploring recently. The Internet of Services provides a mechanism for combining elements of a Future Internet through standardized service interfaces at multiple levels of granularity. Linked Data is a
lightweight mechanism for sharing data at web-scale which we believe can facilitate the management and use of service-based components within global
networks.

Keywords: Linked Data, Internet of Services, Linked Services

1 Introduction

The Future Internet is a fairly recent EU initiative which aims to investigate scientific
and technical areas related to the design and creation of a new global infrastructure.
An overarching goal of the Future Internet is that the new platform should meet
Europe's economic and societal needs. The Internet of Services is seen as a core component of the Future Internet:

> "The Future Internet is polymorphic infrastructure, where the bounda-
> ries between silo systems are changing and blending and where the em-
> phasis is on the integration, interrelationships and interworking of the
> architectural elements through new service-based interfaces". [Frederic
> Gittler, FIA Stockholm]

The Web of Data is a relatively recent effort derived from research on the Semantic
Web [1], whose main objective is to generate a Web exposing and interlinking data
previously enclosed within silos. Like the Semantic Web the Web of Data aims to
extend the current human-readable Web with data formally represented so that software agents are able to process and reason with the information in an automatic and

J. Domingue et al. (Eds.): Future Internet Assembly, LNCS 6656, pp. 351–364, 2011.

flexible way. This effort, however, is based on the simplest form of semantics, RDF(S) [2], and has thus far focused on promoting the publication, sharing and linking of data on the Web.

From a Future Internet perspective a combination of service-orientation and Linked Data provides possibilities for supporting the integration, interrelationship and inter-working of Future Internet components in a partially automated fashion through the extensive use of machine-processable descriptions. From an Internet of Services perspective, Linked Data with its relatively simple formal representations and in-built support for easy access and connectivity provides a set of mechanisms supporting interoperability between services. In fact, the integration between services and Linked Data is increasingly gaining interest within industry and academia. Examples include, for instance, research on linking data from RESTful services by Alarcon et al. [3], work on exposing datasets behind Web APIs as Linked Data by Speiser et al. [4], and Web APIs providing results from the Web of Data like Zemanta[1].

We see that there are possibilities for Linked Data to provide a common 'glue' as services descriptions are shared amongst the different roles involved in the provision, aggregation, hosting and brokering of services. In some sense service descriptions as, and interlinked with, Linked Data is complementary to SAP's Unified Service Description Language[2] [5], within their proposed Internet of Services framework[3], as it provides appropriate means for exposing services and their relationships with providers, products and customers in a rich, yet simple manner which is tailored to its use at Web scale.

In this paper we discuss the relationship between Linked Data and services based on our experiences in a number of projects. Using what we have learnt thus far, at the end of the paper we propose a generalization of Linked Data and service principles for the Future Internet.

2   Linked Data

The Web of Data is based upon four simple principles, known as the Linked Data principles [6], which are:

1. Use URIs (Uniform Resource Identifiers) as names for things.
2. Use HTTP URIs so that people can look up those names.
3. When someone looks up a URI, provide useful information, using standards (RDF*, SPARQL).
4. Include links to other URIs, so that they can discover more things.

---

[1]   http://developer.zemanta.com/
[2]   http://www.internet-of-services.com/index.php?id=288&L=0
[3]   http://www.internet-of-services.com/index.php?id=260&L=0

RDF (Resource Description Framework) is a simple data model for semantically describing resources on the Web. Binary properties interlink terms forming a directed graph. These terms as well as the properties are described by using URIs. Since a property can be a URI, it can again be used as a term interlinked to another property.

SPARQL is a query language for RDF data which supports querying diverse data sources, with the results returned in the form of a variable-binding table, or an RDF graph.

Since the Linked Data principles were outlined in 2006, there has been a large up-take impelled most notably by the Linking Open Data project[4] supported by the W3C Semantic Web Education and Outreach Group.

As of September 2010, the coverage of the domains in the Linked Open Data Cloud is diverse (Figure 1). The cloud now has nearly 25 billion RDF statements and over 400 million links between data sets that cover media, geography, academia, life-sciences and government data sets.

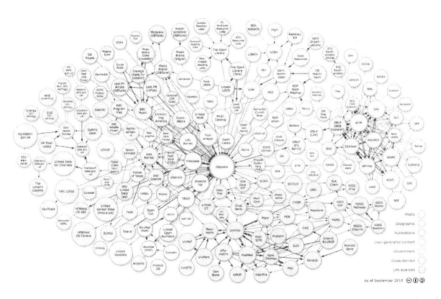

Fig. 1. Linking Open Data cloud diagram as of September 2010, by Richard Cyganiak and Anja Jentzsch[5].

From a government perspective significant impetus to this followed Gordon Brown's announcement when he was UK Prime Minister[6] on making Government data freely available to citizens through a specific Web of Data portal[7] facilitating the creation of a diverse set of citizen-friendly applications.

---

[4] http://esw.w3.org/SweoIG/TaskForces/CommunityProjects/LinkingOpenData

[5] http://lod-cloud.net/

[6] http://www.silicon.com/management/public-sector/2010/03/22/gordon-brown-spends-30m-to-plug-britain-into-semantic-web-39745620/

[7] http://data.gov.uk/

On the corporate side, the BBC has been making use of RDF descriptions for some time. BBC Backstage[8] allows developers to make use of BBC programme data available as RDF. The BBC also made use of scalable RDF repositories for the back-end of the BBC world cup website[9] to facilitate "agile modeling"[10]. This site was very popular during the event receiving over 2 million queries per day.

Other examples of commercial interest include: the acquisition of Metaweb[11] by Google to enhance search, and the release of the OpenGraph[12] API by Facebook. Mark Zuckerberg, Facebook's CEO claimed recently that Open Graph was the "the most transformative thing we've ever done for the Web"[13].

3   Services on the Web

Currently the world of services on the Web is marked by the formation of two main groups of services. On the one hand, "classical" Web services, based on WSDL and SOAP, play a major role in the interoperability within and among enterprises. Web services provide means for the development of open distributed systems, based on decoupled components, by overcoming heterogeneity and enabling the publishing and consuming of functionalities of existing pieces of software. In particular, WSDL is used to provide structured descriptions for services, operations and endpoints, while SOAP is used to wrap the XML messages exchanged between the service consumer and provider. A large number of additional specifications such as WS-Addressing, WS-Messaging and WS-Security complement the stack of technologies.

On the other hand, an increasing number of popular Web and Web 2.0 applications as offered by Facebook, Google, Flickr and Twitter offer easy-to-use, publicly available Web APIs, also referred to as RESTful services (properly when conforming to the REST architectural principles [7]). RESTful services are centred around resources, which are interconnected by hyperlinks and grouped into collections, whose retrieval and manipulation is enabled through a fixed set of operations commonly implemented by using HTTP. In contrast to WSDL-based services, Web APIs build upon a light technology stack relying almost entirely on the use of URIs, for both resource identification and interaction, and HTTP for message transmission.

The take up of both kinds of services is, however, hampered by the amount of manual effort required when manipulating them. Research on semantic Web services [8] has focused on providing semantic descriptions of services so that tasks such as the discovery, negotiation, composition and invocation of Web services can have a higher level of automation. These techniques, originally targeted at WSDL services, have highlighted a number of advantages and are currently being adapted towards lighter and more scalable solutions covering Web APIs as well.

---

[8]   http://backstage.bbc.co.uk/
[9]   http://news.bbc.co.uk/sport1/hi/football/world_cup_2010/default.stm
[10]   http://www.bbc.co.uk/blogs/bbcinternet/2010/07/bbc_world_cup_2010_dynamic_sem.html
[11]   http://www.freebase.com/
[12]   http://developers.facebook.com/docs/opengraph
[13]   http://news.cnet.com/8301-13577_3-20003053-36.html

4   Linked Services

The advent of the Web of Data together with the rise of Web 2.0 technologies and social principles constitute, in our opinion, the final necessary ingredients that will ultimately lead to a widespread adoption of services on the Web. The vision toward the next wave of services, first introduced in [9] and depicted in Figure 1, is based on two simple notions:

1. Publishing service annotations within the Web of Data, and
2. Creating services for the Web of Data, i.e., services that process Linked Data and/or generate Linked Data.

We have since then devoted significant effort to refining the vision [10] and implementing diverse aspects of it such as the annotation of services and the publication of services annotations as Linked Data [11, 12], as well as on wrapping, and openly exposing, existing RESTful services as native Linked Data producers dubbed Linked Open Services [13, 14]. It is worth noting in this respect that these approaches and techniques are different means contributing to the same vision and are not to be considered by any means the only possible approaches. What is essential though is exploiting the complementarity of services and the Web of Data through their integration based on the two notions highlighted above.

As can be seen in Figure 2 there are three main layers that we consider. At the bottom are Legacy Services which are services which may be WSDL-based or Web APIs, for which we provide in essence a Linked Data-oriented view over existing functionality exposed as services. Legacy services could in this way be invoked, either

Fig. 2. Services and the Web of Data

by interpreting their semantic annotations (see Section 4.1) or by invoking dedicated wrappers (see Section 4.2) and RDF information could be obtained on demand. In this way, data from legacy systems, state of the art Web 2.0 sites, or sensors, which do not directly conform to Linked Data principles can easily be made available as Linked Data.

In the second layer are Linked Service descriptions. These are annotations describing various aspects of the service which may include: the inputs and outputs, the functionality, and the non-functional properties. Following Linked Data principles these are given HTTP URIs, are described in terms of lightweight RDFS vocabularies, and are interlinked with existing Web vocabularies. Note that we have already made our descriptions available in the Linked Data Cloud through iServe these are described in more detail in Section 4.1.

The final layer in Figure 2 concerns services which are able to consume RDF data (either natively or via lowering mechanisms), carry out the concrete activity they are responsible for, and return the result, if any, in RDF as well. The invoking system could then store the result obtained or continue with the activity it is carrying out using these newly obtained RDF triples combined with additional sources of data. Such an approach, based on the ideas of semantic spaces, has been sketched for the notion of Linked Open Processes [13]. In a sense, this is similar to the notion of service mashups [15] and RDF mash-ups [16] with the important difference that services are, in this case, RDF-aware and their functionality may range from RDF-specific manipulation functionality up to highly complex processing beyond data fusion that might even have real-life side-effects. The use of services as the core abstraction for constructing Linked Data applications is therefore more generally applicable than that of current data integration oriented mashup solutions.

We expand on the second and third layers in Figure 2 in more detail below.

4.1   Implementing Linked Services with Linked Data-Based Annotations

One thread of our work on Linked Services is based on the use of Linked Data-based descriptions of Linked Services allowing them to be published on the Web of Data and using these annotations for better supporting the discovery, composition and invocation of Linked Services.

Our research there is based on the Minimal Service Model (MSM) [17], originally introduced together with hRESTS [18] and WSMO-Lite [19], and slightly modified for the purposes of this work [12]. In a nutshell, MSM is a simple RDF(S) integration ontology which captures the maximum common denominator between existing conceptual models for services. The best-known approaches to annotating services semantically are OWL-S [20], WSMO [21], SAWSDL [22], and WSMO-Lite for WSDL services, and MicroWSMO [23], and SA-REST for Web APIs. To cater for interoperability, MSM represents essentially the intersection of the structural parts of these formalisms. Additionally, as opposed to most semantic Web services research to date, MSM supports both "classical" WSDL Web services, as well as a procedural view on the increasing number of Web APIs and RESTful services, which appear to be preferred on the Web.

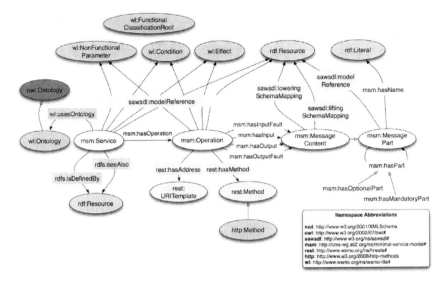

Fig. 3. Conceptual model for services used by iServe

As it can be seen in Figure 3, MSM defines Services, which have a number of *Operations*. Operations in turn have input, output and fault *MessageContent* descriptions. MessageContent may be composed of mandatory or optional *MessageParts*. The addition of message parts extends the earlier definition of the MSM as described in [18]. The SAWSDL, WSMO-Lite and hRESTS vocabularies, depicted in Figure 3 with the `sawsdl`, `wl`, and `rest` namespaces respectively, complete MSM. SAWSDL supports the annotation of WSDL and XML Schema syntactic service descriptions with semantic concepts, but does not specify a particular representation language nor does it provide any specific vocabulary that users should adopt. WSMO-Lite builds upon SAWSDL by extending it with a model specifying the semantics of the particular service annotations. It provides classes for describing non-functional semantics through the concept of *Nonfunctional Parameter*, and functional semantics via the concepts *Condition*, *Effect*, and *Functional Classification Root*. Finally, hRESTS extends the MSM with specific attributes for operations to model information particular to Web APIs, such as a method to indicate the HTTP method used for the invocation.

The practical use of the MSM for service annotation is supported by two tools, namely SWEET [11] and SOWER. The former is a web-based tool that assists users in the creation of semantic annotations of Web APIs, which are typically described solely through an unstructured HTML Web page. SWEET[14] can open any web page and directly insert annotations following the hRESTS/MicroWSMO microformat. It enables the completion of the following key tasks:

---

[14] http://sweet.kmi.open.ac.uk/

- Identification of service properties within the HTML documentation with the help of hRESTS.
- Integrated ontology search for linking semantic information to service properties.
- Adding of semantic annotations and including lifting and lowering mechanisms that handle format transformations.
- Saving of semantically annotated HTML service description, which can be republished on the Web.
- Extraction of RDF service descriptions based on the annotated HTML.

Similarly, the second tool, SOWER, assists users in the annotation of WSDL services and is based in this case on SAWSDL for adding links to semantic descriptions as well as lifting and lowering mechanisms. During the annotation both tools make use of the Web of Data as background knowledge so as to identify and reuse existing vocabularies. Doing so simplifies the annotation and additionally it also leads to service annotations that are potentially more reusable since they are adapted to existing sources of Linked Data.

The annotation tools are both connected to iServe for one click publication. iServe[15], previously introduced in [12], builds upon lessons learnt from research and development on the Web and on service discovery algorithms to provide a generic semantic service registry able to support advanced discovery over both Web APIs and WSDL services described using heterogeneous formalisms. iServe is, to the best of our knowledge, the first system to publish web service descriptions on the Web of Data, as well as the first to provide advanced discovery over Web APIs comparable to that available for WSDL-based services. Thanks to its simplicity, the MSM captures the essence of services in a way that can support service matchmaking and invocation and still remains largely compatible with the RDF mapping of WSDL, with WSMO-based descriptions of Web services, with OWL-S services, and with services annotated according to WSMO-Lite and MicroWSMO.

The essence of the approach followed by iServe is the use of import mechanisms for a wide range of existing service description formalisms to automatically transform them into the MSM. Once the services are transformed, service descriptions are exposed following the Linked Data principles and a range of advanced service analysis and discovery techniques are provided on top. It is worth noting that as service publication is based on Linked Data principles, application developers can easily discover services able to process or provide certain types of data, and other Web systems can seamlessly provide additional data about service descriptions in an incremental and distributed manner through the use of Linked Data principles. One such example is for instance LUF (Linked User Feedback)[16], which links service descriptions with users ratings, tags and comments about services in a separate server. On the basis of these ratings and comments, service recommendation facilities have also been implemented[17].

---

[15] http://iserve.kmi.open.ac.uk/
[16] http://soa4all.isoco.net/luf/about/
[17] http://technologies.kmi.open.ac.uk/soa4all-studio/consumption-platform/rs4all/

In summary, the fundamental objective pursued by iServe is to provide a platform able to publish service annotations, support their analysis, and provide advanced functionality on top like service discovery in a way that would allow people and machines to find and exploit service descriptions easily and conveniently. The simple conceptual model explained earlier is a principal building block to support this as a general model able to abstract away the existing conceptual heterogeneity among service description approaches without introducing considerable complexity from a knowledge acquisition and computational perspectives.

SPICES[18] [24] (Semantic Platform for the Interaction and Consumption of Enriched Services) is a platform for the easy consumption of services based on their semantic descriptions. In particular, SPICES supports both the end-user interaction with services and the invocation process itself, via the generation of appropriate user interfaces. Based on the annotations the user is presented with a set of fields, which must be completed to allow the service to execute, and these fields cover input parameters as well as authentication credentials. By using the provided input and the semantic service description stored in iServe, the service can be automatically invoked through SPICES.

Further tooling covering the composition of services as well as analysis of the execution are also being developed as part of an integrated tool suite called SOA4All Studio[19]. The SOA4All studio is a fully-fledged system that provides extensive support for completing different tasks along the lifecycle of services, enabling the creation of semantic service description, their discovery, composition, invocation and monitoring.

4.2   Services Which Produce and Consume Linked Data

In this section we consider the relationship between service *interactions* and Linked Data; that is, how Linked Data can facilitate the interaction with a service and how the result can contribute to Linked Data. In other words, this section is not about *annotating* service descriptions by means of ontologies and Linked Data, but about how services should be implemented on top of Linked Data in order to become first class citizens of the quickly growing Linking Open Data Cloud. Note that we take a purist view of the type of services which we consider. These services should take RDF as input and the results should be available as RDF; i.e., service consume Linked Data and service produce Linked Data. Although this could be considered restrictive, one main benefit is that everything is instantaneously available in a machine-readable form. Within existing work on Semantic Web Services, considerable effort is often expended in *lifting* from a syntactic description to a semantic representation and lowering from a semantic entity to a syntactic form. Whereas including this information as annotations requires a particular toolset and platform to interpret them, following Linked Data and

---

[18] http://soa4all.isoco.net/spices/about/
[19] http://technologies.kmi.open.ac.uk/soa4all-studio/

REST principles allows for re-exposing the wrappers as RESTful services so that the only required platform to interact with them is the Web (HTTP) itself.

As a general motivation for our case, we consider the status quo of the services offered over the geonames data set, a notable and 'lifelong' member of the Linking Open Data Cloud, which are primarily offered using JSON- and XML-encoded messaging. A simple example is given in Table 1, which depicts an excerpt of a weather report gathered from the station at the Airport in Innsbruck, Austria.

Table 1. Geonames JSON Weather Results Example

```
{"weatherObservation":{
        "stationName":"Innsbruck-Flughafen", "ICAO":"LOWI",
        "countryCode":"AT",
        "lat":47.266666, "lng":11.333333, "elevation":581,
        "clouds":"few clouds", "temperature":"3", … }}
```

While the JSON format is very convenient for consumption in a browser-based client, it conveys neither the result's internal semantics nor its interlinkage with existing data sets. The keys, before each colon, are ambiguous strings that must be understood per API; in Linked Data, on the other hand, geonames itself provides a predicate and values for country codes and the WGS84 vocabulary is widely used for latitude and longitude information. Similarly the value "LOWI" corresponds to a feature found within the geonames dataset (and indeed also within the OurAirports and DBpedia Linked Data sets)[20] but the string value does not convey this interlinkage.

A solution more in keeping with the Linked Data principles, as seen in our version of these services,[21] uses the same languages and technologies in the implementation and description of services, communicated as the Linked Open Service (LOS) principles [14] encouraging the following:

- allowing RDF-encoded messages for input/output;
- reusing URIs from Linked Data source for representing features in input and output messages;
- making explicit the semantic relationship between input and output.

In particular with regard to the last point, we can use predicates from existing vocabularies, such as FOAF's basedNear, to represent the relationship between an input point and the nearest weather station. In order to make the statement of this relationship more useful as Linked Data, the approach of Linked Data Services (LIDS) [25] is to URL-encode the input. For instance, the latitude and longitude and used as query parameters so that the point is represented in a URI forming a new

---

[20] The three identifiers for the Innsbruck Airport resource are
http://sws.geonames.org/6299669/, http://airports.dataincubator.org/airports/LOWI, and
http://dbpedia.org/resource/Innsbruck_Airport, respectively.
[21] http://www.linkedopenservices.org/services/geo/geonames/weather/

resource identifier. This URI is then used as the subject of such a triple, encoding the relationship to the output.

In aligning LOS and LIDS principles, pursued via a Linked Services Wiki[22] and a Linked Data and Services mailing list[23], a URI representing the input is returned using the standard Content-Location HTTP header field. Even in the case of a URL-encoded, LIDS-style input this can be sensible as such a URI will be canonical, whereas a user-encoded input may use variable decimal places for latitude and longitude. LOS has the further advantage that where an input cannot sensibly be URL-encoded, it can first be POSTed as a new resource (Linked Data and Linked Data Services so far concentrate on resource retrieval and therefore primarily the HTTP GET verb), in the standard REST style, and then a resource-oriented service can be offered with respect to it. This can be seen in ontology and query services offered at http://www.linkedopenservices.org/services.

LOS and LIDS also coincide on the idea of refining the general principles of Linked Services communicated in Section 4, of describing accepted/expected messages using SPARQL graph patterns. While this is a design decision, it aims at the greatest familiarity and ease for Linked Data developers. It is not without precedent in semantic service description [26]. The authors of [26] use the SPARQL query language to formulate user goals, and to define the pre- and post-conditions of SAWSDL-based service descriptions, which to some degree, at least conceptually, matches the ideas of our approach of using graph patterns for describing inputs (a pre-condition on the knowledge state prior to service invocation) and outputs (the post-condition of how the knowledge state changes after execution of the service). Although, the use of SPARQL is similar across different proposals, how the patterns are exploited again offers alternative, but complementary views due to LIDS and LOS respectively. On the one hand, atomic user desires can be encoded as a CONSTRUCT query and, under certain restrictions[24], query processing techniques can be used to assemble a set of services whose results can be combined to satisfy the initial user request. On the other hand, where more sophisticated control flow is needed, a process (which we call a Linked Open Process [13]) can be manually created and the graph patterns are used for both the discovery of services, and then also reused in defining the dataflow between services within a process, defined again as SPARQL CONSTRUCT queries. Work is on-going on graph pattern-based discovery and process definition and execution.

---

[22] http://linkedservices.org

[23] http://groups.google.com/group/linkeddataandservices/

[24] Currently that the graph patterns contained in this request, and in the service descriptions, are conjunctive – meaning do not use OPTIONAL or UNION, etc. – and free of FILTERs. etc. [4]

5   Conclusions

In this paper we have outlined how Linked Data provides a mechanism for describing services in a machine readable fashion and enables service descriptions to be seamlessly connected to other Linked Data. We have also described a set of principles for how services should consume and produce Linked Data in order to become first-class Linked Data citizens.

From our work thus far, we see that integrating services with the Web of Data, as depicted before, will give birth to a services ecosystem on top of Linked Data, whereby developers will be able to collaboratively and incrementally construct complex systems exploiting the Web of Data by reusing the results of others. The systematic development of complex applications over Linked Data in a sustainable, efficient, and robust manner shall only be achieved through reuse. We believe that our approach is a particularly suitable abstraction to carry this out at Web scale.

We also believe that Linked Data principles and our extensions can be generalized to the Internet of Services. That is, to scenarios where services sit within a generic Internet platform rather than on the Web. These principles are:

Global unique naming and addressing scheme - services and resources consumed and produced by services should be subject to a global unique naming and addressing scheme. This addressing scheme should be easily resolvable such that software clients are able to access easily underlying descriptions.

Linking – linking between descriptions should be supported to facilitate the reuse of descriptions and to be able to specify relationships.

Service abstraction – building from SOA principles functionality should be encapsulated within services which should have a distinct endpoint available on the Internet, through which they can be invoked using standard protocols.

Machine processability – the descriptions of the services and resources should be machine-processable. RDF(S) achieves this by having an underlying semantics and also with the ability to point to an ontology based description of the schema used. Ideally, the inputs and outputs for services should be machine-processable as well.

Following from the above we believe that the Future Internet will benefit greatly from a coherent approach which integrates service orientation with the principles underlying Linked Data. We are also hopeful that our approach provides a viable starting point for this. More generally, we expect to see lightweight semantics appearing throughout the new global communications platform which is emerging through the Future Internet work and also note that proposals already exist for integrating Linked Data at the network level[25].

Acknowledgements. This work was partly funded by the EU project SOA4All (FP7-215219)[26]. The authors would like to thank the members of the SOA4All project and the members of the STI Conceptual Models for Services Working Group for their interesting feedback on this work.

---

[25] http://socialmedia.net/node/175
[26] http://www.soa4all.eu/

References

1. Berners-Lee, T., Hendler, J., Lassila, O.: The Semantic Web, Scientific American 284(5), May 2001: pages 34-43.
2. Berners-Lee, T., Hendler, J., Lassila, O.: The Semantic Web. Scientific American 284(5), 34–43 (2001)
3. Brickley, D., Guha, D.,, R.V. (eds.): RDF Vocabulary Description Language 1.0: RDF Schema. W3C Recommendation (February 2004), http://www.w3.org/TR/rdf-schema/
4. A.R., Wilde, E.: Linking Data from RESTful Services. In: Workshop on Linked Data on the Web at WWW 2010 (2010)
5. Speiser, S., Harth, A.: Taking the LIDS off Data Silos. In: 6th International Conference on Semantic Systems (I-SEMANTICS) (October 2010)
6. Cardoso, J., Barros, A., May, N., Kylau, U.: Towards a Unified Service Description Language for the Internet of Services: Requirements and First Developments. In: IEEE Int'l Conference on Services Computing, July 2010, pp. 602–609 (2010)
7. Berners-Lee, T.: Linked Data - Design Issues (July 2006), http://www.w3.org/DesignIssues/LinkedData.html
8. Fielding, R.T.: Architectural Styles and the Design of Network-based Software Architectures. PhD Thesis, University of California (2000)
9. McIlraith, S.A., Son, T.C., Zeng, H.: Semantic Web Services. IEEE Intelligent Systems 16(2), 46–53 (2001)
10. Pedrinaci, C., Domingue, J., Krummenacher, R.: Services and the Web of Data: An Unexploited Symbiosis. In: AAAI Spring Symposium "Linked Data Meets Artificial Intelligence", March 2010, AAAI Press, Menlo Park (2010)
11. Pedrinaci, C., Domingue, J.: Toward the Next Wave of Services: Linked Services for the Web of Data. Journal of Universal Computer Science 16(13), 1694–1719 (2010)
12. Maleshkova, M., Pedrinaci, C., Domingue, J.: Supporting the creation of semantic RESTful service descriptions. In: Workshop: Service Matchmaking and Resource Retrieval in the Semantic Web at ISWC (November 2009)
13. Pedrinaci, C., Liu, D., Maleshkova, M., Lambert, D., Kopecky, J., Domingue, J.: iServe: a Linked Services Publishing Platform. In: Workshop: Ontology Repositories and Editors for the Semantic Web at ESWC (June 2010)
14. Krummenacher, R., Norton, B., Marte, A.: Towards Linked Open Services and Processes. In: Future Internet Symposium, October 2010, pp. 68–77.
15. Norton, B., Krummenacher, R.: Consuming Dynamic Linked Data. In: 1st International Workshop on Consuming Linked Data (November 2010)
16. Benslimane, D., Dustdar, S., Sheth, A.: Services Mashups: The New Generation of Web Applications. IEEE Internet Computing 12(5), 13–15 (2008)
17. Phuoc, D.L., Polleres, A., Hauswirth, M., Tummarello, G., Morbidoni, C.: Rapid Prototyping of Semantic Mash-ups Through Semantic Web Pipes. In: 18th Int'l Conference on World Wide Web, April 2009, pp. 581–590 (2009)
18. Maleshkova, M., Kopecky, J., Pedrinaci, C.: Adapting SAWSDL for Semantic Annotations of RESTful Services. In: Workshop: Beyond SAWSDL at OTM, November 2009, pp. 917–926 (2009)

19. Kopecky, J., Gomadam, K., Vitvar, T.: hRESTS: An HTML Microformat for Describing RESTful Web Services. In: IEEE/WIC/ACM Int'l Conference on Web Intelligence and Intelligent Agent Technology, December 2008, pp. 619–625 (2008)
20. Vitvar, T., Kopecký, J., Viskova, J., Fensel, D.: WSMO-lite annotations for web services. In: Bechhofer, S., Hauswirth, M., Hoffmann, J., Koubarakis, M. (eds.) ESWC 2008. LNCS, vol. 5021, pp. 674–689. Springer, Heidelberg (2008)
21. Martin, D., Burstein, M., Hobbs, J., Lassila, O., McDermott, D., McIlraith, S., Narayanan, S., Paolucci, M., Parsia, B., Payne, T.R., Sirin, E., Srinivasan, N., Sycara, K.: OWL-S: Semantic Markup for Web Services. Technical Report, Member Submission, W3C (2004)
22. Fensel, D., Lausen, H., Polleres, A., de Bruijn, J., Stollberg, M., Roman, D., Domingue, J.: Enabling Semantic Web Services - The Web Service Modeling Ontology. Springer, Heidelberg (2006)
23. Farrell, J., Lausen, H.: Semantic Annotations for WSDL and XML Schema. W3C Recommendation (August 2007), http://www.w3.org/TR/sawsdl/
24. Maleshkova, M., Kopecky, J., Pedrinaci, C.: Adapting SAWSDL for Semantic Annotations of RESTful Services. In: Meersman, R., Herrero, P., Dillon, T. (eds.) OTM 2009 Workshops. LNCS, vol. 5872, pp. 917–926. Springer, Heidelberg (2009)
25. Alvaro, G., Martinez, I., Gomez, J.M., Lecue, F., Pedrinaci, C., Villa, M., Di Matteo, G.: Using SPICES for a Better Service Consumption. In: Extended Semantic Web Conference (Posters (June 2010)
26. Speiser, S., Harth, A.: Towards Linked Data Services. In: Int'l Semantic Web Conference (Posters and Demonstrations (November 2010)
27. Iqbal, K., Sbodio, M.L., Peristeras, V., Giuliani, G.: Semantic Service Discovery using SAWSDL and SPARQL. In: 4th Int'l Conference on Semantics, Knowledge and Grid, December 2008, pp. 205–212 (2008)

Part VII:

Future Internet Areas: Content

Introduction

One of the major enablers for the evolution to the Future Internet will be the huge volumes of multimedia content. The new, powerful, low-cost and user friendly capturing devices (e.g. mobile phones, digital cameras, IP networked cameras) supported by new multimedia authoring tools will significantly increase the user generated content. On the other hand, new media sensor networks and tele-immersion applications will further increase the use of automatic generated content. As a result, the Internet as we know it today will be challenged and a (r)evolution towards Media Internet will be initiated.

The Media Internet is defined as the Future Internet variation which supports professional and novice content producers and is at the crossroads of digital multimedia content and Internet technologies. It encompasses two main aspects: Media being delivered through Internet networking technologies (including hybrid technologies) and Media being generated, consumed, shared and experienced on the Web.

The Media Internet is evolving to support novel user experiences such as immersive environments including sensorial experiences beyond video and audio (engaging all the human senses including smell, taste and haptics) that are adaptable to the user, the networks and the provisioned services.

The objective of this section is to offer different views on the processes, techniques and technologies which may pave the way for a Future Media Internet.

First of all, the Future Media Internet should be based on network architectures that can deal with content as a native type, and for this reason the content oriented network architectures for multimedia content delivery will produce a major revolution in the way that content is processed and delivered though the Internet. One particular case concerns content distributed through hybrid and heterogeneous network architectures, e.g. hybrid broadcast and Internet delivery enhancing the immersive experience of the user beyond the classical digital TV interactivity.

Second, enhancing media encoding technologies is required for the Internet with the objective to maintain the overall integrity, and adapt the content to the network, delivery device and user, and also optimize the quality of experience over the Internet.

Third, one of the areas where high investment in research has taken place in recent years is related to the multimedia and multimodal search and retrieval of multimedia objects over the Internet.

Last but not least, collaborative platforms for the experimentation of socially augmented and mixed reality applications are needed to produce advanced applications for the users, and social media including personalization and recommendation, is one of the key orientations of future media technologies. An increasingly large amount of content on the Web, whether multimedia or text is collaboratively generated user content, of which the quality is not always controllable.

In relation to the first point, content oriented network architectures, the paper "Media Ecosystems: A Novel Approach for Content-Awareness in Future Networks", describes a novel architecture for the deployment of a networked "Media Ecosystem" based on a flexible cooperation between providers, operators, and end-users, finally enabling every user – first – to access the offered multimedia services in various contexts, and – second – to share and deliver his/her own audiovisual content dynamically, seamlessly, and transparently to other users. The architecture also relies on autonomous systems to supply users with the necessary infrastructure and a security framework.

Concerning the second point, media encoding technologies for the Internet, the objective of the chapter "Scalable and Adaptable Media Coding Techniques for Future Internet" discusses SVC (Scalable Video Coding) and MDC (Multiple Description Coding) techniques along with the real experience of the authors of SVC/MDC over P2P networks and emphasizes their pertinence in Future Media Internet initiatives in order to decipher potential challenges.

For the third point, multimodal and multimedia search and retrieval in the Future Internet, the chapter "Semantic Context Inference in Multimedia Search" reviews the latest advances in semantic context inference, in which systems exploit the semantic context embedded in multimedia content and its surroundings in order to build a contextual representation scheme. The authors introduce their ideas on how to enable systems to automatically construct semantic context by learning from the available content.

Federico Alvarez, Theodore Zahariadis, Petros Daras, and Henning Müller

Media Ecosystems: A Novel Approach for Content-Awareness in Future Networks

H. Koumaras[1], D. Negru[1], E. Borcoci[2], V. Koumaras[5], C. Troulos[5], Y. Lapid[7],
E. Pallis[8], M. Sidibé[6], A. Pinto[9], G. Gardikis[3], G. Xilouris[3], and C. Timmerer[4]

[1] CNRS LaBRI laboratory, University of Bordeaux, France
koumaras@ieee.org, daniel.negru@labri.fr
[2] Telecommunication Dept., University Politehnica of Bucharest (UPB), Romania
eugen.borcoci@elcom.pub.ro
[3] Institute of Informatics and Telecommunications, NCSR Demokritos, Greece
{gardikis,xilouris}@iit.demokritos.gr
[4] Multimedia Communication, Klagenfurt University, Austria
christian.timmerer@itec.uni-klu.ac.at
[5] PCN, Greece
vkoumaras@pcngreece.com, ktroulos@pcngreece.com
[6] VIOTECH Communications, France
msidibe@viotech.net
[7] Optibase Technologies Ltd, Israel
Yaell@optibase.com
[8] Applied Informatics and Multimedia Dept., TEI of Crete, Greece
pallis@pasiphae.teiher.gr
[9] INESC Porto, Portugal
apinto@inescporto.pt

Abstract. This chapter proposes a novel concept towards the deployment of a networked 'Media Ecosystem'. The proposed solution is based on a flexible cooperation between providers, operators, and end-users, finally enabling every user first to access the offered multimedia services in various contexts, and second to share and deliver his own audiovisual content dynamically, seamlessly, and transparently to other users. Towards this goal, the proposed concept provides content-awareness to the network environment, network- and user context-awareness to the service environment, and adapted services/content to the end user for his best service experience possible, taking the role of a consumer and/or producer.

Keywords: Future Internet, Multimedia Distribution, Content Awareness, Network Awareness, Content/Service Adaptation, Quality of Experience, Quality of Services, Service Composition, Content-Aware Network

1   Introduction

One of the objectives of the future communication networks is the provision of audiovisual content in flexible ways and for different contexts, at various quality standards and associated price levels. The end user (EU) is ultimately interested in getting a

J. Domingue et al. (Eds.): Future Internet Assembly, LNCS 6656, pp. 369–380, 2011.

good Quality of Experience (QoE) at convenient prices. QoE is defined in [15] as "the overall acceptability of an application or service, as perceived subjectively by the end user". Therefore, QoE concepts are considered as important factors in designing multimedia distribution systems. The system capabilities to assure different levels of end-to-end Quality of Services, (i.e. including content production, transport and distribution) together with the possibility to evaluate and react to the QoE level at user terminals, offer to the EU a wide range of potential choices, covering the possibilities of low, medium or high quality levels.

Towards that direction, three factors (technologies) can be considered and can co-operate in an integrated system:

- First, significant advances in digital video encoding and compression techniques are now available. These techniques can achieve high compression ratios by exploiting both spatial and temporal redundancy in video sequences; and provides means for storage, transmission, and provision of very high-volume video data.
- Second, the development of advanced networking technologies in the access and core parts, with QoS assurance is seen. A flexible way of usage – based on virtualised overlays – can offer a strong support for the transportation of multimedia flows.
- Third, the todays' software technologies support the creation and composition of services while being able to take into account information regarding the transport/terminal contexts and adapt the services accordingly.

Bringing together in a synergic way all the above factors, a new "Media Ecosystem" is hence foreseen to arise, gathering a mass of not only existing but also new potential content distributors and media service providers, promoting passive consumers to engaged content creators. Such a "Media Ecosystem" is proposed in ALICANTE FP7 project [1-2]. By analogy to the ecology or market settings, a Media Ecosystem can be defined by inter-working environments, to which various actors belong to and through which they collaborate, in the networked media domain. With this respect, new flexible business models have to be supported by the Media Ecosystem. Finally, such a system can bring breakthrough opportunities in various domains such as communication industry, education, culture and entertainment.

However, the traditional and current layered architectures do not include exchanges of content – and network-based information between the network layers and upper layers. So, the added value of such interactions has not yet been exploited to better support the QoS and QoE requirements that media consumers are placing today. Network neutrality has been the foundational principle of the Internet, albeit today is revisited by service providers, research communities and industry, as a mean for quality provision and profit, to allow sustainable new forms of multimedia communications with an increasing importance in the Future Internet.

This suggests that the emerging approach of Content-Aware Networks (CAN) and Network-Aware services/Applications (NAA) can be a way to overcome the traditional architectures limitations. The network provisions based on CAN concepts, cooperating with powerful media adaptation techniques embedded in the network nodes, can be a foundation for an user-centric approach (i.e. satisfying the different needs of individual users) or service-centric approach (i.e. satisfying the different needs of various service types), that is required for the future services and applications [1-2]. The new CAN/NAA concept no more supposes network neutrality, but more intelligence and a higher degree of coupling to the upper layers are embedded in the network nodes. Based on virtualization, the network can offer enhanced transport and adaptation-capable services.

This chapter will introduce and describe an advanced architecture and new functionalities for efficient cooperation between entities of various environments so as to finally provide the end user with the best and most complete service experience via a Media Ecosystem, aiming to provide content-awareness to the network environment, network- and user context-awareness to the service environment, and adapted services/content to the end user's Environment.

2   Background

Numerous events and studies are currently dedicated to (re)define the directions which the Future Internet development should follow. Among other issues, a higher coupling between application and network layers are investigated, targeting to better performance (for multimedia) but without losing modularity of the architecture. [3-11]. The CAN/NAA approach can naturally lead to a user-centric infrastructure and telecommunication services as described in [3]. The strong orientation of user-centric awareness to services and content is emphasized in [4]. The works [5-6] consider that CAN/NAA can offer a way for evolution of networks beyond IP, while QoE issues related to *-awareness are discussed in [7]. The capability of content-adaptive network awareness is exploited in [8] for joint optimization of video transmission. The architecture can be still richer if to content awareness we add context awareness, [9]. The virtualisation as a powerful tool to overcome the Internet ossification by creating overlays is discussed in [10-11]. Finally, [14] discusses research challenges and open issues when adopting Scalable Video Coding (SVC) as a tool for CANs.

3   System Architecture

The proposed Media Ecosystem in ALICANTE project [1-2], by analogy with the ecology or business counterparts, can be characterized by *inter-working environments* to which the actors belong and through which they collaborate, in the networked media domain. These environments are:

- *User Environment (UE)*, to which the end users belong;
- *Service Environment (SE)*, to which the service and content providers belong;
- *Network Environment (NE)*, to which the network providers belong.

By *Environment*, it is understood a generic and comprehensive name to emphasize a grouping of functions defined around the same functional goal and possibly spanning, vertically, one or more several architectural (sub-)layers. It characterizes a broader scope with respect to the term *layer*. By *Service*, if not specified differently, we understand here *high level services*, as seen at application/service layer.

3.1   Layered Architectural Model

The ALICANTE architecture contains vertically several environments/layers and can be horizontally spanned over multiple network domains.

The *User Environment (UE)* includes all functions related to the discovery, subscription, consumption of the services by the EUs. At the *Service Environment (SE)* the *Service Provider (SP)* entity is the main coordinator. The architecture can support both synchronous communications or publish/subscribed ones. A novel type of service registry with enhanced functionalities allows new services supporting a variety of use case scenarios. Rich service composition in various ways is offered to EUs, opening them the role of SP/CP and manager. User and service mobility is also targeted.

Below the SE there is a new Home-Box (HB) layer to coordinate the actual content delivery to the end user's premises. The HB layer aims at allowing SPs to supply users with advanced context-aware multimedia services in a consistent and interoperable way. It enables uniform access for heterogeneous terminals and supports enhanced Quality of Experience (QoE). At the HB layer, the advanced user context management and monitoring functions provides real-time information on user context and network conditions, allowing better control over multimedia delivery and intelligent adaptation. The assembly of HBs is called layer because the HBs can logically communicate with each other in both client/server style but also in peer-to-peer (P2P) style. The HBs are located at the edges of the *Content Consumer (CC)* premises and have important roles in distribution of media flows, adaptation, routing, media caching, security, etc. Thus, the HB, which can be seen as the evolution of today's Home Gateway, is a physical entity aimed to be deployed inside each user's home.

The *Network Environment (NE)* comprises the virtual CAN layer (on top) and the traditional network infrastructure layer (at the bottom). The virtual CANs can span a single or multiple core network domains having content aware processing capabilities in terms of QoS, monitoring, media flow adaptation, routing/forwarding and security. The goal of the Virtual CAN layer is to offer to higher layers enhanced connectivity services, based on advanced capabilities (1) for network level provisioning of resources in a content-aware fashion and (2) for applying reactive adaptation measures based on media intelligent per flow adaptation. Innovative components, instantiating the CAN are called *Media-Aware Network Elements (MANE)*. They are actually CAN-enabled routers, which together with the associated managers and the other elements of the ecosystem, offer content- and context-aware Quality of Service/Experience, adaptation, security, and monitoring features. The set of MANEs form together a Virtual Content-Aware Network, which will be denoted VCAN or simply CAN where no ambiguity exist.

ALICANTE's advanced concept provides adapted services/content to the end-user for her/his best service experience possible. The adaptation is provided at both the HB and CAN layers, and is managed by the Adaptation Decision-Taking Framework (ADTF) – deployed at both layers. The ADTF is metadata-driven, thus enabling dynamic adaptation based on context information. The adaptation decisions will be taken based on three families of parameters: user preferences, terminal capabilities and network conditions. These parameters are gathered from every environment using dedicated user profile management and/or monitoring entities/subsystems.

The adaptation deployed at the CAN layer will be performed in the Media-Aware Network Elements (MANE): MANEs, which receive feedback messages about the terminal capabilities and channel conditions, can remove the non-required parts from a scalable bitstream before forwarding it. Thus, the loss of important transmission units due to congestion can be avoided and the overall error resilience of the video transmission service can be substantially improved. That is, within the CAN layer only scalable media formats – such as SVC – are delivered adopting a layered-multicast approach which allows the adaptation of scalable media resources by the MANEs implementing the concept of distributed adaptation. In this way, all actors within the media ecosystem will benefit from this approach ranging from providers (i.e., service, content, network) to consumers/end users.

At the border to the user, i.e., the Home-Box, adaptation modules are deployed enabling device-independent access to the SVC-encoded content by providing X-to-SVC and SVC-to-X transcoding functions with X={MPEG-2, MPEG-4 Visual, MPEG-4 AVC, etc.}. An advantage of this approach is the reduction of the load on the network (i.e., no duplicates), making it free for (other) data (e.g., more enhancement layers).

The key innovations of this approach to service/content adaptation are – distributed, self organizing adaptation decision-taking framework; distributed dynamic and intelligent adaptation at HB and CAN layers; efficient, scalable SVC tunnelling and signalling thereof, and as result – high impact on the Quality of Service/Experience (QoS/QoE).

Figure 1 presents a high level view of the general Media Ecosystem layered architecture.

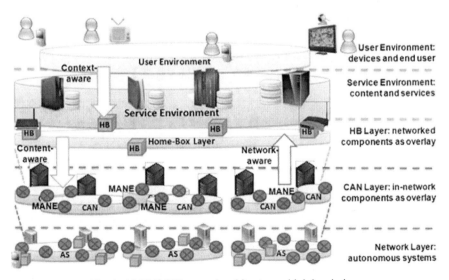

Fig. 1. ALICANTE general architecture – high level view

To preserve each NP independency – which is an important real world need – the ALICANTE solution considers, for each network domain the existence of an Intra-domain Network Resource Manager (NRM). This is responsible for the actual routers configuration in its own network, based on cooperation with the CAN Manager (CANMgr) belonging to the CAN Provider (CANP). Each domain can accommodate several CANs.

The system architecture can be horizontally divided into: *Management, Control* and *Data Planes* (MPl, CPl, DPl), parallel and cooperating (not represented explicitly in the picture). The upper data plane interfaces at the CAN layer and transport the packets between the VCAN layer and the Home-Box layer in both directions. The downloaded packets are especially marked by the application layer, allowing associa-tion with the correct CAN.

3.2   Content-Aware Networks (CAN)

The SPs may request the CANP to create multi-domain VCANs in order to benefit from different purposes (content-aware forwarding, QoS, security, unicast/multicast, etc.). The architecture supports creation of parallel VCANs over the same network infrastructure. Specialization of VCANs may exist (content-type aware), in terms of forwarding, QoS level of guarantees QoS granularity, content adaptation procedures, degree of security, etc. The amount of VCAN resources can be changed during net-work functioning based on monitoring developed at CAN layer and conforming to a contract (SLA) previously concluded between the SP and CANP. From deployment point of view the ALICANTE solution is flexible, specifically an evolutionary ap-proach is possible, i.e. to still keep a single control-management plane, while a more radical approach can also be envisaged towards full virtualization (i.e. independent management and control per VCAN).

The media data flows, are intelligently classified at ingress MANEs, and associated to the appropriate VCANs in order to be processed accordingly based on: (1) protocol header and metadata analysis (in-band information), (2) based on information deliv-ered by the VCAN management, and (3) statistical information in the packet flows. Then, it will assign the flows to the appropriate CANs. Solutions to maximise QoS/QoE can be flexibly applied at the CAN layer in aggregated mode.

Figure 2 depicts an example to illustrate the principle of Internet parallelization based on VCANs, with focus on the classification process performed at ingress MANEs.

In the example given, three autonomous systems (AS) are considered, each having its own NRM and CAN Manager (CANMgr). At request of an SP entity to create VCANs, (this is the Service Manager at SP (SM@SP)) addressed to the CANMgr@AS1, three multi-domain VCANs are created spanning respectively: VCAN1: AS1, AS2; VCAN2: AS1, AS2, AS3; CAN3: AS1, AS3. The dialogue be-tween CANMgr@AS1 and other CAN Managers to chain the VCANs in multi-domain virtual networks is not represented. To simplify the example, it is supposed that the specialisation of these VCANs is performed by their *Meta QoS class* (MQC) of services [12-13].

A MQC is an abstract and flexible definition *which* captures a common set of QoS ranges of parameters spanning several domains. It relies on a worldwide common understanding of application QoS needs. The VCANs can be created assuming that in all AS1, 2, 3 the MQC1, 2, 3 are implemented at network level (supported by Diff-Serv or MPLS). Figure 2 shows the process of VCAN negotiation (action 1 on the figure) and installation in the networks (action 2). Then (action 3) MANE1 is instructed how to classify the data packets, based on information as : *VCAN_Ids, Content description metadata, headers to analyse, QoS class information, policies, PHB – behaviour rules,* etc. obtained from SM@SP via CANMgr@AS1 and In-traNRM@AS1 (actually this is a detail of action 2).

At its turn, the SM@SP instructs the SP/CP servers how to mark the data packets. The information to be used in content aware classification can be: high level protocol headers, content description metadata (optional), VCAN_Id (optional); statistical information extracted from the packets if no other control information is available. The data packets are analysed by the classifier, assigned and forwarded to one of the VCANs for further processing. Special algorithms are needed to reduce the amount of processing of MANE in the data plane based on deep analysis of the first packets of a flow and usage of hashing functions thereafter.

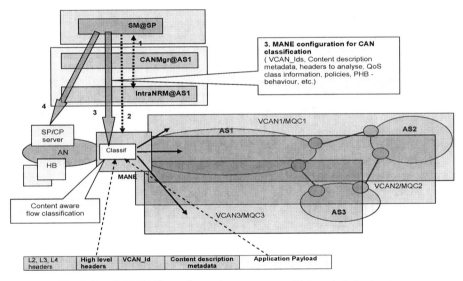

Fig. 2. Parallel VCAN creation and usage on a multi-domain infrastructure

Both traditional routing and forwarding or content/name based style can be used, in different parallel CANs due to MANE intelligence. Scalability is achieved by largely avoiding per-flow signalling in the core part of the network. In the new architecture, MANE also can act as content "caches", depending on their resource capabilities, implementing in such a way content centric capabilities. The architecture can support the client-server communication style and also P2P (between HBs) style.

Additionally, powerful per-flow solutions (adaptation) can be applied in MANEs and HBs, (e.g. for SVC video flows), to maintain, adapt and enhance QoE together with resource management enhancement.

The CAN management is distributed among domains. Each CAN Manager controls one or several CANs deployed in its domain. The CANMgr has interfaces with: i) HB layer and SP layer- to advertise CANs and negotiate their usage by the SP or HBs and to help at the HB layer the establishing of P2P relationships between devices of the HB layer, based on network-related distance information between these devices; ii) to the lower intra-domain network resource managers (Intra-NRM)- in order to negotiate CANs and request their installation; iii) to peer CANMgrs, in order to establish multi-domain CANs. Also CANMgr drives the monitoring at CAN layer. It is supposed that at their turn, the Intra-NRMs have established interconnection agreements at IP level. Technologies like DiffServ or MPLS can be deployed to support CAN virtual networks and offer QoS specific treatment.

3.3   CAN Layer Security

The aim of the security subsystem within the CAN Layer is twofold: 1) data confidentiality, integrity and authenticity; and 2) intelligent and distributed access control policy-based enforcement.

The first objective is characterized by offering, to the Service Provider (SP), a selection of three degrees of security, being: public traffic, secret content, and private communications. In public traffic no security or privacy guarantees are enforced. Secret content addresses content confidentiality and authentication by applying common cryptographic techniques over the packets' payload. Private communications is to be adopted when the confidentiality and authenticity of the entire packets, including headers, are required. The adopted strategy is to evaluate the required end-to-end security along all CAN domains and discretely apply the security mechanisms only where necessary to guarantee the required security level, with respect to the security degree invoked. The evaluation algorithm considers the user flow characteristics, CAN policies and present network conditions. In order to attain the required flexibility, the related security architecture was designed according to the hop-by-hop model [7] on top of the MANEs routers.

The second objective will pursue a content-aware approach that will be enforced by MANE routers over data in motion. Such security enforcement will be done accordingly to policies and filtering rules obtained from the CANMgr. In turn, CANMgr will compute policies and traffic filtering rules by executing security related algorithms over information gathered by the monitoring subsystem. MANE routers will derive filtering rules from packet inspection and will inform the CANMgr about those computed rules. Content-aware security technologies typically perform deep content inspection of data traversing a security element placed in a specific point in the network. The proposed approach differs by being based on MANE routers, which will be used to construct CANs.

An example of a traffic filtering rule could be to drop all traffic matching a set composed of: source IP; source port number; destination IP; destination port number.

An example of a policy (i.e. more generic than a traffic filtering rule) would be to limit the maximum number of active connections for a given source in a period of time. Based on this policy, MANE equipments would be able to derive traffic filtering rules and to enforce them.

MANE's related security functions are then to perform attacks' identification (e.g. port-scan, IP spoofing, replay) and to enforce traffic filtering rules. CANMgr carries out collaborative work with homologous entities in order to implement access control policies definition and distribution, identify large scale attacks' (e.g. network scans based on MANE port-scan information, DDoS), and to trace back attack flow's in order to block the malicious traffic in the proximity of the source.

4   Business Actors and Policy Implications

A Media Ecosystem is comprised of different actors (business and/or operational entities) that interface with each other. The system architecture definition should support flexible ways of collaboration to satisfy both the end-users needs and the Service/Content/Network Providers' objectives. In ALICANTE's novel architectural paradigm, [1], the traditional content consumer can become easily a content/service provider and distributor. In this environment, the main business actors/entities envisaged (as shown in Figure 3) are the following:

Content Consumer (CC) or *End-User (EU)* is an entity (organisation/ individual) which may establish a contract with a Service Provider (SP) for service/content delivery.

Network Provider (NP) is the traditional entity which offers IP connectivity, providing connectivity between network domains/hosts. The NPs own and manage their IP connectivity infrastructures. The NPs may interact with each other to expand content-aware services across a larger geographical span.

Content Provider (CP) gathers/creates, maintains and releases digital content by owning and operating the network hosts (content sources) but it may not necessarily own the distributing network infrastructure.

CAN Provider (CANP) is a new ALICANTE business entity, seen as a virtual layer functionality provider. It is practically an enhanced, virtual NP. The additional CAN functions are actually performed in the network nodes, initiated by CANP but agreed by The CANP offers content-aware network services to the upper layer entities.

Service Provider (SP) is the ultimately ALICANTE business entity which is responsible for the services offered to the end-user and may interact with NPs, and/or CANPs in order to use/expand their service base.

Home-Box (HB) is a new ALICANTE business entity, partially managed by the SP, the NP, and the end-user. The HBs can cooperate with SPs in order to distribute multimedia services (e.g., IPTV) in different modes (e.g. P2P).

Content is offered to the CCs or SPs through quality guarantee schemes such as Service Level Agreements (SLAs). There may be formal business agreements between CPs and NPs for hosting or co-locating CP's content servers in NPs' premises, nevertheless, an individual CC may also be a private CP. SPs aggregate content from multiple CPs and deliver services to CCs with higher quality. SPs may not own a transport infrastructure, but rely on the connectivity services offered by Network Providers (NPs), or CAN Providers (CANP). The SPs are ultimately responsible for the service offered to the CC and may interact with each other in order to expand their service base and use the services of NPs, or CANPs, via appropriate SLAs. In ALI-CANTE a single merged entity SP/CP is considered playing the both roles.

Each of the previously described environments is present in today actual deployments, but there is a profound limitation of collaboration among them. Typically, network providers operate separately from content providers. While content providers create and sell/release content, network providers are in charge of distributing content and ensuring a satisfactory level of QoS for the end users (by appropriating resources to network upgrades etc). User context is not taken into consideration by the Service or Content Provider (SP/CP) delivering the service (content), and therefore they are not able to deliver and adapt the service (content) to the capabilities of the end user equipment. Also, current architectures do not include any information exchange between the network and service layers, thus limiting content and context awareness in service delivery and consequently inhibiting innovation in services and applications. The real challenge and ultimate objective is to find the appropriate means for efficient cooperation between entities of the various environments to provide the end-user with the best service experience while preserving the fundamental principle of network neutrality.

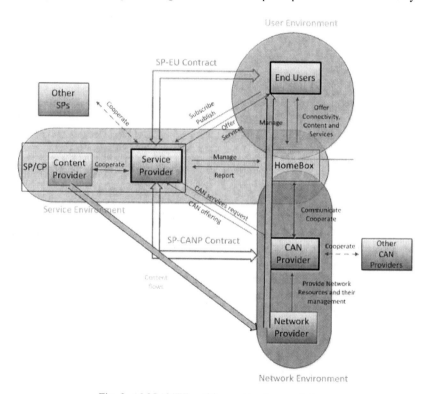

Fig. 3. ALICANTE entities and business relations

The content-aware service delivery platform, discussed previously creates a dynamic market for network and service providers, offering the necessary network and business structures to facilitate collaboration between providers at all levels of the network/content market. This framework motivates business and service innovation in a variety of ways. It allows for revenue (and customer) sharing models between network and content providers, while preserving the upstream revenues from end-users. Also, users may influence service delivery options by requesting specific content on specific priority conditions for a specific period of time (context-awareness).

Moreover, this concept supports the creation of a 2-sided market, with network operators acting as an intermediate between content producers and content consumers, leveraging revenues from both sides. For content providers this opens a new market, characterized by specific business attributes (e.g. QoS guarantees, more intimate relation with their customers, etc.). On the user side, it increases choice and reduces switching costs between content providers (network and content layers are not integrated). In addition, the end user, as a key business actor, can effectively increase the level of choice in content and services by selecting, deploying, controlling and managing easy-to-use, affordable services and applications on service-enabled networks. Eventually the end user will have a choice of service access methods: anywhere, anytime and in any context with the appropriate awareness degree [1].

It also allows competitive content producers to enter and address niche markets. An effective scaling up of the infrastructure across multiple administrative domains (i.e. multiple NPs) could help distinguish competition in the service and network layers. Content and, most importantly, context awareness presents a possible option to address the network neutrality issue which emerges as the key issue in contemporary regulatory agendas in Europe and US. The appropriate implementation would allow management of special services and best-effort services separately. Last and not least, user privacy is a major concern since it directly relates to the consolidation of information sources where user preferences and habits may be retrieved and exploited by third parties.

5   Conclusions

This chapter has presented a novel Media-Ecosystem architecture, which introduces two novel virtual layers on top of the traditional Network layer, i.e. a Content-Aware Network layer (CAN) for network packet processing and a Home-Box layer for the content delivery. The innovative components, proposed to instantiate the CAN are Media-Aware Network Elements (MANE), i.e. CAN-enabled "routers" and associated managers offering together content- and context aware Quality of Service/Experience, security, and monitoring features, in cooperation with the other elements of the ecosystem. The chapter has also indicated the novel business opportunities that are created by the proposed Media-Ecosystem.

Acknowledgement. This work was supported in part by the EC in the context of the ALICANTE project (FP7-ICT-248652).

References

1. FP7 ICT project: MediA Ecosystem Deployment Through Ubiquitous Content-Aware Network Environments. ALICANTE, No248652 (last accessed: March 2011) (2011), http://www.ict-alicante.eu/
2. Borcoci, E., Negru, D., Timmerer, C.: A Novel Architecture for Multimedia Distribution based on Content-Aware Networking. In: CTRQ 2010 Conference Proceedings (2010), http://www.iaria.org
3. Baladrón, C.: User-Centric Future Internet and Telecommunication Services. In: Tselentis, G., et al. (eds.) Towards the Future Internet, pp. 217–226. IOS Press, Amsterdam (2009)
4. Schönwälder, J., et al.: Future Internet = Content + Services + Management. IEEE Communications Magazine 47(7), 27–33 (2009)
5. Zahariadis, T., et al.: Content Adaptation Issues in the Future Internet. In: Tselentis, G., et al. (eds.) Towards the Future Internet, pp. 283–292. IOS Press, Amsterdam (2009)
6. Huszák, Á., Imre, S.: Content-aware Interface Selection Method for Multi-Path Video Streaming in Best-effort Networks. In: Proc. of 16th International Conference on Telecommunications, Marrakech, Morocco, Jul. 2009, pp. 196–201 (2009)
7. Liberal, F., et al.: QoE and *-awareness in the Future Internet. In: Tselentis, G., et al. (eds.) Towards the Future Internet, pp. 293–302. IOS Press, Amsterdam (2009)
8. Martini, M.G., et al.: Content Adaptive Network Aware Joint Optimization of Wireless Video Transmission. IEEE Communications Magazine 45(1), 84–90 (2007)
9. Baker, N.: Context-Aware Systems and Implications for Future Internet. In: Tselentis, G., et al. (eds.) Towards the Future Internet, pp. 335–344. IOS Press, Amsterdam (2009)
10. Anderson, T., et al.: Overcoming the Internet Impasse through Virtualization. Computer 38(4), 34–41 (2005)
11. Chowdhury, N.M., Boutaba, R.: Network Virtualization: State of the Art and Research Challenges. IEEE Communications Magazine 47(7), 20–26 (2009)
12. Levis, P., et al.: The Meta-QoS-Class Concept: a Step Towards Global QoS Interdomain Services. Proc. IEEE, SoftCOM, Oct. 2004 (2004)
13. Paris Flegkas, et.al., Provisioning for Interdomain Quality of Service: the MESCAL Approach. IEEE Communications Magazine (June 2005)
14. Timmerer, C., et al.: Scalable Video Coding in Content-Aware Networks: Research Challenges and Open Issues. In: Proc. International Tyrrhenian Workshop on Digital Communications (ITWDC), Ponza, Italy (September 2010)
15. ITU-T SG12: Definition of Quality of Experience. TD 109rev2 (PLEN/12), Geneva, Switzerland, 16-25 Jan 2007 (2007)

Scalable and Adaptable Media Coding Techniques for Future Internet

Naeem Ramzan and Ebroul Izquierdo

School of Electronic Engineering and Computer Science, Queen Mary University of London,
Mile end, London E1 4NS, United Kingdom
{Naeem.Ramzan, Ebroul.Izquierdo}@elec.qmul.ac.uk

Abstract. High quality multimedia contents can distribute in a flexible, efficient and personalized way through dynamic and heterogeneous environments in Future Internet. Scalable Video Coding (SVC) and Multiple Description Coding (MDC) fulfill these objective thorough P2P distribution techniques. This chapter discusses the SVC and MDC techniques along with the real experience of the authors of SVC/MDC over P2P networks and emphasizes their pertinence in Future Media Internet initiatives in order to decipher potential challenges.

Keywords: Scalable video coding, multiple description coding, P2P distribution.

1   Introduction

Future Media Internet will entail to distribute and dispense high quality multimedia contents in an efficient, supple and personalized way through dynamic and heterogeneous environments. Multimedia content over internet are becoming a well-liked application due to users' growing demand of multimedia content and extraordinary growth of network technologies. A broad assortment of such applications can be found in these days, e.g. as video streaming, video conferencing, surveillance, broadcast, e-learning and storage. In particular for video streaming, over the Internet are becoming popular due to the widespread deployment of broadband access. In customary video streaming techniques the client-server model and the usage of Content Distribution Networks (CDN) along with IP multicast were the most desirable solutions to support media streaming over internet. However, the conventional client/server architecture severely limits the number of simultaneous users for bandwidth intensive video streaming, due to a bandwidth bottleneck at the server side from which all users request the content. In contrast, Peer-to-Peer (P2P) media streaming protocols, motivated by the great success of file sharing applications, have attracted a lot of interest in academic and industrial environments. With respect to conventional approaches, a major advantage in using P2P is that each peer involved in a content delivery contributes with its own resources to the streaming session. However, to provide high quality of service, the video coding/transmission technology needs to be able to cope with varying bandwidth capacities inherent to P2P systems and end-user characteristics

J. Domingue et al. (Eds.): Future Internet Assembly, LNCS 6656, pp. 381–389, 2011.

such as decoding and display capabilities usually tend to be non-homogeneous and dynamic. This means that the content needs to be delivered in different formats simultaneously to different users according to their capabilities and limitations.

In order to handle such obscurity, scalability emerged in the field of video coding in the form of Scalable Video Coding (SVC) [1–4] and Multiple Description Coding (MDC) [5-6]. Both SVC and MDC offers an efficient encoding for applications where content needs to be transmitted to many non-homogeneous clients with different decoding and display capabilities.

Moreover, the bit-rate adaptability inherent in the scalable codec designs provides a natural and efficient way of adaptive content distribution according to changes in network conditions.

In general, a SVC sequence can be adapted in three dimensions, namely, temporal, spatial and quality dimensions, by leaving out parts of the encoded bit-stream, thus reducing the bit-rate and video quality during transmission. By adjusting one or more of the scalability options, the SVC scheme allows flexibility and adaptability of video transmission over resource-constrained networks.

MDC can be considered as an additional way of increasing error resilience and end user adaptation without using intricate error protection methods. The objective of MDC is to generate numerous independent descriptions that can bestow to one or more characteristics of video: spatial or temporal resolution, signal-to-noise ratio. These descriptions can be used to decode the original stream, network congestion or packet loss, which is common in best-effort networks such as the Internet, will not interrupt the reproduction of the stream but will only cause a loss of quality. Descriptions can have the same importance namely "balanced MDC schemes" or they can have different importance "unbalanced MDC schemes". The more descriptions received, the higher the quality of decoded video. MDC combined with path/server diversity offers robust video delivery over unreliable networks and/or in peer-to-peer streaming over multiple multicast trees.

The eventual objective of employing SVC/MDC in Future Internet is to maximize the end-users' quality of experience (QoE) for the delivered multimedia content by selecting an appropriate combination of the temporal, spatial and quality parameters for each client according to the limitation of network and end user devices .

This chapter starts with an overview of SVC and MDC source coding techniques in section 2 and 3. Section 4 describes how to adapt SVC for P2P distribution for Future Internet. MDC over P2P is explained in section 5. Finally, this chapter concludes in section 6.

2   Scalable Video Coding

During the last decade a noteworthy amount of research has been devoted to scalable video coding with the aspire of developing the technology that would offer a low-complexity video adaptation, but preserve the analogous compression efficiency and decoding complexity to those of conventional (non-scalable) video coding systems. This research evolved from two main branches of conventional video coding: 3D wavelet [1] and hybrid video coding [2] techniques. Although some of the earlier video standards, such as H.262 / MPEG-2 [3], H.263+ and MPEG-4 Part 2 included limited support for scalability, the use of scalability in these solutions came at the significant increase in the decoder complexity and / or loss in coding efficiency. The latest video coding standard, H.264 / MPEG-4 AVC [2] provides a fully scalable extension, SVC, which achieves significant compression gain and complexity reduction when scalability is sought, compared to the previous video coding standards.

The scalability is usually required in three different directions (and their combinations). We define these directions of scalability as follows:

- Temporal scalability refers to the possibility of reducing the temporal resolution of encoded video directly from the compressed bit-stream, i.e. number of frames contained in one second of the video.
- Spatial scalability refers to the possibility of reducing the spatial resolution of the encoded video directly from the compressed bit-stream, i.e. number of pixels per spatial region in a video frame.
- Quality scalability, or commonly called SNR (Signal-to-Noise-Ratio) scalability, or fidelity scalability, refers to the possibility of reducing the quality of the encoded video. This is achieved by extraction and decoding of coarsely quantised pixels from the compressed bit-stream.

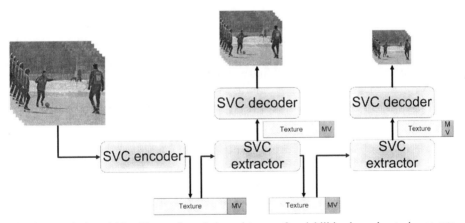

Fig. 1. A typical scalable video coding chain and types of scalabilities by going to lower-rate decoding

An example of basic scalabilities is illustrated in Figure 1, which shows a typical SVC encoding, extraction and decoding chain. The video is encoded at the highest spatio-temporal resolution and quality. After encoding, the video is organised into a scalable bit-stream and the associated bit-stream description is created. This description indicates positions of bit-stream portions that represent various spatio-temporal resolutions and qualities. The encoder is the most complex between the three modules. The compressed video is adapted to a lower spatio-temporal resolution and / or quality by the extractor. The extractor simply parses the bit-stream and decides which portions of the bit-stream to keep and which to discard, according to the input adaptation parameters. An adapted bit-stream is also scalable and thus it can be fed into the extractor again, if further adaptation is required. The extractor represents the least complex part of the chain, as its only role is to provide low-complexity content adaptation without transcoding. Finally, an adapted bit-stream is sent to the decoder, which is capable of decoding any adapted scalable video bit-stream.

2.1   H.264/MPEG-4 SVC

The latest H.264/MPEG-4 AVC standard provides a fully scalable extension, H.264/MPEG-4 SVC, which achieves significant compression gain and complexity reduction when scalability is sought, compared to the previous video coding standards [4]. According to evaluations done by MPEG, SVC based on H.264/MPEG-4 AVC provided significantly better subjective quality than alternative scalable technologies at the time of standardisation. H.264/MPEG-4 SVC reuses the key features of H.264/AVC and also employs some other new techniques to provide scalability and to improve coding efficiency. It provides temporal, spatial and quality scalability with a low increase of bit-rate relative to the single layer H.264/MPEG-4 AVC.

The scalable bit-stream is structured into a base layer and one or several enhancement layers. Temporal scalability can be activated by using hierarchical prediction structures. Spatial scalability is obtained using the multi-layer coding approach. Within each spatial layer, single-layer coding techniques are employed. Moreover, inter-layer prediction mechanisms are utilized to further improve the coding efficiency. Quality scalability is provided using the coarse-grain quality scalability (CGS) and medium-grain quality scalability (MGS). CGS is achieved by requantization of the residual signal in the enhancement layer, while MGS is enabled by distributing the transform coefficients of a slice into different network abstraction layer (NAL) units. All these three scalabilities can be combined into one scalable bit-stream that allows for extraction of different operation points of the video.

2.2   Wavelet-Based SVC (W-SVC)

A enormous amount of research activities are being continued on W-SVC although H.264/MPEG-4 SVC was chosen for standardisation. In the W-SVC, the input video is subjected to a spatio-temporal (ST) decomposition based on a wavelet transform. The rationale of the decomposition is to decorrelate the input video content and offer the basis for spatial and temporal scalability. The ST decomposition results in two distinctive types of data: wavelet coefficients representing of the texture information remaining after the wavelet transform and motion information obtained from motion estimation (ME), which describes spatial displacements between blocks in neighbouring frames. Although the wavelet transform generally performs very well in the task of video content decorrelation, some amount of redundancies still remains between the wavelet coefficients after the decomposition. Moreover, a strong correlation also exists between motion vectors. For these reasons, further compression of the texture and motion vectors is performed. Texture coding is performed in conjunction with so-called embedded quantisation (bit-plane coding) in order to provide the basis for quality scalability. Finally, the resulting data are mapped into the scalable stream in the bit-stream organisation module, which creates a layered representation of the compressed data. This layered representation provides the basis for low-complexity adaptation of the compressed bit-steam.

3 Scalable Multiple Description Coding (SMDC)

SMDC is a source coding technique, which encodes a video into a N (where N≥2) independent decodable sub-bitstreams by exploiting the scalability features of SVC. Each sub-bitstream is called "description". The descriptions can be transmitted through different or independent network paths to reach a destination. The receiver can create a reproduction of the video when at least one of the descriptions is received. The quality of the received video is proportional to the number of descriptions received on time. Thus, the more descriptions are received, the better quality of reconstruction is.

General principles and different approaches for MDC are reviewed in [5]. Approaches for generating multiple descriptions include data partitioning (e.g., even/odd sample or DCT coefficient splitting) [5], multiple description (MD) quantization (e.g.,MD scalar or vector quantization) [6], and multiple description transform coding (e.g., pairwise correlating transform). Another MD encoding approach is based on combinations of video segments (codeblocks, frames, orgroup of pictures) encoded at high and low rates. Different combinations of codeblocks coded at high and low rates have been introduced in [5] for wavelet-based flexible MD encoders.

4 Scalable Video over P2P Network

The proposed system is based on two main modules: scalable video coding and the P2P architecture. In this system, we assume that each peer contains the scalable video coder and the proposed policy of receiving chunk is to make sure that each peer at least receives the base layer of the scalable bit-stream for each group of picture (GOP). Under these circumstances, peers could download different layers from different users, as shown in Figure 2.

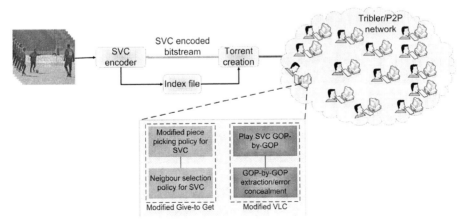

Fig. 2. An example of the proposed system for scalable video coding in P2P network.

In this section, we formulate how the scalable layers are prioritized in our proposed system. First we explain how the video segments or chunks are arranged and prioritized in our proposed system

4.1   Piece Picking Policy

The proposed solution is a variation of the "Give-To-Get" algorithm [8], already implemented in Tribler. Modifications concern the piece picking and neighbour selection policies. Scalable video sequences can be split into GOPs and layers [7] while BitTorrent splits files into pieces. Since there is no correlation between these two divisions, some information is required to map GOPs and layers into pieces and vice versa. This information can be stored inside an index file, which should be transmitted together with the video sequence. Therefore, the first step consists of creating a new torrent that contains both files. It is clear that the index file should have the highest priority and therefore should be downloaded first.

Once the index file is completed, it is opened and information about offsets of different GOPs and layers in the video sequence is extracted. At this point, it is possible to define a sliding window, made of W GOPs and the pre-buffering phase starts. Pieces can only be picked among those inside the window, unless all of them have

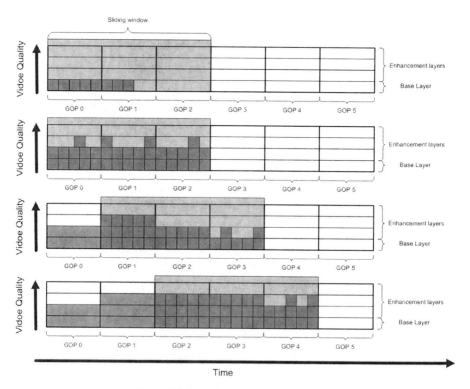

Fig. 3. Sliding window for scalable video

already been downloaded. In the latter case, the piece picking policy will be the same as the original BitTorrent, which is rarest [piece] first.

Inside the window, pieces have different priorities as shown in Figure 3. First of all, a peer will try to download the base layer, then the first enhancement layer and so on. Pieces from the base layer are downloaded in a sequential order, while all the other pieces are downloaded rarest-first (within the same layer).

The window shifts every t(GOP) seconds, where t(GOP) represents the duration of a GOP. The only exception is given by the first shift, which is performed after the pre-buffering, which lasts W * t(GOP) seconds.

Every time the window shifts, two operations are made. First, downloaded pieces are checked, in order to evaluate which layers have been completely downloaded. Second, all pending requests that concern pieces belonging to a GOP that lies before the window are dropped. An important remark is that the window only shifts if at least the base layer has been received, otherwise the system will auto-pause. The detail of modified piece picking policy can be found in [7]. Another issue is the wise choice of the neighbours.

4.2   Neighbour Selection Policy

It is extremely important that at least the base layer of each GOP is received before the window shifts. Occasionally, slow peers in the swarm (or slow neighbours) might delay the receiving of a BT piece, even if the overall download bandwidth is high. This problem is critical if the requested piece belongs to the base layer, as it might force the playback to pause. Therefore, these pieces should be requested from good neighbours. Good neighbours are those peers that own the piece with the highest download rates, which alone could provide the current peer with a transfer rate that is above a certain threshold. During the pre-buffering phase, any piece can be requested from any peer. However, every time the window shifts, the current download rates of all the neighbours are evaluated and the peers are sorted in descending order.

The performance of the proposed framework [7] has been evaluated to transmit wavelet-based SVC encoded video over P2P network as shown in Figure 4.

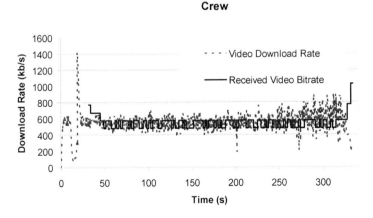

Fig. 4. Received download rate and received video bitrate for Crew CIF sequence

5   Multiple Description Coding over P2P Network

Most of the work on MDC is proposed for wireless applications in which there are issues such as hand-over of a client to another wireless source is present. However, in IP networks, it may be more complicated to have autonomous links among peers. Thus, additional redundancy introduced by using MDC over internet need to be carefully evaluated.

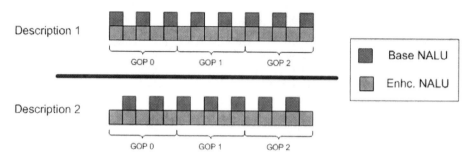

Fig. 5. An example of multiple description using scalable video coding

A simple way to generate multiple descriptions using scalable video coding is to distribute the enhancement layer NAL units to separate descriptions. Thus when both descriptions is received it is possible to have all the enhancement NAL units. Moreover, it is possible to control the redundancy by changing the quality of the base layer as shown in Figure 5.

5.1   Piece Picking Policy

Similar to SVC P2P technique described above, the sliding window is defined for Scalable MDC over P2P. The highest priority is given to the pieces of each description inside the sliding window however the further classification can be enabled with respect to playing. The chunks belong to nearer to the playing time gives higher priority and declared as critical in the sliding window and vice versa as shown in Figure 6. This enhances the overall performance of the system. The main advantage of Scalable

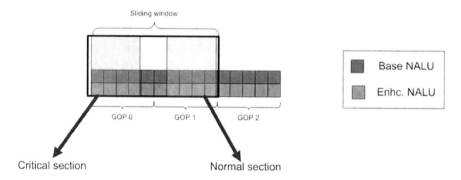

Fig. 6. Sliding window for multiple description of scalable video.

MDC over SVC is that the receiver/client can make a reproduction of the video when any of the description is received at the cost of additional redundancy due to the presence of base layer in each description.

6 Conclusions

This chapter has presented an overview of SVC and MDC with the perspective of content distribution over Future Internet. These coding schemes provide natural robustness and scalability to media streaming over heterogeneous networks. The amalgamation of SVC/MDC and P2P are likely to accomplish some of the Future Media Internet challenges. Tangibly, SVC/MDC over P2P presumes an excellent approach to facilitate future media applications and services, functioning under assorted and vibrant environments while maximizing not only Quality of Service (QoS) but also Quality of Experience (QoE) of the users. At last, we persuade Future Internet initiatives to take into contemplation these techniques when defining new protocols for ground-breaking services and applications.

Acknowledgement. This research has been partially funded by the European Commission under contract FP7-248474 SARACEN.

References

1. Mrak, M., Sprljan, N., Zgaljic, T., Ramzan, N., Wan, S., Izquierdo, E.: Performance evidence of software proposal for Wavelet Video Coding Exploration group, ISO/IEC JTC1/SC29/WG11/ MPEG2006/M13146, 76th MPEG Meeting, Montreux, Switzerland (April 2006)
2. ITU-T and ISO/IEC JTC 1: Advanced video coding for generic audiovisual services, ITU-T Recommendation H.264 and ISO/IEC 14496-10 (MPEG-4 AVC)
3. ITU-T and ISO/IEC JTC 1: Generic coding of moving pictures and associated audio information – Part 2: Video, ITU-T Recommendation H.262 and ISO/IEC 13818-2 (MPEG-2 Video)
4. Schwarz, H., Marpe, D., Wiegand, T.: Overview of the scalable video coding extension of the H.264 / AVC standard. IEEE Transactions on Circuits and Systems for Video Technology 17(9), 1103–1120 (2007)
5. Berkin Abanoz, T., Murat Tekalp, A.: SVC-based scalable multiple description video coding and optimization of encoding configuration. Signal Processing: Image Communication 24(9), 691–701 (2009)
6. Tillo, T., Grangetto, M., Olmo, G.: Redundant slice optimal allocation for H.264 multiple description coding. IEEE Trans. Circuits Syst. Video Technol. 18(1), 59–70 (2008)
7. Asioli, S., Ramzan, N., Izquierdo, E.: A Novel Technique for Efficient Peer-To-Peer Scalable Video Transmission. In: Proc. of European Signal Processing Conference (EUSIPCO-2010), Aalborg, Denmark, August 23-27 (2010)
8. Pouwelse, J.A., Garbacki, P., Wang, J., Bakker, A., Yang, J., Iosup, A., Epema, D.H.J., Reinders, M., van Steen, M.R., Sips, H.J.: Tribler: A social-based based peer to peer system. In: 5th Int'l Workshop on Peer-to-Peer Systems, IPTPS (Feb. 2006)

Semantic Context Inference in Multimedia Search

Qianni Zhang and Ebroul Izquierdo

School of Electronic Engineering and Computer Science
Queen Mary University of London, UK
{qianni.zhang, ebroul.izquierdo}@elec.qmul.ac.uk

Abstract. Multimedia content is usually complex and may contain many semantically meaningful elements interrelated to each other. Therefore to understand the high-level semantic meanings of the content, such interrelations need to be learned and exploited to further improve the search process. We introduce our ideas on how to enable automatic construction of semantic context by learning from the content. Depending on the targeted source of content, representation schemes for its semantic context can be constructed by learning from data. In the target representation scheme, metadata is divided into three levels: low, mid, and high levels. By using the proposed scheme, high-level features are derived out of the mid-level features. In order to explore the hidden interrelationships between mid-level and the high-level terms, a Bayesian network model is built using from a small amount of training data. Semantic inference and reasoning is then performed based on the model to decide the relevance of a video.

Keywords: Multimedia retrieval, context inference, mid-level features, Bayesian network

1 Introduction

In realistic multimedia search scenarios, high-level queries are usually used and search engines are expected to be able to understand underlying semantics in content and match it to the query. Researchers have naturally started thinking of exploiting context for retrieving semantics. Content is usually complex and may contain many semantically meaningful elements interrelated to each other, and such interrelations together form a semantic context for the content. To understand the high-level semantic meanings of the content, these interrelations need to be learned and exploited to further improve the indexing and search process. In latest content-based multimedia retrieval approaches, it is often proposed to build some forms of contextual representation schemes for exploiting the semantic context embedded in multimedia content. Such techniques form a key approach to supporting efficient multimedia content management and search in the Internet. In the literature, such approaches usually incorporates domain knowledge to assist definition of the context representation. However, this kind

J. Domingue et al. (Eds.): Future Internet Assembly, LNCS 6656, pp. 391–400, 2011.

of schemes is often limited as they have to rely on specifically defined domains of knowledge or require specific structure of semantic elements. In this chapter we introduce our ideas on how to enable systems to automatically construct a semantic context representation by learning from the content. Depending on the targeted source of content, which could be online databases, a representation scheme for its semantic context is directly learned from data and will not be restricted to the pre-defined semantic structures in specific application domains.

Another problem hampering bridging the semantic gap is that it is almost impossible to define precise mapping between semantics and visual features since they represent the different understandings about the same content by humans and machines. This problem can be approached by dividing all types of meta-data extracted from multimedia content into three levels - low, mid and high - according to their levels of semantic abstraction and try to define the mapping between them. Low-level features include audio-visual features, the time and place in which content was generated, etc.; mid-level features can be defined as content-related information that involves some degree of semantics but is still not fully understandable by users; while high-level features contains a high degree of semantic reasoning about the meaning or purpose of the content itself. The mid-level features are relatively easy to obtain by analysing the metadata, but usually are not directly useful in real retrieval scenarios. However, they often have strong relationships to high-level queries while these relationships are mostly ignored due to their implicitness. In this way, defining the links between objects on low- and high- levels becomes more tangible through dividing the mapping process in two steps. While linking low-level features to mid-level concepts are relatively easy to solve using the well-defined algorithms in the state-of-the-art, the mapping between mid- and high-level features are still difficult. In the proposed research, in order to explore the hidden interrelationships between mid-level features and the high-level terms, a Bayesian network model is learned from a small amount of training data. Semantic inference and reasoning is then carried out based on the learned model to decide whether a video is relevant to a high-level query. The relevancies are represented by probabilities and videos with high relevancies to a query are annotated with the query terms for future retrieval. The novelty of the proposed approach has two aspects: using a set of mid-level features to reduce the difficulty in matching low-level descriptors and high-level semantic terms; and employing an automatically learned context representation model to assist the mapping between mid- and high-level representations.

Although a full video annotation and retrieval framework using the proposed approach is mentioned in this chapter, the focus will be on the context learning and automatic inference for high-level features. The mid-level features are assumed to be available and are extracted using algorithms with reasonable performance.

The rest of this chapter is organised as follows: Section 2 gives a review on the state-of-the-art techniques on context reasoning for multimedia retrieval task; Section 3 describes the three representation levels of multimedia content and the implemented mid-level features; Section 4 presents the proposed technique for semantic context learning and inference for mid-level to high-level matching; Section 5 shows selected experimental results and the chapter is concluded with Section 6.

2   Related Works

The problem of high-level decision making often consists of reasoning and inference, information fusion, and other common features. Popular techniques related to storing and enforcing high-level information include neural networks, expert systems, statistical association, conditional probability distributions, different kinds of monotonic and non-monotonic, fuzzy logic, decision trees, static and dynamic Bayesian networks, factor graphs, Markov random fields, etc [15,21,9,14,6]. A comprehensive literature review on these topics can be found in [17]. Many object-based image classification frameworks rely on probabilistic models for semantic context modelling and inference [13].

In [16], the authors propose a probabilistic framework to represent the semantics in video indexing and retrieval work. In this approach, the probability density function which denotes the presence of a multimedia object is called a *multiject*. The interaction between the multijects is then modelled using a Bayesian network to form a multinet. The experiments in this research are restricted to a few particularly selected concepts and may not work in general retrieval scenarios. In [5], authors have developed a framework based on semantic visual templates (SVTs), which attempts to abstract the notion of visual semantics relating particular instances in terms of a set of examples representing the location, shape, and motion trajectories of the dominant objects related to that concept. Bayesian relevance feedback was also employed in the retrieval process. In [18], authors concentrate on analysing wildlife videos that capture hunting. Their approach models semantic concepts such as vegetation, terrain, trees, animals, etc. Neural networks are employed in this approach to classify features extracted from video blobs for their classification task. A coupled HMM approach is presented in [3] for detecting complex actions in Tai Chi movies, a very interesting application. In literature, the Bayesian approach is mainly used in region or object-based image retrieval systems [11,10,12], in which the object's likelihood can be calculated from the conditional probability of feature vectors. Some systems use probabilistic reasoning to exploit the relationship between objects and their features. In [4], the authors propose a content-based semi-automatic annotation procedure for providing images with semantic labels. This approach uses Bayes point machines to give images a confidence level for each trained semantic label. This vector of confidence labels can be exploited to rank relevant images in case of a keyword search. Some other systems employ Bayesian approach in scene classification e.g., *indoor/outdoor* or *city view/landscape* [20,1,2]. As we have already stated, this kind of classification has been restricted to mutually exclusive categories, and so is only suitable for images that have one dominant concept. But in more realistic scenarios in image classification and retrieval, images are complex and usually consist of many semantically meaningful objects. Therefore the relationships between semantically meaningful concepts of the objects cannot be ignored and need to be explored to a greater degree.

3 Three-Level Multimedia Representation

The aim of multimedia retrieval techniques is to elicit, store and retrieve the audio- and imagery-based information content in multimedia. Taking images as an example of various digital multimedia content types, the search for a desired image from a repository might involve many image attributes including a particular combination of colour, texture or shape features, a specific type of object; a particular type of event or scene; named individuals, locations or events; subjectively associated emotions; metadata such as who created the image, where and when. According to these attributes, the types of query in a multimedia retrieval scenario can be categorised into three levels of increasing complexity corresponding to the used features [7]:

"*Level 1*: Primitive features such as colour, texture, shape, sound or the spatial location of image elements."

"*Level 2*: Derived features involving some degree of logical inference about the identity of the objects depicted in the image."

"*Level 3*: Abstract attributes involving a significant amount of high-level reasoning about the meaning and purpose of the objects or scenes depicted."

The first level of primitive features can be directly derived from the visual content, and does not require employing any knowledge base. It is relatively easy to generate this type of features but they are mostly restricted to applications about retrieving strong visual characters, such as trademark registration, identification of drawings in a design archive or colour matching of fashion accessories. The second and third levels are more commonly demanded in real-world scenarios as semantic image retrieval. However, the semantic gap lying between levels 1 and 2 hampers the progress of multimedia retrieval area [19]. The semantic gap is commonly defined as "the discrepancy between low-level features or content descriptors that can be computed automatically by current machines and algorithms, and the richness, and subjectivity of semantics in high-level human interpretations of audiovisual media". To bridge the semantic gap has become the most challenging task in semantic multimedia retrieval. Thus in this paper, by dividing this task into two steps: matching *Level 1* (low-level features) to *Level 2* (mid-level features) and matching *Level 2* (mid-level features) to *Level 3* (high-level features), we hope to minimise the difficulty in the task and try to find feasible solutions.

4 Semantic Inference for Video Annotation and Retrieval

The proposed semantic context learning and inference approach analyses the inter-relationships between the high-level queried concepts and mid-level features by constructing a model of these relationships automatically and use it to infer high-level terms out of mid-level features in future search.

Figure 1 shows the work flow of this approach. There are two processes in the work flow, the learning process and the inference process. In the learning process which is usually carried out off-line. First, several mid-level features are extracted using any specifically designed classifiers. A subset of the database randomly selected for training purpose is then manually annotated on the high-level query concept. Then extracted mid-level features and the manual annotations on the training subset is used to derive the semantic context model. In this research Bayesian network is used for modelling the semantic context involving mid-level features and the particular query concept. The learning process concerns learning of both the network structure and probability tables of nodes.

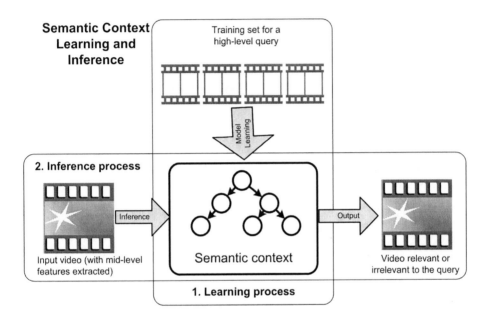

Fig. 1. Semantic inference work flow

One important feature in this module is that the Bayesian network model is constructed automatically using a learning approach based on $K2$ algorithm [8], which is basically a greedy search technique. It is able to automatically learn the structure of a Bayesian model based on a given training dataset. In this algorithm, a Bayesian network is created by starting with an empty network and iteratively adding a directed arc to a given node from each parent node. The iteration is terminated when no more possible additions could increase a score which is calculated as:

$$K2score = \prod_{j=1}^{p} \frac{\Gamma\left(\sum_k a_{ijk}\right)}{\Gamma\left(\sum_k a_{ijk} + \sum_k b_{ijk}\right)} \prod_{k=1}^{q_i} \frac{\Gamma\left(a_{ijk} + b_{ijk}\right)}{\Gamma\left(a_{ijk}\right)}$$

where i, j and k are respectively the index of the child node, the index of the parents of the child node, and the index of the possible values of the child node. p is the number of different instantiations of parent nodes, q_i is the number of possible values of the i_{th} child node, b is the number of times that the child node has the value of the k_{th} index value of the node, and a is the number of times that the parents and the child correlate positively in discrete cases. This selection criterion is basically a measure of how well the given graph correlates to the data. Due to the scope of this paper, we give only a brief introduction to $K2$ algorithm here. If the reader is interested in more details about this algorithm, please refer to [8].

Then in the inference stage, when an un-annotated data item is present, the Bayesian network model derived from the training stage conducts automatic semantic inferences for the high-level query. The Bayesian network in this case represent the semantic context involving the mid-level features of this video and their underlying links to the high-level query.

Assume that, for dataset D, a total number of n pre-defined mid-level features are available: $M = \{m_i | i = 1, 2, ..., n\}$. For the high-level query q in the same dataset D, each m_i are supposed to be more or less interrelated to q, and M together form the semantic context of D and can be used as the 'semantic evidence' for carrying out context reasoning for q. Consider an example case in which five mid-level features are used, $n = 5$, then the joint probability that q exists in a piece of multimedia content considering the semantic context can be calculated as:

$$
\begin{aligned}
P(M, q) &= P(q) \cdot P(M|q) = P(q) \cdot P(m_1, m_2, ..., m_5|q)] \\
&= P(q) \cdot P(m_1|q) \cdot P(m_2, m_3, m_4, m_5|m_1, q) \\
&= P(q) \cdot P(m_1|q) \cdot P(m_2|m_1, q) \cdot P(m_3, m_4, m_5|m_1, m_2, q) \\
&= P(q) \cdot P(m_1|q) \cdot P(m_2|m_1, q) \cdot P(m_3|m_1, m_2, q) \\
&\quad \cdot P(m_4, m_5|m_1, m_2, m_3, q) \\
&= P(q) \cdot P(m_1|q) \cdot P(m_2|m_1, q) \cdot P(m_3|m_1, m_2, q) \\
&\quad \cdot P(m_4|m_1, m_2, m_3, q) \cdot P(m_5|m_1, m_2, m_3, m_4, q)
\end{aligned}
$$

Assuming the components of M are independent to each other, the joint probability takes the form of

$$
P(M, q) = P(q, m_1, m_2, ..., m_5) = P(q) \cdot P(m_1|q) \cdot P(m_2|q).... \cdot P(m_5|q)
$$

A similar expansion can be obtained for an arbitrary number of classes n:

$$
P(M, q) = P(q, m_1, m_2, ..., m_n) = P(q) \cdot \prod_{i=1}^{n} P(m_i|q)
$$

5   Experiments

The experiments were carried out on the good sized un-edited video database. Videos were segmented into shots and each shot was treated as an independent multimedia element for analysis and annotation. Several mid-level features were selected by observing the content and they belong to *Level 2* of the three-level representation scheme as described in Section 3. These mid-level features include: 'Regular shapes', '2D flat-zone', 'Human faces', 'Flashlight', 'Text', 'Vegetation', 'Ship', 'Music', 'Speech'.

On top of that, two high-level queries have been carefully selected considering those commonly exist in the database with reasonable proportions and have relatively rich connections to the mid-level features: *City view from helicopter* (CVH) and *Football stadium*.

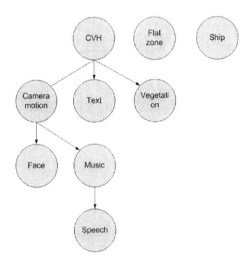

Fig. 2. Bayesian model for CVH

For the sake of illustration, an example of the learned model structures is shown in Figure 5 considering the case for *City view from helicopter* and some examples of probability distribution (PD) tables are given in Table 1.

In an attempt to read the networks manually, one may note that these learned structures and parameters are sometimes not meaningful in the understanding of human-beings. This is because they only reveal the underlying relationships between the query concept node and feature nodes from the machine's perspective, as the overall aim in the proposed semantic inference approach is to build a context network from available metadata rather than those containing full human intelligence in brains. The considered mid-level features and high-level queries are merely a small subset in the specific scenario due to the lack of rich subjects in the used video content. However, the experiment carried out based on this specific scenario shows that suitable high-level queries can be obtained purely based on semantic reasoning on a small subset of mid-level features.

The training processes for each semantic inference model were performed on a randomly selected subset of less than 10% of the whole dataset and this process took only a few seconds on a PC with Pentium D CPU 3.40GHZ and 2.00GB of RAM.

Ground-truth annotations of the two high-level terms have been made available on the whole experiment datasets. Based on these ground-truth annotations, we managed to obtain statistical evaluations on the results to show the performance of the proposed approach, as given in Table 2.

Table 1. Examples of PD tables for CVH

Camera motion		
CVH	True	False
True	0.169	0.831
False	0.002	0.998
Text		
CVH	True	False
True	0.101	0.899
False	0.236	0.764
Vegetation		
CVH	True	False
True	0.764	0.236
False	0.966	0.034

Table 2. Evaluation results

City view from helicopter			
TP	TN	FP	FN
10	2891	4	63
Precision	Recall	Accuracy	ROC area
71.40%	13.70%	97.74%	63.1%
Football stadium			
TP	TN	FP	FN
4	2950	3	11
Precision	Recall	Accuracy	ROC area
57.10%	26.70%	99.53%	97.1%

As it can be observed from Table 2, the detection performance of the proposed approach is reasonably good, given the limited accuracy of the mid-level feature extractors and the abstractness and sparse distribution of the query terms throughout the dataset.

6   Conclusions

In this chapter an approach for semantic context learning and inference has been presented. The basic idea was to organise the representations for multimedia content in three semantic levels, by adding a mid-level feature category between low-level visual features and high-level semantic terms. Thus, the mapping between low-level content and high-level semantics could be achieved in two steps, which were easier to achieve than a single step. This research has concentrated on the second step, high-level semantic extraction based on the semantic context involving a number of mid-level features. This was achieved by automatically building a context model for the relationships between the mid-level features and the query. Modelling and inference in this case were carried out using the $K2$ algorithm. The proposed approach was tested on a large size video dataset. The obtained results have shown that this approach was capable of extracting very abstract semantic terms that were scarcely distributed in the database.

Acknowledgments. The research that leads to this chapter was partially supported by the European Commission under the contracts FP6-045189 RUSHES and FP7-247688 3DLife.

References

1. Boutell, M., Luo, J.: Beyond pixels: Exploiting camera metadata for photo classification. Pattern recognition 38(6), 935–946 (2005)
2. Bradshaw, B.: Semantic based image retrieval: a probabilistic approach. In: Proceedings of the eighth ACM international conference on Multimedia, pp. 167–176 (2000)
3. Brand, M., Oliver, N., Pentland, A.: Coupled hidden markov models for complex action recognition. In: IEEE Computer Society Conference on Computer Vision and Pattern Recognition, pp. 994–999 (1997)
4. Chang, E., Goh, K., Sychay, G., Wu, G.: Cbsa: content-based soft annotation for multimodal image retrieval using bayes point machines. IEEE Transactions on Circuits and Systems for Video Technology 13(1), 26–38 (2003)
5. Chen, W., Chang, S.F.: Generating semantic visual templates for video databases. In: 2000 IEEE International Conference on Multimedia and Expo, 2000. ICME 2000, vol. 3 (2000)
6. De Jong, F.M.G., Westerveld, T., De Vries, A.P.: Multimedia search without visual analysis: the value of linguistic and contextual information. IEEE Transactions on Circuits and Systems for Video Technology 17(3), 365–371 (2007)
7. Eakins, J.P., Graham, M.E.: Content-based image retrieval: A report to the jisc technology applications programme. Tech. rep., Institute for Image Data Research, University of Northumbria at Newcastle (1999), http://www.jisc.ac.uk/uploaded_documents/jtap-039.doc
8. Cooper, G.F., Herskovits, E.: A bayesian method for the induction of probabilistic networks from data. Machine learning 9(4), 309–347 (1992)

9. Fan, J., Gao, Y., Luo, H., Jain, R.: Mining multilevel image semantics via hierarchical classification. IEEE Transactions on Multimedia 10(2), 167–187 (2008)
10. Fei-Fei, L., Fergus, R., Perona, P.: A bayesian approach to unsupervised one-shot learning of object categories. In: Proc. ICCV, vol. (2003)
11. Fergus, R., Perona, P., A., Zisserman, o.: Object class recognition by unsupervised scale-invariant learning. In: IEEE Computer Society Conference on Computer Vision and Pattern Recognition, vol. 2 (2003)
12. Hoiem, D., Sukthankar, R., Schneiderman, H., Huston, L.: Object-based image retrieval using the statistical structure of images. In: IEEE Computer Society Conference on Computer Vision and Pattern Recognition, vol. 2 (2004)
13. Kherfi, M.L., Ziou, D.: Image collection organization and its application to indexing, browsing, summarization, and semantic retrieval. IEEE Transactions on multimedia 9(4), 893–900 (2007)
14. Koskela, M., Smeaton, A.F., Laaksonen, J.: Measuring concept similarities in multimedia ontologies: Analysis and evaluations. IEEE Transactions on Multimedia 9(5), 912–922 (2007)
15. Lavrenko, V., Feng, S., Manmatha, R.: Statistical models for automatic video annotation and retrieval. In: IEEE International Conference on Acoustics, Speech, and Signal Processing ICASSP'04, vol. 3, IEEE Computer Society Press, Los Alamitos (2004)
16. Naphade, M.R., Huang, T.S.: A probabilistic framework for semantic video indexing, filtering, and retrieval. IEEE Transactions on Multimedia 3(1), 141–151 (2001)
17. Naphade, M.R., Huang, T.S.: Extracting semantics from audio-visual content: the final frontier in multimedia retrieval. IEEE Transactions on Neural Networks 13(4), 793–810 (2002)
18. Qian, R., Haering, N., Sezan, I.: A computational approach to semantic event detection. In: Proceedings of the IEEE Computer Society Conference on Computer Vision and Pattern Recognition, vol. 1, pp. 200–206 (1999)
19. Smeulders, A.W.M., Worring, M., Santini, S., Gupta, A., Jain, R.: Content-based image retrieval at the end of the early years. IEEE Transactions on pattern analysis and machine intelligence 22(12), 1349–1380 (2000)
20. Vailaya, A., Figueiredo, M.A.T., Jain, A.K., Zhang, H.J.: Image classification for content-based indexing. IEEE Transactions on Image Processing 10(1), 117–130 (2001)
21. Zhu, X., Wu, X., Elmagarmid, A.K., Feng, Z., Wu, L.: Video data mining: Semantic indexing and event detection from the association perspective. IEEE Transactions on Knowledge and Data engineering, 665–677 (2005)

Part VIII:

Future Internet Applications

Introduction

The Future Internet is grounded in the technological infrastructure for advanced networks and applications. It constitutes a complex and dynamic system and societal phenomenon; it comprises the processes of innovation, shaping and the actual use of these technologies and infrastructures in private and public organisations, in different sectors of the economy including the service sectors, and in social networks. Research on the Future Internet therefore includes the development, piloting and validation of high-value applications in domains such as healthcare, energy, transport, utilities, manufacturing and finance. Increasingly, research and innovation on the Future Internet such as envisaged in the Future Internet PPP programme forms part of a diverse, dynamic and increasingly open Future Internet innovation-ecosystem, where different stakeholders such as researchers, businesses, government actors and user communities are brought together to interact and engage in networked and collaborative innovation.

In the field of Future Internet application areas, several research and innovation topics are emerging for the next years. In particular, there is a need to explore the opportunities provided by Future Internet technologies in various business and societal sectors and how these opportunities could be realized through open innovation models.

One of the key developments is towards smart enterprises and collaborative enterprise networks. Enterprises of the future are envisioned to be ever more open, creative and sustainable; they will become "smart enterprises". Innovation lies at the core of smart enterprises and includes not only products, services and processes but also the organizational model and full set of relations that comprise the enterprise's value network. The Future Internet should provide enterprises a new set of capabilities, enabling them to innovate through flexibility and diversity in experimenting with new business values, models, structures and arrangements. Combinations of Future Internet technologies are needed to deliver maximum value and these combinations require the federation and integration of appropriate software building blocks. A new generation of enterprise systems comprising applications and services are expected to emerge, fine-tuned to the needs of enterprise users by leveraging a basic infrastructure of utility-like software services.

High-value Future Internet applications are also foreseen in the domain of living, healthcare, and energy. "Smart Living" is one of the areas where the focus lies clearly on the human user, and encompasses the combination of technologies in areas such as smart content, personal networks and ubiquitous services, to provide the user a simpler, easier and enriched life across many domains including home life, education and learning, working, and assisted living. Much interest is in "smart health" applications, including the provision of assisted living services for the elderly and handicapped, and also to increase the efficiency and effectiveness of the health value chain e.g. through enabling access to and sharing of patient data, secure data exchange between healthcare actors, and applications for remote and collaborative diagnosis, cure and care.

Another high-potential area also is that of "smart energy" systems. Two main topics of interest can be distinguished. The first topic concerns the resources of telecom operators and service providers such as networks, switching, computing and data centers which are prominent targets for energy efficiency. The second includes solutions allowing for energy management and reduction of the over-all energy consumption. One of the key developments in this respect is the use of advanced communication and computing infrastructure as part of the Smart Grid. Related topics include the cost-effective deployment of supporting infrastructures such as sensors and meters, advanced metering infrastructure, integration of power grid and ICT networks, and context-aware monitoring and control of energy consumption.

A final emerging area of high interest is "Smart Cities". Cities and urban areas of today are complex ecosystems, where ensuring quality of life is an important concern. In such urban environments, people, companies and public authorities experience specific needs and demands regarding domains such as healthcare, media, energy and the environment, safety, and public services. These domains are increasingly enabled and facilitated by Internet-based applications and infrastructures based on common platforms. Therefore, cities and urban environments are facing challenges to maintain and upgrade the required infrastructures and establish efficient, effective, open and participative innovation processes to jointly create the innovative applications that meet the demands of their citizens. In this context, cities and urban areas represent a critical mass when it comes to shaping the demand for advanced Internet-based services. The "living labs" approach which comprises open and user driven innovation in large-scale real-life settings opens up a promising opportunity to enrich the experimentally-driven research approach as currently adopted in the Future Internet community.

The four chapters in the Application Areas part of this book illustrate the developments and opportunities mentioned. The first chapter "Future Internet Enterprise Systems: a Flexible Architectural Approach for Innovation" discusses how emerging paradigms, such as Cloud Computing and Software-as-a-Service are opening up a significant transformation process for enterprise systems. This transformation arises from commoditization of the traditional enterprise system functions and is accelerated by new and innovative development methods and architectures of Future Internet Enterprise Systems. The chapter foresees a rich, complex, articulated digital world reflecting the real business world, where computational elements referred to as Future Internet Enterprise Resources will directly act and evolve according to what exists in the real world.

The chapter "Renewable Energy Provisioning for ICT Services in a Future Internet" discusses the GreenStar Network (GSN), of the first worldwide initiatives for provisioning ICT services that are entirely based on renewable energy such as solar, wind and hydroelectricity across Canada and around the world. GSN is developed to dynamically transport user services to be processed in data centers built in proximity to green energy sources, thereby reducing greenhouse gas emissions of ICT equipments. While current approaches mainly focus on reducing energy consumption at the micro-level through energy efficiency improvements, the proposed approach is much broader because it focuses on greenhouse gas emission reductions at the macro-level and focuses on heavy computing services dedicated to data centers powered completely by green energy, from a large abundant reserve of natural resources in Canada, Europe and the US.

The third chapter "Smart Cities and Future Internet: towards Cooperation Frameworks for Open Innovation" elaborates the concept of "smart cities" as environments of open and user driven innovation for experimenting and validating Future Internet-enabled services. The chapter describes how the living labs concept has started to fulfill a role in the development of cities towards becoming "smart". In order to exploit the opportunities of services enabled by the Future Internet for smart cities, there is a need to clarify the way how living lab innovation methods, user communities and Future Internet experimentation approaches and testbed facilities constitute a common set of resources. These common resources can be made accessible and shared in open innovation environments, to achieve ambitious city development goals. This approach requires sustainable partnerships and cooperation strategies among the main stakeholders.

The fourth chapter "Smart Cities at the forefront of the Future Internet" presents an example of city-scale platform architecture for utilizing innovative Internet of Things technologies to enhance the quality of life of citizens. It provides some results of generic implementations based on the Ubiquitous Sensor Network (USN) model. The referenced platform model fulfils basic federation principles at two different levels: the infrastructure level, where it provides a common ground for heterogeneous Internet of Things facilities that are interworking; and at the service level, where the platform can be used to interconnect with different Internet of Services testbeds, helping to bridge the existing gap between the two levels.

Hans Schaffers, Man-Sze Li, and Anastasius Gavras

Future Internet Enterprise Systems:
A Flexible Architectural Approach for Innovation

Daniela Angelucci, Michele Missikoff, and Francesco Taglino

Istituto di Analisi dei Sistemi ed Informatica "A. Ruberti"
Viale Manzoni 30, I-00185 Rome, Italy
daniela.angelucci@gmail.com, michele.missikoff@iasi.cnr.it,
francesco.taglino@iasi.cnr.it

Abstract. In recent years, the evolution of infrastructures and technologies carried out by emerging paradigms, such as Cloud Computing, Future Internet and SaaS (Software-as-a-Service), is leading the area of enterprise systems to a progressive, significant transformation process. This evolution is characterized by two aspects: a progressive commoditization of the traditional ES functions, with the 'usual' management and planning of resources, while the challenge is shifted toward the support to enterprise innovation. This process will be accelerated by the advent of FInES (Future Internet Enterprise System) research initiatives, where different scientific disciplines converge, together with empirical practices, engineering techniques and technological solutions. All together they aim at revisiting the development methods and architectures of the Future Enterprise Systems, according to the different articulations that Future Internet Systems (FIS) are assuming, to achieve the Future Internet Enterprise Systems (FInES). In particular, this paper foresees a progressive implementation of a rich, complex, articulated digital world that reflects the real business world, where computational elements, referred to as FInER (Future Internet Enterprise Resources), will directly act and evolve according to what exists in the real world.

Keywords: Future Internet, Future Enterprise Systems, component-based software engineering, COTS, SOA, MAS, smart objects, FInES, FInER.

1 Introduction

In recent years, software development methods and technologies have markedly evolved, with the advent of SOA [15], MDA [16], Ontologies and Semantic Web, to name a few. But there are still a number of open issues that require further research and yet new solutions. One crucial issue is the excessive time and cost required to develop enterprise systems (ES), even if one adopts a customisable pre-built application platform, e.g., an ERP solution.

This paper explores some emerging ideas concerning a new generation of Internet-based enterprise systems, along the line of what has been indicated in the FInES

J. Domingue et al. (Eds.): Future Internet Assembly, LNCS 6656, pp. 407–418, 2011.

(Future Internet Enterprise Systems) Research Roadmap[1], a study carried out in the context of the European Commission, and in particular the FInES Cluster of the D4 Unit: Internet of Things and Enterprise Environments (DG InfSo). The report claims that we are close to a significant transformation in the enterprise systems, where (i) the way they are developed, and (ii) their architectures, will undergo a progressive paradigm shift. Such paradigm shift is primarily motivated by the need to repositioning the role of enterprise systems that, since their inception, have been conceived to support the management and planning of enterprise resources. Payroll, inventory management, and accounting have been the first application areas. Then, ES progressively expanded their functions and aims, but the underlying philosophy remained the same: supporting the value production in the day by day business, optimising operations and the use of resources, with some look ahead capabilities (i.e., planning). In the recent period there has been a clear movement towards a progressive commoditization of such traditional ES functions. This movement is further facilitated by the evolution of infrastructures and technologies, starting from Cloud Computing and Future Internet, and, on top of those, the Software-as-a-Service (SaaS) paradigm that is progressively providing new ways of conceiving and realising enterprises software applications.

In essence, while enterprise management and planning services will be increasingly available from the 'cloud', in a commoditised form, the future business needs (and challenges) are progressively shifting towards the support to enterprise innovation. But also innovation cannot remain as it used to be: Future Internet, Web 2.0, Semantic Web, Cloud Computing, SaaS, Social Media, and similar emerging forms of distributed, open computing will push forward new forms of innovation such as, and in particular, Open Innovation [3]. The quest for continuous, systematic business innovation requires (i) ES capable of shifting the focus to ideas generation and innovation support, and (ii) new agile architectures, capable of (instantly) adjusting to the continuous change required to enterprises. New business requirements that current software engineering practices do not seem to meet. Therefore we need to orientate the research towards new ES architectures and development paradigms, when the role of ICT experts will be substantially reduced. To this end, we wish to propose three grand research challenges.

The first grand research challenge (GRC) implies for ICT people to surrender the mastership of ES development, handing it over to business experts. To this end, the ICT domain needs to push forward the implementation of future ES **development environments**, specifically conceived to be directly used by business experts. Such development environments will be based on an evolution of MDA, being able to separate the specification and development of the (i) strategic business logic from the (ii) specific business operations and, finally, their (iii) actual implementation. A central role will be played by enterprise system Business Process Engineering, for the above point (i), and a new vision, based on a new family of reusable components, in the implementation of enterprise operations (and related services) automation, for the last two points. Reusable components mash-up techniques, and advanced graphical user

---

[1] http://cordis.europa.eu/fp7/ict/enet/documents/task-forces/research-roadmap/

interfaces will foster new development environments conceived for business experts to directly intervene in the development process.

The second grand research challenge concerns the **architecture** of the Future Internet Enterprise Systems (FInES) that need to deeply change with respect to what we have today. A new paradigm is somehow already emerging nowadays, pushed by the new solutions offered in the Future Internet Systems (FIS) field. In particular, we may mention: the Internet of Services (IoS), Internet of Things (IoT) and *smart objects*, Internet of Knowledge (IoK), Internet of People (IoP). But these solutions need to further evolve towards a better characterisation in the business direction, allowing different aspects of the business reality (functions, objects, actors, etc.) to acquire their networked identity, together with a clear and precise definition (i.e., science based) of their (information) structure, capabilities, and mutual relationships. In fact, what is missing today is a unifying vision of the disparate business aspects and entities of an enterprise, supported by an adequate theory, able to propose new technological paradigms. The idea is that all possible entities within (and outside) an enterprise will have a digital image (a sort of 'avatar') that has been referred to as Future Internet Enterprise Resource (FInER) in the FInES Research Roadmap. So, the second grand research challenge consists in conceiving new, highly modular, flexible FInERs for the FInES architectures to be based on.

Finally, there is a third grand research challenge, that of shifting the focus of the attention from the management and planning of business and enterprise resources to **enterprise innovation**. This GRC requires, again, a strategic synergy between ICT and business experts. Together, they need to cooperate in developing a new breed of services, tools, software packages, interfaces and user interaction solutions that are not available at the present time. A new family of ICT solutions aimed at supporting the conception, design, implementation and deployment of enterprise innovation, including assessment of impact and risks.

In this paper, we intend to further elaborate on these challenges. In particular on the first and the second GRC that concern the development of new FInESs capable of offering to the business experts the possibility of directly governing the development of software architectures. This will be possible if such software architectures will correspond to the enterprise architectures, and will be composed by elements tightly coupled with business entities. The achievement of this objective relies on a number of ICT solutions that are already emerging: from Cloud Computing to Social Media, to Service-oriented Computing, from Business Process Engineering to semantic technologies and mash-up. An exhaustive analysis of the mentioned technologies is outside the scope of this paper, below we will briefly survey some of them.

The rest of the paper is organised as follows. Next Section II provides an overview of the evolution of the main technologies that, in our vision, will support the advent of the FInES. Section III presents the main characteristics of a FInES innovation-oriented architecture. Section IV introduces the new component-oriented approach based on the notion of a FInER, seen as the new frontier to software components aimed at achieving agile system architectures. Section V provides some conclusions and a few lines that will guide our future work.

2   A Long March towards Component-Based Enterprise Systems

FInES represents a new generation of enterprise systems aimed at supporting continuous, open innovation. Innovation implies continuous, often deep changes in the enterprise; such changes must be mirrored by the enterprise systems: if the latter are too complex, rigid, difficult to evolve, they will represent a hindering factor for innovation. A key approach to system flexibility and evolvability is represented by highly componentized architectures.

Traditionally, the software engineering community has devoted great attention to design approaches, methods and tools, supporting the idea that large software systems can be created starting from independent, reusable collections of pre-existing software components.

This technical area is often referred to as Component Based Software Engineering (CBSE). The basic idea of software componentization is quite the same as software modularization, but mainly focused on reuse. CBSE distinguishes the process of "component development" from that of "system development with components" [9].

CBSE laid the groundwork for the Object Oriented Programming (OOP) paradigm that in a short time imposed itself over the pre-existing modular software development techniques. OOP aims at developing applications and software systems that provide a high level of data abstraction and modularity (using technologies such as COM, .NET, EJB and J2EE).

Another approach to componentization is that of the Multi Agent Systems (MAS), which is based on the development of *autonomous, heterogeneous, interacting* software agents. Agents mark a fundamental difference from conventional software modules in that they are inherently autonomous and endowed with advanced communication capability [10].

On the other side, the spread of the Internet technologies and the rising of new communication paradigms, has encouraged the development of loosely coupled and highly interoperable software architectures through the spread of the Service-Oriented approach, and the consequent proliferation of Service-Oriented Architectures (SOA). SOA is an architectural approach whose goal is to achieve loose coupling among interacting software services, i.e., units of work performed by software applications, typically communicating over the Internet [11].

In general, a SOA will be implemented starting from a collection of components (e-services) of two different sorts. Some services will have a 'technical' nature, conceived to the specific needs of ICT people; some other will have a 'business' nature, reflecting the needs of the enterprise. Furthermore, the very same notion of an e-service is an abstraction that often hides the entity (or agent) that in the real world provides such a service. Such an issue may seem trivial to ICT people (they need a given computation to take place; where it is performed or who is taking care of it is inconsequential). Conversely, for business people, services are not generated 'in the air': there is an active entity (a person, an organization, a computer, a robot, etc.) that provides the services, with a given cost and time (not to mention SLA, etc.), associated to it.

In summary, Web services were essentially introduced as a computation resource, transforming a given input to produce the desired output, originally without the need to have a persistent memory and an evident state. Such a notion of ICT service is very specific and, as such, not always suited when we consider business services, where states, memories, and even the pre-existing history of the entity providing the business service, are important.

Our aim to achieve an agile system architecture made up of FInERs put its basis upon the spread of the Cloud Computing philosophy, but revising and applying it into the specific context of developing new FInESs, where business expert can directly manage a new generation enterprise software architectures. Cloud Computing represents an innovative way to architect and remotely manage computing resources: this approach aims at delivering scalable IT resources over the Internet, as opposed to hosting and operating those resources (i.e. applications, services and the infrastructure on which they operate) locally. It refers to both the applications delivered as services over the Internet and the hardware and system software in the datacenters that provide those services [12]. Cloud Computing may be considered the basic support for a brand new business reality where FInERs can easily be searched, composed and executed by a business expert. FInERs will implement a cloud-oriented way of designing, organizing and implementing the enterprises of the future.

In conclusion, for decades component technologies have been developed with an ICT approach, to ease software development processes. Conversely, we propose to base a FInES architecture on building blocks based on business components. In addressing a component-based architecture for FInES, we felt the need for a new component-oriented philosophy that should be closer to, and make good sense for, business people. The idea of a FinER goes in this direction. From an ICT perspective, the notion of a FInER integrates many characteristics of the OOP, MAS, and SOA just illustrated, plus the key notion of *smart object*. But the key issue is that a FInER can exist only if it represents a business entity and is recognized as such by business people.

3   Guidelines for a FInES Architecture

With the idea of a continuous innovation process going on in parallel to the everyday business activities, a FInES needs to integrate the two levels: doing business and pushing forward innovation, constantly evolving along a loop similar to that represented in Fig.1. From the architectural point of view, a FInES is seen as a federation of systems relaying on two major infrastructures: one for the advanced management of knowledge and the other for semantic interoperability. The figure reports four systems identified by the same numerals reported in the FInES Research Roadmap, but the prefix here is the letter 'S' to indicate the *systems* and 'I' to indicate the *infrastructures* (instead of 'RC' for research challenge used in the FInES Research Roadmap).

The macro-architecture is briefly described below along the lines of what is reported in the cited research roadmap (where additional details can be found).

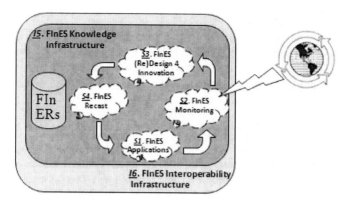

Fig. 1. FInES Macro-architecture

S1 – FInES Open Application System

This system is devoted to the business operations for the day by day business and value production, with planning capabilities. In terms of functionalities it is similar to an ERP as we know it today, but its architecture is inherently different, since it is built by business experts by using Enterprise Systems/Architectures (including Business Process) Engineering methods and tools starting from a repository of FInERs, the new sort of computational enterprise components just introduced (see below for more details).

S2 – FInES Open Monitoring System

This system is dedicated to the constant monitoring and assessment of the activities of S1, to keep under control the health of the enterprise, its performances, both internally (HR, resources, productivity, targets, etc.) and with respect to the external world (markets, competitors, natural resources, new technologies, etc.). S2 is able to signal when and where an innovation intervention is necessary, or even suitable.

S3 – FInES Re-design System

This system is mainly used by business experts who, once identified the area(s) where it is necessary / suitable to intervene, proceed in designing the interventions on the enterprise and, correspondingly, on FInES. This task is achieved by using a platform with a rich set of tools necessary to support the business experts in their redesign activities that are, and will remain, largely a 'brain intensive' job.

S4 – FInES Recast System

This system has the critical task of implementing the new specifications identified and released by S3. The activities of S4 can be seen roughly divided in two phases. The first phase is focused on the identification and acquisition of the new components to be used (e.g., the new FInERs). The second phase is the actual deployment of the new FInES. Updating a FInES is a particularly delicate job, since the changes need to be achieved without stopping the business activities.

Finally, the proposed architecture relies on two key infrastructures: the FInES **Knowledge Infrastructure** (I5-FKI) aimed at the management of the information and knowledge distributed within the different FInERs, maintaining a complex, networked structure, conceived as an evolution of the Linked Open Data[2] of today; and the FInES **Interoperability Infrastructure** (I6-FII), supporting the smooth communication among the great variety of components, services, tools, platforms, resources, (produced by different providers) that compose a FInES.

4   The New Frontier for ES Components: The FInER Approach

In FInES architectures we intend to push the component-oriented engineering to an extreme, applied such an approach both horizontally (i.e., for different classes of applications and services) and vertically (using sub-parts at different levels of granularity). But the most relevant aspect is the large role played by a new sort of components: FInERs, i.e., computational units representing enterprise entities. They are recognised by business people as constituent parts of the enterprise, and therefore easily manipulated by them. A FInER has also a computational nature, characterised by 5 aspects, as described below.

A FInER can be a concrete, tangible entity, such as a drilling machine or an automobile, or intangible, such as a training course, a business process, or a marketing strategy. A complex organization, such as an enterprise, is itself a FInER. FInERs are conceived to interact and cooperate among themselves, in a more or less tight way, depending on cases (alike to real world business entities do).

To be more precise, we define a FInER as a 5-tuple:

$$F = (FID, GR, M, B, N)$$

The defining elements can greatly vary, depending on the complexity of the enterprise entity represented. In general we have:

FID: FInER identifier. This is a unique identifier defined according to a precise, universally accepted standard (e.g., according to IPv6, URI[3], or ENS[4]).

GR: Graphical Representation. This can vary from a simple GIF to a 3D model, to a JPEG representation, including possibly a video.

M: Memory. The FInER memory again can be simply a ROM with basic info (e.g., its sort, date of production, etc.) to complex knowledge about its components, properties and the history of its lifecycle.

B: Behaviour. The description of the functional capability of the FInER. It structured on the line of the profile (IOPE: Input, Output, Processing, Effects) and model (essentially, a set of workflows) of OWL-S [13]. Besides, it may have self-monitoring capabilities.

N: Networking. The specification of all the possible interactions the FInER can achieve, with the protocols for issuing (as client) or responding (as server) to request messages. It is structured according to the grounding of OWL-S.

---

[2] http://esw.w3.org/SweoIG/TaskForces/CommunityProjects/LinkingOpenData
[3] Universal Resource Identifier, see Berners-Lee et al., 2005: http://gbiv.com/protocols/uri/rfc/rfc3986.html
[4] ENS: Entity Name System, proposed by the OKKAM project (www.okkam.org)

As a first level, 5 FInERs categories have been identified:

- **Enterprise**, being the 'key assembly' in our work.
- **Public Administration**, seen in its interactions with the enterprise.
- **People**, a special class of FInERs for which avatars are mandatory.
- **Tangible entity**, from computers to aircrafts, to buildings and furniture.
- **Intangible entity**, for which a digital image is mandatory.

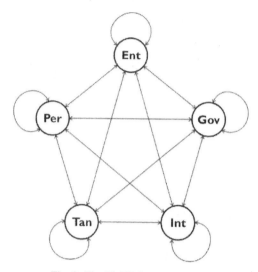

Fig. 2. The FInER Pentagone

All these FInERs will freely interact and cooperate, according to what happens for their real world counterparts. A complete interaction scheme is sketchily represented by the FInERs Pentagon in Fig. 2.

The proposed approach foresees a progressive implementation of a rich, complex, articulated digital world where computational elements, i.e. FInERs, will largely reflect what exists in the real (analogical) world. As a consequence, we expect that the FInER world will include all possible objects, creatures, entities, both simple and complex, animated and inanimate, tangible and intangible, that can be found in the real world.

As previously said, the FInER entities will have a unique identity and will be constantly connected (transparently, in a wired or wireless mode) to the Internet, to reach other FInERs, but at the same time to be themselves reachable anytime, anywhere, by any other FInER.

5   The FInES Approach to Design and Runtime Operations

As anticipated, the key platforms in the adoption of the FInES approach are the (re)design and development, and the runtime execution platforms.

5.1   A Business-Driven FInES Develpment Platform

In order to put the business experts at the centre of the ES development process, we foresee a platform where FInERs are visualised and directly manipulated by the user, while they reside in the Cloud and are reached through the Internet. On the FInES development environment (see Fig. 3), FInERs are visually represented in a 3D space that models the enterprise reality (i.e., a Virtual Enteprise Reality) where the user can navigate and manage changes. At a lower level, simpler FInERs will be aggregated to form more complex ones. The composition will take place in a partial automatic and bottom-up way, since many FInERs have autonomic capabilities and know already how to aggregate and connect to each other. Then, business experts supervise and complete the work. This approach represents a marked discontinuity with the past, since a FInES will be directly engineered by business experts and not by IT specialists. In fact, business experts will be able to select, manipulate, and compose FInERs at best, since they know better than IT specialists what the different business entities (represented by FInERs) are, which characteristics they have, how they can be connected one another to cooperate for achieving successful business undertakings. Future Internet will play a central role in supporting the discovery of the needed FInERs that often will be virtually acquired (in case of intangible assets), since they will be positioned in different parts of the enterprise or in the Cloud, depending on the cases.

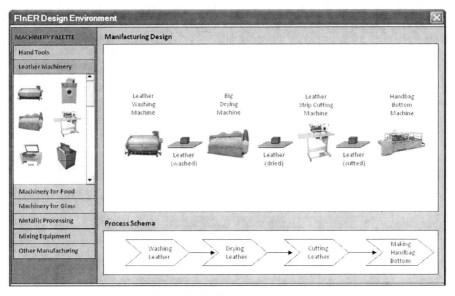

Fig. 3. FInES design environment

5.2   A Cloud-Based Architecture for FInERs Runtime

Once a FInES has been assembled (or re-casted, see Fig. 1), a runtime environment will recognise, connect, and support the execution and collaboration of the FInER components. Since FInERs will be developed by different suppliers, using different approaches and techniques, one of the key aspects here is represented by the availability of a number of Smart Interoperability Enablers capable of recognising and transmitting events and data generated during FInERs' operations. There is not a centralised database, the information will stay by the business entity to which they pertain or in the Cloud. A similar interface, representing a Virtual Enterprise Reality, will be made available to the users during business operations to navigate in the enterprise and see how the operations evolve.

The computational resources of a FInES are maintained in the Computing Cloud, and are recursively linked to compose complex FInERs starting from simpler ones. Fig. 4 reports a three levels macro-architecture where the top level is represented by the real world, with the enterprise and the actual business resources. Below we distinguish *High Level* and *Low Level* and FInERs. At the bottom level, there are atomic and simple FInER aggregations that do not represent yet consistent business entities. E.g, a simple production machine, with its components, elementary operations belonging to one or more business process, a teaching textbook. The High Level FInERs are representatives of business entities that have a clear business identity and are aimed at the reaching a business goal. For instance, a production chain, a training course, a company department. (A formal definition of Low and High Level FInERs is difficult to be achieved, much depends on the business perspective, the industrial sector, etc.).

The platform is mainly activated by business events that are generated in the enterprise (but also the external world, in general) and trigger the execution of the FInERs that correspond to the business entities that need to respond to the triggering events. It is also possible to have pre-planned events that are generated by a FInER (e.g., the achievement of a specific operation or the reaching of a deadline) and propagate over the FInES according to a pre-defined semantics (e.g., a specific business process). In principle, all the FInERs are hosted in the Cloud.

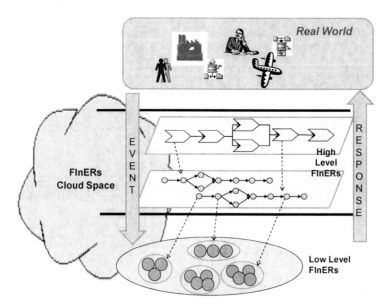

Fig. 4. FInES Runtime Environment

The runtime architecture of Fig. 4 is described in a sketchy way, aiming to highlight the main issues represented by (i) the highly modular structure, (ii) the mirroring of the real world business entities, (iii) the full control of the architecture by the business experts.

6   Conclusions

At the beginning of the 80s, the SUN had a visionary catchphrase, summarised in the sentence 'The Network is the Computer'. As it happens with early intuitions, it took too long to happen and now only few remember this foretelling. Actually, current ICT achievements show that this is going to be fully achieved in a short term to go. As a next prophecy we propose "the Enterprise is the Computer", meaning that an enterprise, with all its FInERs deployed and operational, will enjoy a fully distributed computing power, where computation will be directly performed by enterprise components, mainly positioned in the enterprise itself of in the Cloud (typically, in case of intangible entities). This approach represents a disruptive change, from both a technological point of view and a business perspective. In particular, in the latter, business people will be involved in building and maintaining large scale computing solutions simply interacting with a familiar (though technologically enhanced) business reality. Along this line, the (in)famous business / IT alignment problem will simply disappear, since the gap has been solved at its roots.

We are aware that the vision presented in this paper is a long term one, however, we believe that it will advance progressively and in the next years we will see an increasing integration of technologies that today are still loosely connected (e.g., IoT, IoS, Multi-Agent Systems, Cloud Computing, Autonomic Systems) and, in parallel, some key areas of the enterprise that will start to benefit of the FInES approach.

Acknowledgment. We acknowledge the FInES Cluster (European Commission), and in particular the FInES Research Roadmap Task Force for their contribution in producing much of the material reported in this paper.

References

1. Bouckaert, S., De Poorter, E., Latré, B., et al.: Strategies and Challenges for Interconnecting Wireless Mesh and Wireless Sensor Networks. Wireless Personal Communications 53(3) (2010)
2. Buxmann, P., Hess, T., Ruggaber, R.: - Internet of Services. Business & Information Systems Engineering 1(5), 341–342 (2009)
3. Chesbrough, H.: Open Innovation: The new Imperative for Creating and Profiting from Technology. Harvard Business School Press (2003)

4. Luftmann, J.N., Papp, R., Brier, T.: Enablers and Inhibitors of Business-IT-Alignment. Communications of AIS 1(11) (1999)
5. Mansell, R.E.: Introduction to Volume II: Knowledge, Economics and Organization. In: Mansell (ed.), The Information Society, Critical Concepts in Sociology, Routledge (2009)
6. Cordis.lu: Proposition, Informal Study Group on Value. Retrieved May 29, 2010, from Value Proposition for Enterprise Interoperability Report (2009), http://cordis.europa.eu/fp7/ict/enet/ei-isg_en.html
7. Sykes, D., Heaven, W., Magee, J., Kramer, J.: From goals to components: a combined approach to self-management. In: Proceedings of the 2008 international workshop on Software engineering for adaptive and self-managing systems (2008)
8. Villa, F., Athanasiadis, I.A., Rizzoli, A.E.: Modelling with knowledge: A review of emerging semantic approaches to environmental modeling. Environmental Modelling & Software 24(5) (2009)
9. Crnkovic, I., Larsson, S., Chaudron, M.: Component-based Development Process and Component Lifecycle. In: 27th International Conference on Information Technology Interfaces (ITI), Cavtat, Croatia, IEEE, Los Alamitos (2005)
10. Nierstrasz, O., Gibbs, S., Tsichritzis, D.: Component-oriented software development, Special issue on alaysis and modeling in software development, pp. 160–165 (1992)
11. Petritsch, H.: Service-Oriented Architecture (SOA) vs. Component Based Architecture, white paper, TU Wien (2006), http://whitepapers.techrepublic.com.com
12. Armbrust, M., et al.: Above the Clouds: A Berkley View of Cloud Computing, EECS-2009-28 (2009)
13. Martin, D., et al.: Bringing Semantics to Web Services with OWL-S. In: Proc. Of WWW Conference (2007)
14. Clark, D., et al.: Newarch project: Future-generation internet architecture. Tech Rep. MIT Laboratory for Computer Science (2003), http://www.isi.edu/newarch/
15. Tselentis, G., et al. (eds.): Towards the Future Internet- Emerging Trends from European Research. IOS Press, Amsterdam (2010)
16. Papazoglou, M.P.: Web Services: Principles and Technology. Prentice-Hall, Englewood Cliffs (2007)
17. Mellor, S.J., Scott, K., Uhl, A., Weise, D.: Model-driven architecture. In: Bruel, J.-M., Bellahsène, Z., et al. (eds.) OOIS 2002. LNCS, vol. 2426, p. 290. Springer, Heidelberg (2002)

Renewable Energy Provisioning for ICT Services in a Future Internet

Kim Khoa Nguyen[1], Mohamed Cheriet[1], Mathieu Lemay[2], Bill St. Arnaud[3],
Victor Reijs[4], Andrew Mackarel[4], Pau Minoves[5], Alin Pastrama[6], and
Ward Van Heddeghem[7]

[1] Ecole de Technologie Superieure, University of Quebec, Canada
kim.nguyen@synchromedia.ca, Mohamed.Cheriet@etsmtl.ca
[2] Inocybe Technologies, Canada
mlemay@inocybe.ca
[3] St. Arnaud Walker & Associates, Canada
bill.st.arnaud@gmail.com
[4] HEAnet, Ireland
{victor.reijs, andrew.mackarel}@heanet.ie
[5] i2CAT Foundation, Spain
pau.minoves@i2cat.net
[6] Nordunet, Iceland
alin@nordu.net
[7] Interdisciplinary institute for BroadBand Technology, Belgium.
ward.vanheddeghem@intec.ugent.be

Abstract. As one of the first worldwide initiatives provisioning ICT (Information and Communication Technologies) services entirely based on renewable energy such as solar, wind and hydroelectricity across Canada and around the world, the GreenStar Network (GSN) is developed to dynamically transport user services to be processed in data centers built in proximity to green energy sources, reducing GHG (Greenhouse Gas) emissions of ICT equipments. Regarding the current approach, which focuses mainly in reducing energy consumption at the micro-level through energy efficiency improvements, the overall energy consumption will eventually increase due to the growing demand from new services and users, resulting in an increase in GHG emissions. Based on the cooperation between Mantychore FP7 and the GSN, our approach is, therefore, much broader and more appropriate because it focuses on GHG emission reductions at the macro-level. Whilst energy efficiency techniques are still encouraged at low-end client equipments, the heaviest computing services are dedicated to virtual data centers powered completely by green energy from a large abundant reserve of natural resources, particularly in northern countries.

Keywords: Green Star Network, Mantychore FP7, green ICT, Future Internet

1   Introduction

Nowadays, reducing greenhouse gas (GHG) emissions is becoming one of the most challenging research topics in Information and Communication Technologies (ICT)

J. Domingue et al. (Eds.): Future Internet Assembly, LNCS 6656, pp. 419–429, 2011.

because of the alarming growth of indirect GHG emissions resulting from the overwhelming utilization of ICT electrical devices [1]. The current approach when dealing with the ICT GHG problem is improving energy efficiency, aimed at reducing energy consumption at the micro level. Research projects following this direction have focused on micro-processor design, computer design, power-on-demand architectures and virtual machine consolidation techniques. However, a micro-level energy efficiency approach will likely lead to an overall increase in energy consumption due to the Khazzoom–Brookes postulate (also known as Jevons paradox) [2], which states that *"energy efficiency improvements that, on the broadest considerations, are economically justified at the micro level, lead to higher levels of energy consumption at the macro level"*. Therefore, we believe that reducing GHG emissions at the macro level is a more appropriate solution. Large ICT companies, like Microsoft which consumes up to 27MW of energy at any given time [1], have built their data centers near green power sources. Unfortunately, many computing centers are not so close to green energy sources. Thus, green energy distributed network is an emerging technology, given that losses incurred in energy transmission over power utility infrastructures are much higher than those caused by data transmission, which makes relocating a data center near a renewable energy source a more efficient solution than trying to bring the energy to an existing location.

The GreenStar Network (GSN) project [3] is one of the first worldwide initiatives aimed at providing ICT services based entirely on renewable energy sources such as solar, wind and hydroelectricity across Canada and around the world. The network can transport user service applications to be processed in data centers built in proximity to green energy sources, thus GHG emissions of ICT equipments are reduced to minimal. Whilst energy efficiency techniques are still encouraged at low-end client equipments (e.g., such as hand-held devices, home PCs), the heaviest computing services will be dedicated to data centers powered completely by green energy. This is enabled thanks to a large abundant reserve of natural green energy resources in Canada, Europe and USA. The carbon credit saving that we refer to in this paper is the emission due to the operation of the network; the GHG emission during the production phase of the equipments used in the network and in the server farms is not considered since no special equipment is deployed in the GSN.

In order to move virtualized data centers towards network nodes powered by green energy sources distributed in such a multi-domain network, particularly between Europe and North America domains, the GSN is based on a flexible routing platform provided by the Mantychore FP7 project [8], which collaborates with the GSN project to enhance the carbon footprint exchange standard for ICT services. This collaboration enables research on the feasibility of powering e-Infrastructures in multiple domains worldwide with renewable energy sources. Management and technical policies will be developed to leverage virtualization, which helps to migrate virtual infrastructure resources from one site to another based on power availability. This will facilitate use of renewable energy within the GSN providing an Infrastructure as a Service (IaaS) management tool. By integrating connectivity to parts of the European National Research and Education Network (NREN) infrastructures with the GSN network this develops competencies to understand how a set of green nodes (where each one is powered by a different renewable energy source) could be integrated into an everyday network. Energy considerations are taken before moving virtual services without suffering connectivity interruptions. The influence of physical location in that relocation is also addressed, such as weather prediction, estimation of solar power generation. Energy produced from renewable sources has the potential to become one of the industries in each of these countries. In addition to the GSN-Mantychore project, a number of other Green ICT projects are also conducted in parallel in the FP7 framework [16].

The main objective of the GSN/Mantychore liaison is to create a pilot and a testbed environment from which to derive best practices and guidelines to follow when building low carbon networks. Core nodes are linked by an underlying high speed optical network having up to 1,000 Gbit/s bandwidth capacity provided by CANARIE. Note that optical networks have a modest increase in power consumption, especially with new 100Gbit/s, in comparison to electronic equipments such as routers and aggregators [4]. The migration of virtual data centers over network nodes is indeed a result of a convergence of server and network virtualizations as virtual infrastructure management. The GSN as a network architecture is built with multiple layers, resulting in a large number of resources to be managed. Virtualized management has therefore been proposed for service delivery regardless of the physical location of the infrastructure which is determined by resource providers. This allows complex underlying services to remain hidden inside the infrastructure provider. Resources are allocated according to user requirements; hence high utilization and optimization levels can be achieved. During the service, the user monitors and controls resources as if he was the owner, allowing the user to run their application in a virtual infrastructure powered by green energy sources.

2   Provisioning of ICT Services over Mantychore FP7 and GSN with Renewable Energy

In the European NREN community connectivity services are provisioned on a manual basis with some effort now focusing towards automating the service setup and operation. Rising energy costs, working in an austerity based environment which has dynamically changing business requirements has raised the focus of the community to control some characteristics of these connectivity services, so that users can change some of the service characteristics without having to renegotiate with the service provider.

The Mantychore FP7 project has evolved from previous research projects MANTICORE and MANTICORE II [8][9]. The initial MANTICORE project goal was to implement a proof of concept based on the idea that routers and an IP network can be setup as a Service (IPNaaS, as a management Layer 3 network). MANTICORE II continued in the steps of its predecessor to implement stable and robust software while running trials on a range of network equipment.

The Mantychore FP7 project allows the NRENs to provide a complete, flexible network service that offers research communities the ability to create an IP network under their control, where they can configure:

a) Layer 1, Optical links. Users will be able to get access control over optical devices like optical switches, to configure important properties of its cards and ports. Mantychore integrates the Argia framework [10] which provides complete control of optical resources.

b) Layer 2, Ethernet and MPLS. Users will be able to get control over Ethernet and MPLS (Layer 2.5) switches to configure different services. In this aspect, Mantychore will integrate the Ether project [6] and its capabilities for the management of Ethernet and MPLS resources.

c) Layer 3, Mantychore FP7 suite includes set of features for: i)Configuration and creation of virtual networks, ii) Configuration of physical interfaces, iii) Support of routing protocols, both internal (RIP, OSPF) and external (BGP), iv) Support of QoS and firewall services, v) Creation, modification and deletion of resources (interfaces, routers) both physical and logical, and vi) Support of IPv6. It allows the configuration of IPv6 in interfaces, routing protocols, networks.

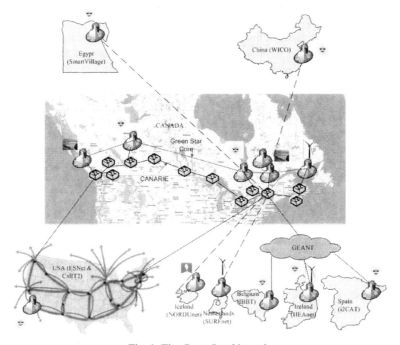

Fig. 1. The GreenStar Nework

Figure 1 shows the connection plan of the GSN. The Canadian section of the GSN has the largest deployment of six GSN nodes powered by sun, wind and hydroelectricity. It is connected to the European green nodes in Ireland (HEAnet), Iceland (NORDUnet), Spain (i2CAT), Belgium (IBBT), the Netherlands (SURFnet), and some other nodes in other parts of the world such as in China (WiCo), Egypt (Smart Village) and USA (ESNet).

One of the key objectives of the liaison between Mantychore FP7 and GSN projects is to enable renewable energy provisioning for NRENs. Building competency using renewable energy resources is vital for any NREN with such an abundance of natural power generation resources at their backdoor and has been targeted as a potential major industry for the future. HEAnet in Ireland [11], where the share of electricity generated from renewable energy sources in 2009 was 14.4%, have connected two GSN nodes via the GEANT Plus service and its own NREN network to a GSN solar powered node in the South East of Ireland and to a wind powered grid supplied location in the North East of the country. NORDUnet [12], which links Scandinavian countries having the highest proportion of renewable energy sources in Europe, houses a GSN Node at a data center in Reykjavík (Iceland) and also contributes to the GSN/Mantychore controller interface development. In Spain [13], where 12.5% of energy comes from renewable energy sources (mostly solar and wind), i2CAT is leading the Mantychore development as well as actively defining the interface for GSN/Mantychore and they will setup a solar powered node in Lleida (Spain). There will be also a GSN node provided by the Interdisciplinary institute for BroadBand Technology (IBBT) network in Belgium [15]. IBBT is involved in several research projects on controlling energy usage and setting up dynamic architectures and reducing CO2 emission and they will setup a solar powered node in Gent in order to perform power related experiments.

3   Architecture of a Zero-Carbon Network

We now focus on the design and management of a GSN network which provides zero-carbon ICT services. The only difference between the GSN and a regular network is that the former one is able to transport ICT services to data centers powered

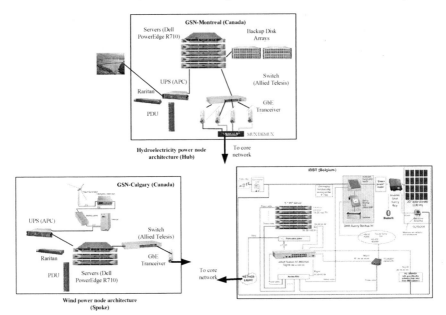

Fig. 2. Architecture of green nodes (hydro, wind and solar types)

by green energy and adjust the network to the needs controlled by software. The cost of producing and maintaining network elements, such as routers and servers, is not considered, because no special hardware equipment is used in the GSN.

Figure 2 illustrates the architectures of a hydroelectricity and two green nodes, one is powered by solar energy and the other is powered by wind. The solar panels are grouped in bundles of 9 or 10 panels, each panel generates a power of 220-230W. The wind turbine system is a 15kW generator. After being accumulated in a battery bank, electrical energy is treated by an inverter/charger in order to produce an appropriate output current for computing and networking devices. User applications are running on multiple Dell PowerEdge R710 systems, hosted by a rack mount structure in an outdoor climate-controlled enclosure. The air conditioning and heating elements are powered by green energy at solar and wind nodes; they are connected to regular the power grid at hydro nodes. The PDUs (Power Distribution Unit), provided with power monitoring features, measures electrical current and voltage. Within each node, servers are linked by a local network, which is then connected to the core network through GE transceivers. Data flows are transferred among GSN nodes over dedicated circuits (like light paths or P2P links), tunnels over Internet or logical IP networks.

The Montreal GSN node plays a role of a manager (hub node) that opportunistically sets up required connectivity for Layer 1 and Layer 2 using dynamic services, then pushes Virtual Machines (VMs) or software virtual routers from the hub to a sun or wind node (spoke node) when power is available. VMs will be pulled back to the hub node when power dwindles. In such a case, the spoke node may switch over to grid energy for running other services if it is required. However, GSN services are powered entirely by green energy. The VMs are used to run user applications, particularly heavy-computing services. Based on this testbed network, experiments and research are performed targeting cloud management algorithms and optimization of the intermittently-available renewable energy sources.

The cloud management solution developed in order to run the GSN enables the control of a large number of devices of different layers. With power monitoring and control capabilities, the proposed solution aims at distributing user-oriented services

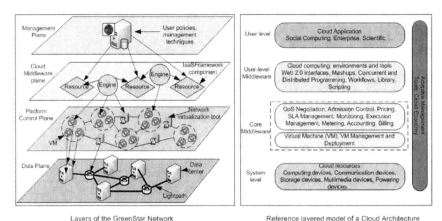

Fig. 3. Layered GSN and Cloud Computing Architectures

regardless of the underlying infrastructure. Such a management approach is essential for data center migration across a wide-area network, because the migration must be achieved in a timely manner and transparently to service users. The proposed web-based cloud management solution is based on the IaaS concept, which is a new software platform specific for dealing with the delivery of computing infrastructure [5].

Figure 3 compares the layered architecture of the GSN with a general architecture of a cloud comprising four layers. The GSN Data plane corresponds to the System level, including massive physical resources, such as storage servers and application servers linked by controlled circuits (i.e., lightpaths). The Platform Control plane corresponds to the Core Middleware layer, implementing the platform level services that provide running environment enabling cloud computing and networking capabilities to GSN services. The Cloud Middleware plane corresponds to the User-level Middleware, providing Platform as a Service capabilities based on IaaS Framework components [5]. The top Management plane or User level focuses on application services by making use of services provided by the lower layer services.

4   Virtual Data Center Migration

In the GSN project, we are interested in moving a virtual data center from one node to another. Such a migration is required for large-scale applications running on multiple servers with a high density connection local network. The migration involves four steps: i) Setting up a new environment (i.e., a new data center) for hosting the application with required configurations, ii) Configuring network connection, iii) Moving VMs and their running state information through this high speed connection to the new location, and iv) Turning off computing resources at the original node. Indeed, solutions for the migration of simple applications have been provided by many ICT operators in the market. However, large scale data centers require arbitrarily setting their complex working environments when being moved. This results in a reconfiguration of a large number of servers and network devices in a multi-domain environment.

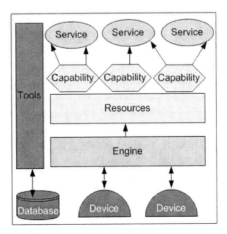

Fig. 4. IaaS Framework Architecture Overview

In our experiments with an online interactive application like Geochronos [7] each VM migration requires 32Mbps bandwidth in order to keep the service live during the migration, thus a 10 Gbit/s link between two data centers can transport more than 300 VMs in parallel. Given that each VM occupies one processor and that each server has up to 16 processors, 20 servers can be moved in parallel. If each VM consumes 4GByte memory space, the time required for such a migration is 1000sec.

The migration of data centers among GSN nodes is based on cloud management. The whole network is considered as a set of clouds of computing resources, which is managed using the IaaS Framework [5]. The IaaS Framework include four main components: i) IaaS Engine used to create model and devices interactions abstractions, ii) IaaS Resource used to build web services interfaces for manageable resources, iii) IaaS Service serves as a broker which controls and assigns tasks to each VM, and iv) IaaS Tool provides various tools and utilities that can be used by the three previous components (Figure 4).

The Engine component is positioned at the lowest level of the architecture and maintains interfaces with physical devices. It uses services provided by protocols and transport layers in order to achieve communications. Each engine has a state machine, which parses commands and decides to perform appropriate actions. The GSN management is achieved by three types of engines: i) Computing engine is responsible for managing VMs, ii) Power engine takes care of power monitoring and control and iii) Network engine controls network devices. The engines allow GSN users to quantify the power consumption of their service. Engines notify upper layers by triggering events. The Resource component serves as an intermediate layer between Engine and Service. It provides Service with different capabilities. Capabilities can contribute to a resources Business, Presentation or Data Access Tier. The Tool component provides additional services, such as persistence, which are shared by other components.

Based on the J2EE/OSGi platform, the IaaS Framework is designed in such a modular manner that each module can be used independently from others. OSGi (Open Services Gateway initiative) is a Java framework for remotely deployed service applications, which provides high reliability, collaboration, large scale distribution and wide-range of device usage. With an IaaS Framework based solution, the GSN can easily be extended to cover different layers and technologies.

Through a Web interface, users may determine GHG emission boundaries based on information providing VM power and their energy sources, and then take actions in order to reduce GHG emissions. The project is therefore ISO 14064 compliant. Indeed, cloud management has been addressed in many other research projects, e.g., [14], however, the IaaS Framework is chosen for the GSN because it is an open platform and converges server and network virtualizations. Whilst most of cloud management solutions in the market focus particularly on computing resources, IaaS Framework components can be used to build network virtualized tools [6][10], which provides for a flexible set of data flows among data centers. The ability of incorporating third-party power control components is also an advantage of the IaaS Framework.

5   Federated Network

GSN takes advantage of the virtualization to link virtual resources together to span multiple cloud and substrate types. The key issue is how to describe, package, and deploy such multi-domain cloud applications. An orchestration middleware is built to federate clouds across domains, coordinate user registration, resource allocation, stitching, launch, monitoring, and adaptation for multi-domain cloud applications. Such a tool also requires solutions for identity, authorization, monitoring, and re-source policy specification and enforcement. An extensible plug-in architecture has been developed in order to enable these solutions to evolve over time and leverage and interoperate with software outside of the GSN.

Along with the participation of international nodes, there is an increasing need of support for dynamic circuits on GSN, which is a multi-layer network, including inter-domain links that span more than one network. Technologies to virtualize networks are continuing to advance beyond the VPN tunneling available to the early cloud network efforts. Some recently built multi-layer networks, like GENI, offer direct control of the network substrate to instantiate isolated virtual pipes, which may appear as VLANs, MPLS tunnels, or VPLS services at the network edge.

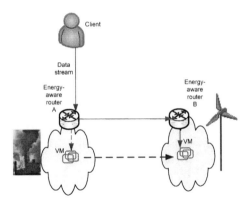

Fig. 5. Energy-aware routing

In the proposed new energy-ware routing scheme based on Mantychore support, the client will contact firstly an energy-aware router in order to get an appropriate VM for his service. The router will look for a VM which is optimal in terms of GHG emission, i.e., the one which is powered by a green energy source. If no VM is found, a migration process would be triggered in order to move a VM to a greener data center. The process is as follows (Figure 5): i) Copy VM memory between old and new locations, ii) VM sends an ARP, iii) Router B receives the ARP and sends the message to the client, iv) New routing entry is installed in router B for the VM, and v) New routing entry is added in router A.

In our design, the GSN is provided with a component called the Federation Stitcher which is responsible for establishing connection among domains, and forwarding user requests to appropriate data centers. The big picture of the GSN network management solution is shown in Figure 6. The heart of the network is the programmable Federation Stitcher, which accepts connections from service users through Internet. This point is powered by green sustainable energy, i.e., hydroelectricity. It links user requests to appropriate services provided by data centers distributed across the network. Each data center is represented by a virtual instance, including virtual servers and virtual routers and/or virtual switches interconnecting the servers. Such a virtual data center can be hosted by any physical network node, according to the power availability. There is a domain controller within each data center or a set of data centers sharing the same network architecture/policy. User requests will be forwarded by the Federation Stitcher to the appropriate domain controller. When a VM or a data center is migrated, the new location will be registered with the Federation Stitcher then user requests are tunneled to the new domain controller.

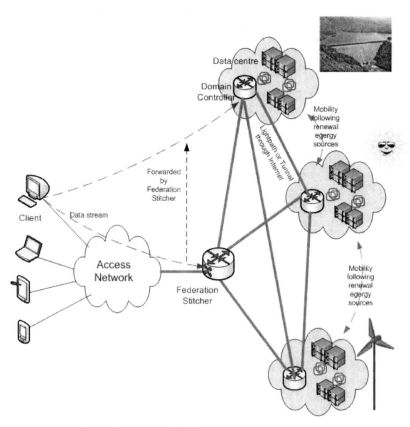

Fig. 6. Overview of GSN network management solution

6   Conclusion

In this chapter, we have presented a prototype of a Future Internet powered only by green energy sources. As a result of the cooperation between Europe and North America researchers, the GreenStar Network is a promising model to deal with GHG reporting and carbon tax issues for large ICT organizations. Based on the Mantychore FP7 project, a number of techniques have been developed in order to provision renewable energy for ICT services worldwide. Virtualization techniques are shown to be the most appropriate solution to manage such a network and to migrate data centers following green energy source availability, such as solar and wind.

Our future work includes research on the quality of services hosted by the GSN and a scalable resource management.

Acknowledgments. The authors thank all partners for their contribution in the GSN and Mantychore FP7 projects.

References

1. The Climate Group: SMART2020: Enabling the low carbon economy in the information age. Report on behalf of the Global eSustainability Initiative, GeSI (2008)
2. Saunders, H.: The Khazzoom-Brookes Postulate and Neoclassical Growth. The Energy J 13(4) (1992)
3. The GreenStar Network Project, http://www.greenstarnetwork.com/
4. Figuerola, S., Lemay, M., Reijs, V., Savoie, M., St. Arnaud, B.: Converged Optical Network Infrastructures in Support of Future Internet and Grid Services Using IaaS to Reduce GHG Emissions. J. of Lightwave Technology 27(12) (2009)
5. Lemay, M.: An Introduction to IaaS Framework. (8/2008), http://www.iaasframework.com/
6. Figuerola, S., Lemay, M.: Infrastructure Services for Optical Networks. J. of Optical Communications and Networking 1(2) (2009)
7. Kiddle, C.: GeoChronos: A Platform for Earth Observation Scientists. OpenGridForum 28, (3/2010)
8. Grasa, E., Hesselbach, X., Figuerola, S., Reijs, V., Wilson, D., Uzé, J.M., Fischer, L., de Miguel, T.: The MANTICORE Project: Providing users with a Logical IP Network Service. TERENA Networking Conference (5/2008)
9. Grasa, E., et al.: MANTICORE II: IP Network as a Service Pilots at HEAnet, NORDUnet and RedIRIS. TERENA Networking Conference (6/2010)
10. Grasa, E., Figuerola, S., Forns, A., Junyent, G., Mambretti, J.: Extending the Argia software with a dynamic optical multicast service to support high performance digital media. Optical Switching and Networking 6(2) (2009)
11. HEAnet website, http://www.heanet.ie/
12. NORDUnet website, http://www.nordu.net
13. Moth, J.: GN3 Study of Environmental Impact Inventory of Greenhouse Gas Emissions and Removals – NORDUnet (9/2010)
14. OpenNEbula Project, http://www.opennebula.org/
15. IBBT Website, http://www.ibbt.be/
16. Reservoir FP7, http://www.reservoir-fp7.eu/

Smart Cities and the Future Internet: Towards Cooperation Frameworks for Open Innovation

Hans Schaffers[1], Nicos Komninos[2], Marc Pallot[3], Brigitte Trousse[3],
Michael Nilsson[4], Alvaro Oliveira[5]

[1] ESoCE Net
hschaffers@esoce.net
[2] Urenio, Aristotle University of Thessaloniki
komninos@urenio.org
[3] INRIA Sophia Antipolis,
marc.pallot@inria.fr, brigitte.trousse@inria.fr
[4] CDT Luleå University of Technology
michael.nilsson@cdt.ltu.se
[5] Alfamicro Lda
alvaro.oliveira@alfamicro.pt

Abstract. Cities nowadays face complex challenges to meet objectives regarding socio-economic development and quality of life. The concept of "smart cities" is a response to these challenges. This paper explores "smart cities" as environments of open and user-driven innovation for experimenting and validating Future Internet-enabled services. Based on an analysis of the current landscape of smart city pilot programmes, Future Internet experimentally-driven research and projects in the domain of Living Labs, common resources regarding research and innovation can be identified that can be shared in open innovation environments. Effectively sharing these common resources for the purpose of establishing urban and regional innovation ecosystems requires sustainable partnerships and cooperation strategies among the main stakeholders.

Keywords: Smart Cities, Future Internet, Collaboration, Innovation Ecosystems, User Co-Creation, Living Labs, Resource Sharing

1   Introduction

The concept of "smart cities" has attracted considerable attention in the context of urban development policies. The Internet and broadband network technologies as enablers of e-services become more and more important for urban development while cities are increasingly assuming a critical role as drivers of innovation in areas such as health, inclusion, environment and business [1]. Therefore the issue arises of how cities, surrounding regions and rural areas can evolve towards sustainable open and user-driven innovation ecosystems to boost Future Internet research and experimentation for user-driven services and how they can accelerate the cycle of research, inno-

J. Domingue et al. (Eds.): Future Internet Assembly, LNCS 6656, pp. 431–446, 2011.

vation and adoption in real-life environments. This paper pays particular attention to collaboration frameworks which integrate elements such as Future Internet testbeds and Living Lab environments that establish and foster such innovation ecosystems.

The point of departure is the definition which states that a city may be called 'smart' "when investments in human and social capital and traditional (transport) and modern (ICT) communication infrastructure fuel sustainable economic growth and a high quality of life, with a wise management of natural resources, through participatory government" [2]. This holistic definition nicely balances different economic and social demands as well as the needs implied in urban development, while also encompassing peripheral and less developed cities. It also emphasises the process of economic recovery for welfare and well-being purposes. Secondly, this characterisation implicitly builds upon the role of the Internet and Web 2.0 as potential enablers of urban welfare creation through social participation, for addressing hot societal challenges, such as energy efficiency, environment and health.

Whereas until now the role of cities and regions in ICT-based innovation mostly focused on deploying broadband infrastructure [3], the stimulation of ICT-based applications enhancing citizens' quality of life is now becoming a key priority. As a next step, the potential role of cities as innovation environments is gaining recognition [4]. The current European Commission programmes FP7-ICT and CIP ICT-PSP stimulate experimentation into the smart cities concept as piloting user-driven open innovation environments. The implicit aim of such initiatives is to mobilise cities and urban areas as well as rural and regional environments as agents for change, and as environments of "democratic innovation" [5]. Increasingly, cities and urban areas are considered not only as the object of innovation but also as innovation ecosystems empowering the collective intelligence and co-creation capabilities of user/citizen communities for designing innovative living and working scenarios.

Partnerships and clear cooperation strategies among main stakeholders are needed in order to share research and innovation resources such as experimental technology platforms, emerging ICT tools, methodologies and know-how, and user communities

Table 1. Three perspectives shaping the landscape of Future Internet and City Development

	Future Internet Research	Cities and Urban Development	User-Driven Innovation Ecosystems
Actors	Researchers ICT companies National and EU actors	City policy actors Citizen platforms Business associations	Living Lab managers, citizens, governments, enterprises, researchers as co-creators
Priorities	Future Internet technical challenges (e.g. routing, scaling, mobility)	Urban development Essential infrastructures Business creation	User-driven open innovation Engagement of citizens
Resources	Experimental facilities Pilot environments Technologies	Urban policy framework Organisational assets Development plans	Living lab facilities: methodologies & tools, physical infrastructures
Policies	Creation of advanced and testbed facilities Federated cooperation Experimental research	City policies to stimulate innovation, business and urban development Innovative procurement	User-driven innovation projects Open, collaborative innovation

for experimentation on Future Internet technologies and e-service applications. Common, shared research and innovation resources as well as cooperation models providing access to such resources will constitute the future backbone of urban innovation environments for exploiting the opportunities provided by Future Internet technologies. Three perspectives are addressed in this paper in order to explore the conditions for rising to this challenge (see Table 1).

The first perspective of Future Internet research and experimentation represents a technology-oriented and longer term contribution to urban innovation ecosystems. Cities and urban areas provide a potentially attractive testing and validating environment. However, a wide gap exists between the technology orientation of Future Internet research and the needs and ambitions of cities. Hence, the second perspective is comprised of city and urban development policies. City policy-makers, citizens and enterprises are primarily interested in concrete and short-term solutions, benefiting business creation, stimulation of SMEs and social participation. While many cities have initiated ICT innovation programmes to stimulate business and societal applications, scaling-up of pilot projects to large-scale, real-life deployment is nowadays crucial. Therefore, a third perspective is the concept of open and user-driven innovation ecosystems, which are close to the interests and needs of cities and their stakeholders, including citizens and businesses, and which may bridge the gap between short-term city development priorities and longer term technological research and experimentation.

A key challenge is the development of cooperation frameworks and synergy linkages between Future internet research, urban development policies and open user-driven innovation. Elements of such frameworks include sharing of and access to diverse sets of knowledge resources and experimentation facilities; using innovative procurement policies to align technology development and societal challenges; and establishing open innovation models to create sustainable cooperation. The concept of open and user-driven innovation looks well positioned to serve as a mediating, exploratory and participative playground combining Future Internet push and urban policy pull in demand-driven cycles of experimentation and innovation. Living Lab-driven innovation ecosystems may evolve to constitute the core of "4P" (Public-Private-People-Partnership) ecosystems providing opportunities to citizens and businesses to co-create, explore, experiment and validate innovative scenarios based on technology platforms such as Future Internet experimental facilities involving SMEs and large companies as well as stakeholders from different disciplines.

This paper is structured as follows. Section 2 addresses challenges for cities to exploit the opportunities of the Future Internet and of Living Lab-innovation ecosystems. How methodologies of Future Internet experimentation and Living Labs could constitute the innovation ecosystems of smart cities is discussed in section 3. Initial examples of such ecosystems and related collaboration models are presented in section 4. Finally, section 5 presents conclusions and an outlook.

2   City and Urban Development Challenges

In the early 1990s the phrase "smart city" was coined to signify how urban development was turning towards technology, innovation and globalisation [6]. The World Foundation for Smart Communities advocated the use of information technology to meet the challenges of cities within a global knowledge economy [7]. However, the more recent interest in smart cities can be attributed to the strong concern for sustainability, and to the rise of new Internet technologies, such as mobile devices (e.g. smart phones), the semantic web, cloud computing, and the Internet of Things (IoT) promoting real world user interfaces.

The concept of smart cities seen from the perspective of technologies and components has some specific properties within the wider cyber, digital, smart, intelligent cities literatures. It focuses on the latest advancements in mobile and pervasive computing, wireless networks, middleware and agent technologies as they become embedded into the physical spaces of cities. The emphasis on smart embedded devices represents a distinctive characteristic of smart cities compared to intelligent cities, which create territorial innovation systems combining knowledge-intensive activities, institutions for cooperation and learning, and web-based applications of collective intelligence [8, 9].

Box: A New Spatiality of Cities - Multiple Concepts

Cyber cities, from cyberspace, cybernetics, governance and control spaces based on information feedback, city governance; but also meaning the negative / dark sides of cyberspace, cybercrime, tracking, identification, military control over cities.

Digital cities, from digital representation of cities, virtual cities, digital metaphor of cities, cities of avatars, second life cities, simulation (sim) city.

Intelligent cities, from the new intelligence of cities, collective intelligence of citizens, distributed intelligence, crowdsourcing, online collaboration, broadband for innovation, social capital of cities, collaborative learning and innovation, people-driven innovation.

Smart cities, from smart phones, mobile devices, sensors, embedded systems, smart environments, smart meters, and instrumentation sustaining the intelligence of cities.

It is anticipated that smart city solutions, with the help of instrumentation and interconnection of mobile devices, sensors and actuators allowing real-world urban data to be collected and analysed, will improve the ability to forecast and manage urban flows and push the collective intelligence of cities forward [10]. Smart and intelligent cities have this modernisation potential because they are not events in the cybersphere, but integrated social, physical, institutional, and digital spaces, in which digital components improve the functioning of socio-economic activities, and the management of physical infrastructures of cities, while also enhancing the problem-solving capacities of urban communities.

The most urgent challenge of smart city environments is to address the problems and development priorities of cities within a global and innovation-led world. A recent public consultation held by the European Commission [11] on the major urban and regional development challenges in the EU has identified three main priorities for the future cohesion policy after 2013. It appears that competitiveness will remain at the heart of cohesion policy, in particular, research, innovation, and upgrading of skills to promote the knowledge economy. Active labour market policy is a top priority to sustain employment, strengthen social cohesion and reduce the risk of poverty. Other hot societal issues are sustainable development, reducing greenhouse gases emissions and improving the energy efficiency of urban infrastructure. Smart city solutions are expected to deal with these challenges, sustain the innovation economy and wealth of cities, maintain employment and fight against poverty through employment generation, the optimisation of energy and water usage and savings, and by offering safer cities. However, to achieve these goals, city authorities have to undertake initiatives and strategies that create the physical-digital environment of smart cities, actualising useful applications and e-services, and assuring the long-term sustainability of smart cities through viable business models.

The first task that cities must address in becoming smart is to create a rich environment of broadband networks that support digital applications. This includes: (1) the development of broadband infrastructure combining cable, optical fibre, and wireless networks, offering high connectivity and bandwidth to citizens and organisations located in the city, (2) the enrichment of the physical space and infrastructures of cities with embedded systems, smart devices, sensors, and actuators, offering real-time data management, alerts, and information processing, and (3) the creation of applications enabling data collection and processing, web-based collaboration, and actualisation of the collective intelligence of citizens. The latest developments in cloud computing and the emerging Internet of Things, open data, semantic web, and future media technologies have much to offer. These technologies can assure economies of scale in infrastructure, standardisation of applications, and turn-key solutions for software as a service, which dramatically decrease the development costs while accelerating the learning curve for operating smart cities.

The second task consists of initiating large-scale participatory innovation processes for the creation of applications that will run and improve every sector of activity, city cluster, and infrastructure. All city economic activities and utilities can be seen as innovation ecosystems in which citizens and organisations participate in the development,

Innovation Economy
- 1- Intelligent city clusters: manufacturing, business services, health, tourism
- 2- Intelligent city districts: CBD, techno park, mall, university campus, port area, airport city
- 3- New companies creation / intelligent incubators

City Infrastructure and Utilities
- 4- Smart transport, mobility and parking
- 5- Broadband, DSL, FTTH, wi-fi, embedded systems
- 6- Energy saving / smart grid
- 7- Environment monitoring, real time alert, safety

Governance
- 8- Government services to citizens
- 9- Decision making / participation / direct democracy
- 10- Monitoring & measurement: The city as database

Fig. 1. Smart city key application areas

supply and consumption of goods and services. Fig. 1 presents three key domains of potential smart city applications in the fields of innovation economy, infrastructure and utilities, and governance.

Future media research and technologies offer a series of solutions that might work in parallel with the Internet of Things and embedded systems, providing new opportunities for content management [12, 13]. Media Internet technologies are at the crossroads of digital multimedia content and Internet technologies, which encompasses media being delivered through Internet networking technologies, and media being generated, consumed, shared and experienced on the web. Technologies, such as content and context fusion, immersive multi-sensory environments, location-based content dependent on user location and context, augmented reality applications, open and federated platforms for content storage and distribution, provide the ground for new e-services within the innovation ecosystems of cities (see Table 2).

Table 2. Media Internet technologies and components for Smart Cities

Solutions and RTD challenges	Short term (2014)	Mid term (2018)	Longer term (2022)
Content management tools	Media Internet technologies	Scalable multimedia compression and transmission	Immersive multimedia
Collaboration tools	Crowd-based location content; augmented reality tools	Content and context fusion technologies	Intelligent content objects; large scale ontologies and semantic content
Cloud services and software components	City-based clouds	Open and federated content platforms	Cloud-based fully connected city
Smart systems based on Internet of Things	Smart power management Portable systems	Smart systems enabling integrated solutions e.g. health and care	Software agents and advanced sensor fusion; telepresence

Demand for e-services in the domains outlined in Fig. 1 is increasing, but not at a disruptive pace. There is a critical gap between software applications and the provision of e-services in terms of sustainability and financial viability. Not all applications are turned into e-services. Those that succeed in bridging the gap rely on successful business models that turn technological capabilities into innovations, secure a continuous flow of data and information, and offer useful services. It is here that the third task for city authorities comes into play, that of creating business models that sustain the long-term operation of smart cities. To date, the environment for applications and their business models has been very complex, with limited solutions available 'off the shelf', a lot of experimentation, and many failures. Cities currently face a problem of standardisation of the main building blocks of smart / intelligent cities in terms of applications, business models, and services. Standardisation would dramatically reduce the development and maintenance costs of e-services due to cooperation, exchange and sharing of resources among localities. Open source communities may also substantially contribute to the exchange of good practices and open solutions.

The current research on smart cities is partly guided by the above priorities of contemporary urban development and city governance. Large companies in the ICT sector, such as IBM, Cisco, Microsoft, are strongly involved in and are contributing to shaping the research agenda. EU research within the context of the FP7 and CIP programmes also aims at stimulating a wider uptake of innovative ICT-based services for smart cities, linking smart cities with user-driven innovation, future Internet technologies, and experimental facilities for exploring new applications and innovative services.

Technology push is still dominant in the actual research agenda. A recent Forrester survey states that smart city solutions are currently more vendor push than city government pull based. However, the survey points out that, "smart city solutions must start with the city not the smart" [14]. The positive impact of available smart city solutions on European cities has not yet been demonstrated, nor have the necessary funding mechanisms and business models for their sustainability been developed. Creating the market constitutes the first priority. Innovation ecosystems for smart cities have to be defined, in terms of applications, services, financial engineering and partnerships. This will help cities to secure funding, identify revenue streams, broker public-private partnerships, and open public data up to developers as well as user communities. As the major challenge facing European cities is to secure high living standards through the innovation economy, smart cities must instrument new ways to enhance local innovation ecosystems and the knowledge economy overall.

3   Future Internet Experimentation and Living Labs Interfaces

In exploring the role of Future Internet experimentation facilities in benefiting urban development as we move towards smart cities, we will succinctly summarise the role of experimental facilities and the experimentation process, as well as the potential role of the 'Living Labs' concept in enriching experimentally-driven research on the Future Internet. Within the context of the now emerging FIRE portfolio [15], the potential exists to support new classes of users and experiments combining heterogeneous technologies that represent key aspects of the Future Internet. The considerable obstacles of complexity and unfamiliarity that are faced when trying to explore the effects of new applications that bring future users the increasing power of the Future Internet have not yet been overcome. Issues that are being dealt with in the attempt of FIRE projects to move closer to the goal of a federated testbed facility, and which are also important in collaborating with smart city and Living Labs activities, are authentication and access to facilities; security and privacy as well as IPR protection; operation and research monitoring as well as experiment control; and the issue of defining and monitoring experiments in large-scale usage settings.

The portfolio of FIRE experimentation projects shows that users in such FIRE projects are mostly academic and industry researchers. End-user involvement and end user experimentation is beyond the current scope of FIRE, although some interesting initiatives in that respect have started such as the Smart Santander project (services and applications for Internet of Things in the city), the TEFIS project (platform for managing experimental facilities, among which Living Labs) and the ELLIOT project (co-creation of wellbeing, logistics and environment IoT-based services).

A comparison of the role of users in FIRE facilities projects compared to Living Labs is presented in Table 3. Importantly, FIRE projects typically involve users in assessing the impacts of technologies in socio-economic terms, whereas Living Labs projects aim to engage users in the innovation process itself. Also, the predominant approach of FIRE facilities is controlled experimentation, whereas Living Labs engage users in the actual innovation process (co-creation). The European Commission has voiced its support for stronger user orientation in the Future Internet facilities projects; not only users in terms of academic and industry researchers who will use these facilities for their research projects, but also end-users. Emphasis is on involving communities of end-users at an early stage of development to assess the impacts of technological changes, and possibly engage them in co-creative activities.

Table 3. User Role in FIRE and Living Labs

	Future Internet Experiments	**Living Labs Innovation**
Approach	Controlled experiments Observing large-scale deployment and usage patterns Federated testbeds	Both controlled and natural situation experiments User co-creation via Living Labs methodologies, action research Open, cooperative innovation
Object of testing	Technologies, services, architectures, platforms, system requirements; impacts	Validation of user ideas, prototype applications and solutions. Testing as joint validation activity
Scale of testing	Large-scale mainly	From small to large scale
Stakeholders	FI Researchers (ICT industry & academia)	IT multidisciplinary researchers, End-users, enterprises (large & SMEs)
Objective	Facilities to support research Impact assessment of tested solutions	Support the process of user-driven innovation as co-creation

In order to explore the opportunities and interfaces, we will now take a further look at Living Labs. The Web 2.0 era has pushed cities to consider the Internet, including mobile networks, as a participative tool for engaging citizens and tourists. Many initiatives have been launched by cities, such as Wikicity in Rome stemming from MIT's Senseable City Lab which studies the impact of new technologies on cities, Real-Time City Copenhagen, and Visible City Amsterdam. This collection of initiatives already looks like a "networked Living Lab" of cities for investigating and anticipating how digital technologies affect people as well as how citizens are "shaping" those technologies to change the way people are living and working.

Apart from the diversity of research streams and related topics for designing alternatives of the Internet of tomorrow, it becomes increasingly challenging to design open infrastructures that efficiently support emerging events and citizens' changing needs. Such infrastructure also creates many opportunities for innovative services such as green services, mobility services, wellbeing services, and playable city services based on real-time digital data representing digital traces of human activity and their context in the urban space. Environmental sensors measure parameters such as air quality, temperature or noise levels; telecommunication networks reflect connectivity and the location of their users; transportation networks digitally manage the mobility of people and vehicles as well as products in the city, just to give a few examples. Today, it is becoming increasingly relevant to explore ways in which such data streams can become tools for people taking decisions within the city. Promising applications and services seem to be emerging from user co-creation processes.

Recent paradigms, such as open innovation and open business models [16], Web 2.0 [17] as well as Living Labs [18], a concept originating from the work of William Mitchell at MIT and currently considered as user-driven open innovation ecosystems, promote a more proactive and co-creative role of users in the research and innovation process. Within the territorial context of cities, rural areas and regions, the main goal of Living Labs is to involve communities of users at an early stage of the innovation process. The confrontation of technology push and application pull in a Living Lab enables the emergence of breakthrough ideas, concepts and scenarios leading to adoptable innovative solutions. Some of the methodologies used in Living Labs innovation projects demonstrate a potential interface with FIRE experimentation approaches. In [19], a useful classification is elaborated of different platforms for testing and experimentation including testbeds, prototyping projects, field trials, societal pilots and Living Labs. In [20] a landscape of user engagement approaches is presented. Methodologies for Living Labs organisation, phased development and process management integrated with user experiments within an action research setting have been developed and implemented in [21].

Altogether, Future Internet experimental facilities, Living Labs and Urban development programmes form an innovation ecosystem consisting of users and citizens, ICT companies, research scientists and policy-makers. In contrast with a testbed, a Living Lab constitutes a "4P" (Public, Private and People Partnership) ecosystem that provides opportunities to users/citizens to co-create innovative scenarios based on technology platforms such as Future Internet technology environments involving large enterprises and SMEs as well as academia from different disciplines. It appears that Future Internet testbeds could be enabling the co-creation of innovative scenarios by users/citizens contributing with their own content or building new applications that would mash-up with the city's open, public data.

4   Emerging Smart City Innovation Ecosystems

As Table 4 illustrates, several FP7-ICT projects are devoted to research and experimentation on the Future Internet and the Internet of Things within cities, such as Smart Santander and, within the IoT cluster, ELLIOT. The CIP ICT-PSP programme has initiated several pilot projects dedicated to smart cities and Living Labs, some with a clear Future Internet dimension (Apollon, Periphèria, and to a less extent too, Open Cities and EPIC). Among the earlier projects with interesting aspects on the interface of Living Labs and Future Internet is C@R (FP6).

The Smart Santander project proposes an experimental research facility based on sensor networks which will eventually include more than 20,000 sensors, considered as IoT devices. The architecture supports a secure and open platform of heterogeneous technologies. The project is intended to use user-driven innovation methods for designing and implementing 'use cases'. Bus tracking and air quality (EKOBUS: a map of sensor data available on smart phone) as well as urban waste management are two of the use cases from the Smart Santander project.

Table 4. Examples of Living Lab Initiatives Related to Smart Cities, Rural Areas and Regions

Cities and urban areas	• Smart Santander (FP7-ICT, 2010). Internet services and sensor network in the city. www.smartsantander.eu
	• ELLIOT (FP7-ICT, 2010). Experimental Living Lab for Internet of Things. Three Living Labs are involved. http://www.elliot-project.eu/
	• Periphèria (CIP ICT-PSP, 2010). Internet of Things in Smart City. www.peripheria.eu
	• Open Cities (CIP ICT-PSP, 2010). Public sector services.
	• EPIC (CIP ICT-PSP, 2010). Platforms for intelligent cities.
	• Apollon (CIP ICT-PSP, 2010). Domain-specific Pilots of Living Labs in cross-border networks, targeting city areas. www.apollon-pilot.eu
Villages in rural areas and regions	• Collaboration@Rural – C@R (FP6-ICT, 2006-2010). Six Living Labs in Rural areas using a common service platform. www.c-rural.eu
	• Networking for Communications Challenges Communities (N4C). Extending Internet access to remote regions. www.n4c.eu
	• MedLab (Interreg IVc). Living Labs and Regional Development.

The ELLIOT project (Experiential Living Lab for the Internet of Things) represents a clear example of Living Labs and Future Internet interaction, elaborating three IoT use cases in three different Living Labs. The first use case is dedicated to co-creation by users of green services in the areas of air quality and ambient noise pollution with innovative devices such as the "green watch" (http://www.lamontreverte.org/en/) and customised sensors being used by citizens. The second one addresses wellbeing services in connection with a hospital and the third focuses on logistic services in product development facilities with professional users. Its goal is to investigate evidence of the social dynamics of the Living Lab approach for the purpose of ensuring a wide and rapid spread of innovative solutions through socio-emotional intelligence mechanisms.

The green services use case takes place in the context of the ICT Usage Lab and within the Urban Community of Nice - Cote d'Azur (NCA). This use case involves local stakeholders, such as the regional institution for air measurement quality (Atmo PACA), the local research institute providing the IoT-based green service portal and managing the experiments (INRIA/AxIS), the Internet Foundation for the New Generation (FING) facilitating user workshops, and a local SME providing data access from electric cars equipped with air quality sensors (VULog) and a citizen IT platform (a regional Internet space for citizens in the NCA area). The objectives of the IoT-based green services use case are twofold: to investigate experiential learning of the IoT in an open and environmental data context, and to facilitate the co-creation of green services based on environmental data obtained via sensors. Various environmental sensors will be used, such as fixed sensors from Atmo PACA in the NCA area, fixed Arduino-assembled sensors by citizens, mobile sensors, such as citizen-wired green watches or sensors installed on electric vehicles. The backbone of the green services use case is an IoT-based service portal which addresses three main IoT-related portal services by allowing the user: 1) to participate in the collection of environmental data; 2) to participate in the co-creation of services based on environmental data; and 3) to access services based on environmental data, such as accessing and/or visualising environmental data in real time. Three complementary approaches have already been identified as relevant for the green services use case: participatory/user-centred design methods; diary studies for IoT experience analysis, and coupling quantitative and qualitative approaches for portal usage analysis. In this context of an open innovation and Living Lab innovation eco-system, focus groups involving stakeholders and/or citizen may be run either online or face-to-face.

The Perifèria project is among the Smart Cities portfolio of seven projects recently launched in the European Commission ICT Policy Support Programme. Their aim is to develop smart cities infrastructures and services in real-life urban environments in Europe. Actually, the Perifèria project forms a bridge between the Smart Cities portfolio of projects and the Internet of Things European Research Cluster (IERC) and can therefore be taken as a model of Smart Cities and Future Internet integration. At the core of Perifèria lies the role of Living Labs in constituting a bridge between Future Internet technology push and Smart City application pull, re-focusing the attention on "People in Places" to situate the human-centric approach within physical urban settings. People in Places becomes the context and the situation – including the relational situations between people and between people and spaces, infrastructures, services, etc. – in which the integration of Future Internet infrastructures and services occurs as part of a "discovery-driven" process. The Cloud is considered to be a resource environment that is dynamically configured (run-time) to bring together testbeds, applets, services, and whatever is relevant, available and configured for integration at the moment that the social interaction of People in Places calls for those services.

Participation is at the heart of this bottom-up approach to Future Internet technology integration, whereby Future Internet research adopts a "competitive offer" stance to prove its added value to users. Platform and service convergence is promoted by the use of serious games that engage citizens and users in the process of discovering the potential of Future Internet technologies and the possible sustainable scenarios that can be built upon them. Serious gaming thus constitutes a mechanism to enhance participation and transform individual and collective behaviour by working directly on the social norms that shape them; in addition, they constitute a monitoring and governance platform for increasing self-awareness of the changes brought about by the adoption of Future Internet technologies. Perifèria has identified five archetypal urban settings: (1) the Smart Neighbourhood where media-based social interaction occurs; (2) the Smart Street where new mobility behaviours develop; (3) the Smart Square where participatory civic decisions are taken; (4) the Smart Museum and Park where natural and cultural heritage feed learning; and (5) the Smart City Hall where mobile e-government services are delivered.

As an example (see Fig. 2), the City of Genova is experimenting with the Smart Museum and Park arena, with to the aim of blending the fruition of the city's natural and cultural heritage with safety and security in urban spaces. This approach draws on and integrates Future Internet technologies (such as augmented reality services for the appreciation of cultural heritage) with networks of video-cameras used to monitor public spaces. In addition, the integration of these services occurs in the Living Lab context where citizens contribute both to the definition and prioritisation of the cultural heritage in their city and also to an exploration of the privacy and security issues that are central to the acceptance and success of Future Internet services for the safety of urban environments.

Fig. 2. Genoa smart city experiments on Smart Museum and Smart Park

This example illustrates the central role of users and citizens in defining the services that make up a Smart City as well as the new sustainable lifestyles and workstyles made possible by Future Internet technologies. In addition, it shows how the Future Internet is a mixture of technologies and paradigms with overlapping implementation time-frames. While the deployment of IPv6 networks may be a medium-term effort, other Future Internet paradigms such as cloud services and camera and sensor networks can be considered as already operational. The discovery-driven arena settings in Perifèria are guiding the development of Living Lab-convergent service platforms that bring these technologies together into integrated, dynamic co-creation environments that make up a Smart City.

These projects examples provide initial examples of collaboration models in smart city innovation ecosystems, governing the sharing and common use of resources such as testing facilities, user groups and experimentation methodologies. Two different layers of collaboration can be distinguished. The first layer is collaboration *within* the innovation process, which is understood as ongoing interaction between research, technology and applications development and validation and utilisation in practice. Cases mentioned above such as ELLIOT, SmartSantander and Perifèria constitute typical arenas where potential orchestrations of these interactions are explored. Still, many issues need to be clarified such as how the different research and innovation resources in a network, such as specific testing facilities, tools, data and user groups, can be made accessible and adaptable to specific demands of any research and innovation projects.

The second layer concerns collaboration at the territorial level, driven by urban and regional development policies aiming at strengthening the urban innovation systems through creating effective conditions for sustainable innovation. This layer builds on Michael Porter's concept of "national competitive advantage" [22] which borrows the 'national systems of innovation' thinking, which was originally developed by Chris Freeman. Following this thinking, the "urban value creation system" can be considered as being shaped by four determinants: 1) physical and immaterial infrastructure, 2) networks and collaboration, 3) entrepreneurial climate and business networks, 4) demand for services and availability of advanced end-users (see Fig. 3). Additionally, the value creation system in its conceptualisation by Michael Porter is affected by policy interventions aimed at stimulating the building of networks, the creation of public-private partnerships, and the enhancement of innovative conditions.

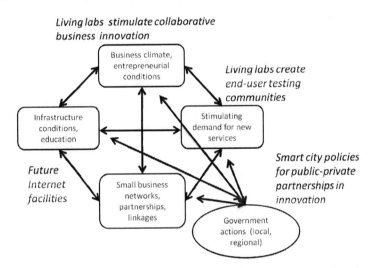

Fig. 3. Conceptualisation of smart city value creation and innovation system (based on Porter)

The challenge in this layer is to create a collaborative approach to innovation ecosystems based on sustainable partnerships among the main stakeholders from business, research, policy and citizen groups and achieve an alignment of local, regional and European policy levels and resources. The ELLIOT project is an example of a Future Internet research and innovation project embedded in regional and even national innovation policy. From the perspective of smart cities, managing innovation at the level of urban innovation ecosystems becomes a task of managing the portfolio of resources and fostering fruitful interlinkages. Smart city innovation ecosystem management aims to manage the portfolio of "innovation assets" made up of the different facilities and resources, by creating partnerships among actors that govern these assets, by fostering knowledge and information flows, and by providing open access to resources made available to users and developers.

5   Conclusions and Outlook

In this paper we explored the concept of "smart cities" as environments of open and user driven innovation for experimenting and validating Future Internet-enabled services. Smart cities are enabled by advanced ICT infrastructure contributed to by current Future Internet research and experimentation. Such infrastructure is one of the key determinants of the welfare of cities. Other determinants of the welfare of cities will be important as well: the infrastructure for education and innovation, the networks between businesses and governments, the existence of demanding citizens and businesses to push for innovation and the quality of services. Here we see a clear analogy to Porter's concept of national competitive advantage: the welfare potential of cities and urban areas.

The Living Labs concept represents a powerful view of how user-driven open innovation ecosystems could be organised. As a concept applied to smart cities it embodies open business models of collaboration between citizens, enterprises and local governments, and the willingness of all parties -including citizens and SMEs- to engage actively in innovation. The Living Lab concept should be considered also as a methodology, a model for organising specific innovation programmes and innovation projects and conducting innovation experiments. Whereas the last aspect has gained most attention, both levels and their interaction are important: *shaping* and *operating* the innovation ecosystem.

Based on an analysis of challenges of smart cities on the one hand and current projects in the domain of Future Internet research and Living Labs on the other, common resources for research and innovation can be identified, such as testbeds, Living Lab facilities, user communities, technologies and know-how, data, and innovation methods. Such common resources potentially can be shared in open innovation environments. Two layers of collaboration were distinguished that govern the sharing of these resources. One layer focuses on the actual resources within the Future Internet research and innovation process, the second layer addresses the urban innovation system. Several projects discussed in this paper provide evidence of collaboration models for sharing resources at both layers, e.g. the use of Living Lab facilities and methods in experimenting on Future Internet technologies, and the use of Living Lab methodologies for implementing innovation policies of cities.

The potential types and structures of these collaboration frameworks and the concrete issues to be resolved in sharing of research and innovation resources, such as governance, ownership, access, transferability and interoperability need further examination and also need development and piloting in future pilot projects. The current experimentation and innovation approaches used in some of the FIRE and Living Lab projects should be studied more closely in order to develop concrete examples of resource sharing opportunities. Initial examples of resource sharing appear in making user communities available for joint use with Future Internet facilities (e.g. the TEFIS project), and in making accessible Future Internet facilities for developing and validating IoT-based service concepts and applications through Living Labs approaches for smart cities (e.g. the SmartSantander and ELLIOT projects).

The Future Internet constitutes both a key technology domain and a complex societal phenomenon. Effective, user driven processes of innovation, shaping and application of

Future Internet technologies in business and society are crucial for achieving socio-economic benefits. A key requirement emphasised in this paper is how, within an environment of open innovation in smart cities and governed by cooperation frameworks, the diverse set of resources or assets that constitutes the "engine" of ongoing research and innovation cycles can be made open accessible for users and developers.

References

1. Kroes, N.: European Commissioner for Digital agenda, The critical role of cities in making the Digital Agenda a reality. Closing speech to Global Cities Dialogue Spring Summit of Mayors Brussels, 28 May (2010)
2. Caragliu, A., Del Bo, C., Nijkamp, P.: Smart cities in Europe. Series Research Memoranda 0048. VU University Amsterdam, Faculty of Economics, Business Administration and Econometrics (2009)
3. Eurocities: Strategic Policy Paper on Broadband in Cities (2010)
4. Eurocities: Cities and Innovation in Europe. Discussion paper (2010)
5. Von Hippel, E.: Democratizing Innovation. The MIT Press, Cambridge (2005)
6. Gibson, D.V., Kozmetsky, G., Smilor, R.W. (eds.): The Technopolis Phenomenon: Smart Cities, Fast Systems, Global Networks. Rowman & Littlefield, New York (1992)
7. WFSC: Smart Communities, http://www.smartcommunities.org/about.htm
8. Komninos, N.: Intelligent Cities: Innovation, knowledge systems and digital spaces. Taylor & Francis, London and New York (2002)
9. Komninos, N.: Intelligent Cities and Globalisation of Innovation Networks. Routledge, London and New York (2008)
10. Chen-Ritzo, C.H., Harrison, C., Paraszczak, J., Parr, F.: Instrumenting the Planet. IBM Journal of Research & Development 53(3), 338–353 (2009)
11. European Commission: Growing Regions, Growing Europe: Fifth progress report on economic and social cohesion. European Commission COM(2008) 371 final (2008)
12. European Commission: Future Media Internet: Research challenges and road ahead. DG Information Society and Media, Luxembourg, Publications Office of the European Union (2010)
13. European Commission: Future Media Networks: Research challenges 2010. DG Information Society and Media, Luxembourg, Publications Office of the European Union (2010)
14. Belissent, J.: Getting clever about smart cities: New opportunities require new business models. Forrester for Ventor Strategy Professionals (2010)
15. European Commission, DG INFSO: Future Internet Research and Experimentation (September 2010)
16. Chesbrough, H.W.: Open Innovation: The New Imperative for Creating and Profiting from Technology. Harvard Business School Press, Boston (2003)
17. O'Reilly, T., Battelle, J.: Web Squared: Web 2.0 Five Years On. Special Report, Web 2.0 Summit, Co-produced by O'Reilly & Techweb (2009)
18. European Commission, DG INFSO: Advancing and Applying Living Lab Methodologies (2010)
19. Ballon, P., Pierson, J., Delaere, S., et al.: Test and Experimentation Platforms for Broadband Innovation. IBBT/VUB-SMIT Report (2005)

20. Pallot, M., Trousse, B., Senach, B., Scapin, D.: Living Lab Research Landscape: From User Centred Design and User Experience Towards User Co-creation. Position Paper, First Living Labs Summer School (http://www-sop.inria.fr/llss2010/), Paris (August. 2010)
21. Schaffers, H., Garcia Guzmán, J., Navarro, M., Merz, C. (eds.): Living Labs for Rural Development. TRAGSA, Madrid (2010), http://www.c-rural.eu
22. Porter, M.: The Competitive Advantage of Nations. Free Press, New York (1990)

Smart Cities at the Forefront of the Future Internet

José M. Hernández-Muñoz[1], Jesús Bernat Vercher[1], Luis Muñoz[2], José A. Galache[2], Mirko Presser[3], Luis A. Hernández Gómez[4], and Jan Pettersson[5]

[1] Telefonica I+D, Madrid, Spain
{jmhm, bernat}@tid.es
[2] University of Cantabria, Santander, Spain
{luis, jgalache}@tlmat.unican.es
[3] Alexandra Institute, Aahrus, Denmark
mirko.presser@alexandra.dk
[4] Universidad Politécnica Madrid, Spain
luisalfonso.hernandez@upm.es
[5] Centre for Distance-Spanning Technology, Skellefteå, Sweden
jan.pettersson@ltu.se

Abstract. Smart cities have been recently pointed out by M2M experts as an emerging market with enormous potential, which is expected to drive the digital economy forward in the coming years. However, most of the current city and urban developments are based on vertical ICT solutions leading to an unsustainable sea of systems and market islands. In this work we discuss how the recent vision of the Future Internet (FI), and its particular components, Internet of Things (IoT) and Internet of Services (IoS), can become building blocks to progress towards a unified urban-scale ICT platform transforming a Smart City into an open innovation platform. Moreover, we present some results of generic implementations based on the ITU-T's Ubiquitous Sensor Network (USN) model. The referenced platform model fulfills basic principles of open, federated and trusted platforms (FOTs) at two different levels: the infrastructure level (IoT to support the complexity of heterogeneous sensors deployed in urban spaces), and at the service level (IoS as a suit of open and standardized enablers to facilitate the composition of interoperable smart city services). We also discuss the need of infrastructures at the European level for a realistic large-scale experimentally-driven research, and present main principles of the unique-in-the-world experimental test facility under development within the SmartSantander EU project.

Keywords: Smart Cities, Sensor and Actuator Networks, Internet of Things, Internet of Services, Ubiquitous Sensor Networks, Open, Federated and Trusted innovation platforms, Future Internet.

1 Introduction

At a holistic level, cities are 'systems of systems', and this could stand as the simplest definition for the term. However, one of the most well-known definitions was provided by the EU project 'European Smart Cities' [1]. Under this work, six dimensions

J. Domingue et al. (Eds.): Future Internet Assembly, LNCS 6656, pp. 447–462, 2011.

of 'smartness' were identified (economy, people, governance, mobility, environment, and living).

As the upsurge of information and communication technologies (ICT) has become the nervous system of all modern economies, making cities smarter is usually achieved through the use of ICT intensive solutions. In fact, ICT is already at the heart of many current models for urban development: revamping their critical infrastructure and enabling new ways of city transport management, traffic control or environmental pollution monitoring. The extensive use of ICT is also empowering the development of essential services for health, security, police and fire departments, governance and delivery of public services.

Nevertheless, the main concern with respect to most of these solutions is that its own commercial approach is leading to an unmanageable and unsustainable sea of systems and market islands. From the point of view of the European Commission, there is a need to reach to a high level agreement at an industrial level to overcome this increasing market fragmentation, which prevents solutions of becoming more efficient, scalable and suitable for supporting new generations of services that are not even envisaged nowadays.

Consequently, the successful development of the Smart Cities paradigm will "require a unified ICT infrastructure to allow a sustainable economic growth" [2], and this unified ICT platform must be suitable to "model, measure, optimize, control, and monitor complex interdependent systems of dense urban life" [3]. Therefore in the design of urban-scale ICT platforms, three main core functionalities can be identified:

- *Urban Communications Abstraction.* One of the most urgent demands for sustainable urban ICT developments is to solve the inefficient use (i.e. duplications) of existing or new communication infrastructures. Due to the broad set of heterogeneous urban scenarios, there will be also a pronounced heterogeneity of the underlying communication layers. So far, through communications abstraction, urban-scale ICT platforms will allow unified communications regardless the different network standards and will enable data transfer services agnostic to the underlying connection protocol. Furthermore, a major challenge in future urban spaces will be how to manage the increasing number of heterogeneous and geographically dispersed machines, sensors and actuators intensively deployed everywhere in the city.
- *Unified Urban Information Models.* Also related to the huge amount of heterogeneous information generated at urban scale, a unified ICT platform should be built on top of a unified model so that data and information could be shared among different applications and services at global urban levels. This will relay on the articulation of different enriched semantic descriptions, enabling the development of information processing services involving different urban resources and entities of interest. Specific information management policies should also be addressed to ensure the required level of security and privacy of information.
- *Open Urban Services Development.* Together with unified communications and information, a key functionality of urban ICT Platforms should be to guarantee interoperability at both the application and service levels. Only through open, easy-to-use, and flexible interfaces the different agents involved (public administrations, enterprises, and citizens) will be able to conceive new innovative solutions to interact

with and manage all aspects of urban life in a cost-effective way. This will provide the necessary innovation-enabling capabilities for attracting public and private investments to create products and services which have not yet been envisioned, a crucial aspect for SmartCities to become future engines of a productive and profitable economy.

Once major challenges of unified urban-scale ICT platforms are identified, it is clear that the future development of Smart Cities will be only achievable in conjunction with a technological leap in the underlying ICT infrastructure. In this work we advocate that this technological leap can be done by considering Smart Cities at the forefront of the recent vision of the Future Internet (FI). Although there is no universally accepted definition of the Future Internet, it can be approached as "a socio-technical system comprising Internet-accessible information and services, coupled to the physical environment and human behavior, and supporting smart applications of societal importance" [4]. Thus the FI can transform a Smart City into an open innovation platform supporting vertical domain of business applications built upon horizontal enabling technologies. The most relevant basic FI pillars [11] for a Smart City environment are the following:

- The Internet of Things (IoT): defined as a global network infrastructure based on standard and interoperable communication protocols where physical and virtual "things" are seamlessly integrated into the information network [5].
- The Internet of Services (IoS): flexible, open and standardized enablers that facilitate the harmonization of various applications into interoperable services as well as the use of semantics for the understanding, combination and processing of data and information from different service provides, sources and formats.
- The Internet of People (IoP): envisaged as people becoming part of ubiquitous intelligent networks having the potential to seamlessly connect, interact and exchange information about themselves and their social context and environment.

At this point, it is important to highlight a bidirectional relationship between the FI and Smart Cities: as if, in the one direction, FI can offer solutions to many challenges that Smart Cities face; on the other direction, Smart Cities can provide an excellent experimental environment for the development, experimentation and testing of common FI service enablers required to achieve 'smartness' in a variety of application domains [6]. To fully develop the Smart City paradigm at a wide geographical scope, a better understanding and insight on issues like: required capacity, scalability, interoperability, and stimulation of faster development of new and innovative applications is required. This knowledge must be taken into account to influence the specification of the FI architecture design. The availability of such infrastructures is expected to stimulate the development of new services and applications by various types of users, and to help gathering a more realistic assessment of users' perspective by means of acceptability tests. To this later extent, close to the IoP vision, the Living Labs network [7] based on the user-driven approach is of main relevance, although in this paper we also advocate the need of large, open and federated experimental facilities able to support experimental research in order to gain real feedback at a city scale [8].

The rest of the paper is organized as follows: Section 2 discusses how major components of the Future Internet, namely IoT and IoS, can be essential building blocks in future Smart Cities open innovation platforms. Several technical details related to the development of next generation urban IoT platforms are outlined in Section 3. Section 4 discusses the need for realistic urban-scale open and federated experimental facilities, and presents most relevant current initiatives, with special attention to the SmartSantander EU Project. Finally, conclusions and future challenges are given in Section 5.

2   IoT and IoS as ICT Building Blocks for Smart Cities

In the analysis from Forrester Research [9] on the role that ICT will play in creating the foundation for Smart Cities, a smart city is described as one that "uses information and communications technologies to make the critical infrastructure components and services of a city — administration, education, healthcare, public safety, real estate, transportation and utilities — more aware, interactive and efficient. "According to this approach, frequent in research reports, several key ICT technologies can be identified for their benefits on different city "systems":

- Transportation: sensors can be used to manage the mobility needs with an appropriate Intelligent Transport System (ITS) that takes care of congestion, predicts the arrival of trains, buses or other public transportation options; managing parking space availability, expired meters, reserved lanes, etc.
- ICT can be also used for environmental and energy monitoring: sensors that detect when trash pick-ups are needed, or notify authorities about landfill toxicity; energy consumption and emissions monitoring across sectors to improve accountability in the use of energy and carbon, etc.
- Building management: smart meters and monitoring devices can help monitor and manage water consumption, heating, air-conditioning, lighting and physical security. This can allow the development of smart utilities grids with bidirectional flow in a distributed generation scheme requiring real-time exchange of information.
- Healthcare: telemedicine, electronic records, and health information exchanges in remote assistance and medical surveillance for disabled or elderly people.
- Public Safety and Security: sensor-activated video surveillance systems; location-aware enhanced security systems; estimation and risk prevention systems (e.g. sensitivity to pollution, extreme summer heating).
- Remote working and e-commerce services for businesses, entertainment and communications for individuals. Advanced location based services, social networking and collaborative crowd-sourcing collecting citizens' generated data.

By analyzing these different Smart Cities application scenarios, together with the need of a broadband communication infrastructure that is becoming, or starting to be considered, the 4th utility (after electricity, gas and water), two major ICT building blocks of a Smart City can be identified among the main pillars that the FI provides:

- Recent advances in Sensors and Actuator Networks (SAN) are stimulating massive sensor networks deployments, particularly for the previously described urban application areas. Therefore IoT, essential to the FI, can be invaluable to provide the necessary technological support to manage in a homogeneous and sustainable way the huge amount of sensor and devices connected to the Smart City infrastructure.
- In a complementary vein, only an open and easy-to-use service enablement suite, that allows the efficient orchestration and reuse of applications, can foster new solutions and services to meet the needs of cities and their inhabitants. In this context, IoS evolution must be undoubtedly correlated with IoT advances. Otherwise, a number of future Smart City services will never have an opportunity to be conceived due to the lack of the required links to the real world.

So far, it may be relevant to consider both the benefits and challenges of implementing IoT and IoS at the city scale.

Starting with the benefits of IoT technologies, they are two-fold: on the one hand they can increase the efficiency, accuracy and effectiveness in operation and management of the city's complex ecosystem and, on the other, they can provide the necessary support for new innovative applications and services (the city as an Open Innovation Platform).

In that sense, the FI PPP promoted by the EC [10][11] seeks for the cooperation among the main European stakeholders in order to develop cross-domain Next Generation (NG) IoT platforms suitable to different usage areas and open business models to improve market dynamics by involving third parties in the value chain (SMEs).

Some of the essential functionalities identified as required for NG IoT platforms comprise the support for horizontality, verticality, heterogeneity, mobility, scalability, as well as security, privacy, and trust [12][13]. Cross-domain NG IoT platforms may foster the creation of new services taking advantage of the increasing levels of efficiency attained by the re-use of deployed infrastructures.

Considering now the IoS, it must be stressed that it is widely recognized (see for example [12]) that the real impact of future IoT developments is heavily tied to the parallel evolution of the IoS. So, a Smart City could only become a true open innovation platform through the proper harmonization of IoS and IoT. There can be a long list of potential benefits for Smart Cities' services relaying on the same basic sensed information and a suite of application enablers (i.e. from sensor data processing applications, to enablers for accessing multimedia mobile communications or social networks, etc.). Thus the integration of innovative principles and philosophy of IoS will engage collective end-user intelligence from Web 2.0 and Telco 2.0 models that will drive the next wave of value creation at urban scales, a key aspect typically missing in other technologically-driven initiatives.

The technological challenge of developing the IoS has been assumed at EU level, and actions are being initiated to overcome the undesirable dissociation between technological and service infrastructures [13]. Of particular relevance for Smart Cities scenarios can be the relatively new evolving concept of a Global Service Delivery Platform GSDP under the FOT (Federated, Open and Trusted) platform model [14]. As Figure 1 illustrates, the GSDP vision can represent one single point of access to a federation/network of interoperable urban platforms (including both experimental and deployed IoT platforms). In that way, an increasing number of Smart Cities' services could be searched, discovered and composed (following Web 2.0/Telco2.0 principles and including QoS, trust, security, and privacy) in a standard, easy and flexible way. Now that a number of different approaches towards future GSDP are being addressed in several EU research projects such as SOA4ALL, SLA@SOI, MASTER, NEXOF-RA, etc. (as stated in the 2008 FIA meeting: "One project on its own cannot develop a GSDP" [20] [15]), the Smart Cities can represent an extraordinary rich ecosystem to promote the generation of massive deployments of city-scale applications and services for a large number of activity sectors. Furthermore this will enable future urban models of convergent IT/Telecom/Content services, Machine-to-Machine (M2M) services, or entirely new service delivery models simultaneously involving virtual and real worlds.

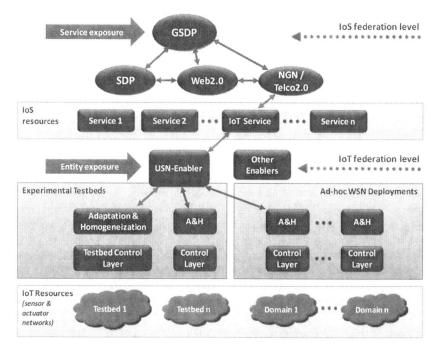

Fig. 1. Global Service Delivery Platform (GSDP) integrating IoT / IoS building blocks

3   Developing Urban IoT Platforms

At present, some works have been reported of practical implementations in order to develop IoT platforms inspired by the Ubiquitous Sensor Networks concept from the ITU-T USN Standardization Group [21]. Some research teams have already initiated activities in this line, but there are currently very few references to them in the literature. ITU's USN concept envisions a *"technological framework for ambient sensor networks not as simple sets of interconnected networks but as intelligent information infrastructures"*. The concept translates directly into cities, as they can be considered as one multi-dimensional eco-system, where data is binding the different dimensions, as most aspects are closely related (e.g. environment and traffic, both of them to health, etc.).

3.1   USN Functionalities

The main goal of a USN platform is to provide an infrastructure that allows the integration of heterogeneous and geographically disperse sensor networks into a common technological ground where services can be developed in a cost efficient manner. Consequently, at urban-scale, a USN platform can represent an invaluable infrastructure to have access and manage the huge amount of sensor and devices connected in Smart City environments. Through a set of basic functionalities it will support different types of Smart City services in multiple application areas:

- Sensor Discovery: this functionality will provide services and applications information about all the registered sensors in the city. In that way, a particular service interested in finding information (such as available parking places in a given area) will have access to efficient look-up mechanisms based on the information they provide.
- Observation Storage: many Smart City services will rely on continuously generated sensor data (for example for energy monitoring, video surveillance or traffic control). This functionality will provide a repository where observations / sensors' data are stored to allow later retrieval or processing, to extract information from data by applying semantic annotation and data linkage techniques.
- Publish-Subscribe-Notify: in other cases, services rely on some specific events happening in the city (such as traffic jams or extreme pollution situations). The platform will allow services to subscribe not just to the observations provided by the sensors, but also to complex conditions involving also other sensors and previous observations and measurements.
- Homogeneous Remote Execution capabilities: This functionality allows executing tasks in the sensor/actuator nodes, so city services could either change sensor configuration parameters (i.e. the sensibility of a critical sensor) or to call actuator commands (as, for example, closing a water pipe).

Another important set of capabilities provided by a USN platform is related to the need for a homogeneous representation and access to heterogeneous urban information, as it was discussed in Section 1. In this sense, the main basic principles covered by USN platforms are:

- Unified information modeling: The information should be provided to the Smart City services using a unified information model, regardless of the particular information model used by the sensor technologies deployed through the city infrastructure. This principle should be applied both to the sensor descriptions and to observations.
- Unified communication protocol: given the extension of an urban area, several standards can co-exist to communicate sensors and sensor networks (ZigBee, 6LowPan, ISA-100.11.a, xDSL, GPRS, etc.). Services should be agnostic to the communication protocol used. The platform should provide access to the information regardless the particular underlying communication protocol used.

- Horizontally layered approach: The platform should also be built following a layered approach, so services and networks are decoupled in order to evolve independently [22]. This capability will allow a seamless link between IoT and IoS, as discussed in Section 2.

Also relevant will be the definition of open APIs, so that USN platforms could provide support for third-party's agents interested in the deployment of different Smart City services, thus allowing federation with different service creation environments and different business processes.

3.2   USN Architecture for Urban IoT Platforms

While the new wave of Next Generation IoT platforms are expected to be defined by initiatives and projects like IoT-A [23], the IERC cluster [24] or the emerging PPP IoT Core Platform Working Group discussion [25], multiple different approaches for First Generation IoT-platforms are currently being implemented. In essence, many of them are realizations of the described ITU-T's model. For reference on the current state of the technology, this Section describes a practical USN platform implementation (more details can be found in [22]), integrated into the Next Generation Networks Infrastructures [35], as one of the most remarkable currently reported solutions for advanced IoT platforms. As shown in Figure 2, a functional specialization of the building blocks has been applied in this work.

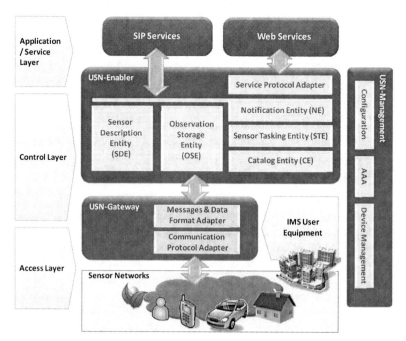

Fig. 2. High-Level Architecture of a USN IoT Platform

As sketched in the figure, the USN platform is based on two components, the *USN-Enabler* (that interfaces services) and the *USN-Gateways* (that interacts with Sensor networks). This approach is inspired by the Open Geospatial Consortium (OGC) Sensor Web Enablement (SWE) activity [26]. Its goal is the creation of the foundational components to enable the *Sensor Web* concept, where services will be capable to access any type of sensors through the web. This has been reflected by a set of standards used in the platform (SensorML, Observation & Measurements, Sensor Observation Service, Sensor Planning Service, Sensor Alert Service and Web Notification Service [26]). Besides the SWE influence, the USN-Enabler relays on existing specifications from the OMA Service Environment (OSE) [27] enablers (such as presence, call conferencing, transcoding, billing, etc.). Especially important has been the Presence SIMPLE Specification (for publish and subscribe mechanisms to sensor information) and XML Document Management, also known as XDM (for XML information modeling in Service Enablers).

The **USN-Gateway** represents a logical entity acting as data producers to the USN-Enabler that implements two main adaptation procedures to integrate physical or logical Sensor and Actuator Networks (SANs):

- Communication Protocol Adaptation. As a connection point between two networks (sensors networks deployed throughout the city and the core IP communication network), the main responsibility is to provide independence from the communication protocol used by the sensor networks.
- Sensor Data Format Adaptation. This functionality is intended to provide USN-Enabler both SensorML (meta-information) and O&M (observation & measurements) data from specific SANs data (i.e. ZigBee).

Adaptation and Homogenization are two key requirements for the USN Platform aiming at its integration with different Smart Cities' testbeds and experimental deployments. They are also essential requirements for a successful seamless integration, and the proper basement for the new heterogeneous sensor network infrastructures needed to enable an evolving FI based on the IoT and IoS paradigms.

Functionalities required to support services are offered both in synchronous and asynchronous mode by the **USN-Enabler** through the following entities:

- The *Sensor Description Entity* (SDE) is responsible for keeping all the information of the different city sensors registered in the USN-Platform. It uses the SensorML language.
- The *Observation Storage Entity* (OSE) is responsible for storing all the information that the different sensors have provided to the platform related to the different Smart City resources and Entities of Interest. This information is stored using the OGC® O&M language.
- The *Notification Entity* (NE) is the interface with any sensor data consumer that require filtering or information processing over urban-generated data. The main functionalities provided by this entity are the subscription (receive the filter that will be applied), the analysis of the filters (analyze the filter condition) and the notification (notify the application when the condition of the filter occurs).

- The *Sensor Tasking Entity* (STE) allows services to perform requests operations to the sensor network, like for example a request to gather data, without the need to wait for an answer. The service will receive an immediate response and, when the desired data gets available it will receive the corresponding alert. This is mainly used for configuration and for calling actuators.
- The *Service Protocol Adapter* (SPA) provides protocol adaptation between the Web Services and SIP requests and responses.
- The *Catalogue and Location Entity* (CLE) provides mechanisms in a distributed environment to discover which of the different instances of the entities is the one performing the request a user might be interested in. For example, in an architecture where several Sensor Description Entities (SDEs) exist, a client might be interested in a particular sensor deployed in a particular city zone. The client should interrogate the CLE to know which particular existing SDEs in the requested urban area contain the information needed.

4   The Need of Urban Scale Experimental Facilities

Experimentally-driven research is becoming more and more important in ICT research. When designing heterogeneous large scale systems, difficulties arise in modeling the diversity, complexity and environmental conditions to create a realistic simulation environment. The consequence is clear: simulation results can only give very limited information about the feasibility of an algorithm or a protocol in the field.

In many cases, due to practical and outside plant constraints, a number of issues arise at the implementation phase, compromising the viability of new services and applications. Most of these problems are related to scalability aspects and performance degradation. The level of maturity achieved at the networking level, despite the fact that they can be further improved, foresees an increasing necessity of additional research activity at the sensor and context information management level [17]. Nevertheless, FI research is no longer ending at the simulation stage. Advances in sensor networking technologies need field validation at large scales, also posing new requirements on the experimentation facilities. Besides, new cross-layer mechanisms should be introduced to abstract the networking level from the higher ones, so new services and information management activities can be performed over heterogeneous networking technologies. This increasing demand to move from network experimentation towards service provisioning requirements does not just apply to the Smart Cities field, but also in a more generic way it is common to most FI experimentation areas.

4.1   Smart Cities as Open Innovation Platforms

To perform reliable large scale experimentation, the need of a city scale testbed emerges. Setting such an experimental facility into a city context has several reasons: the first one is the extent to which the necessary infrastructure of a Smart City will rely on technologies of the IoT. The resulting scale and heterogeneity of the environment makes it an ideal environment for enabling the above mentioned broad range of experimentation needs. Furthermore, a city can serve as an excellent catalyst for IoT research, as it forms a very dense techno-social eco-system. Cities can act as invaluable source of challenging functional and non-functional requirements from a variety of problem and application domains (such as vertical solutions for the environment control and safety, horizontal application to test network layers, content delivery networks, etc.). They provide the necessary critical mass of experimental businesses and end-users that are required for testing of IoT as well as other Future Internet technologies for market adoption. This new smart city model can serve as an excellent incubator for the development of a diverse set of highly innovative services and applications [18].

For all these reasons, systems' research in ICT needs more powerful and realistic tools, which can only be provided by large-scale experimental facilities. At a urban scale, this approach yields to a rising importance of non-technical practical issues, like interworking at the organizational level, or even administrative and political constraints. There are very few initiatives addressing the creation of such smart city environments. Some examples are Oulu in Finland [28], Cambridge, Massachusetts [29], or Friedrichshafen, Germany [30]. Most recent and interesting initiatives are Sense Smart City in Skellefteå, Sweden [31], and SmartSantander [32] at a European level. The first one is a Swedish project designed to conduct research, create new business opportunities and sustainably increase ICT research and innovation capability with specific objective to make urban cities/areas "smarter". The project will generate new and better ICT solutions that instrument urban areas to gather and combine information (energy, traffic, weather, events, activities, needs and opinions) continuously as well as "on-demand". This will enable city environments to become "smarter", as more adaptive and supportive environment, for people as well as organizations.

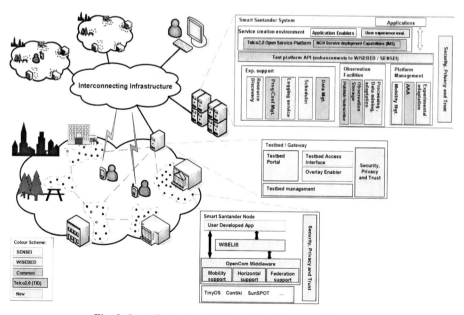

Fig. 3. SmartSantander: A city-scale platform architecture

4.2   SmartSantander Experimental Research Facility

At urban scale, SmartSantander [32] represents the most challenging reference nowa-days, and aims at creating a unique-in-the-world European experimental test facility for the research and experimentation of architectures, key enabling technologies, services and applications for the IoT. The facility will allow large-scale experimenta-tion and testing in a real-world environment. The infrastructure will be mainly de-ployed in Santander in the North of Spain, with nodes in Guildford, UK; Lübeck, Germany; Belgrade, Serbia; Aahrus, Denmark and Melbourne, Australia. Apart from providing state of the art functionalities, the facility will count with advanced capa-bilities to offer support for experimentation of different FI components. Some of these additional functionalities that are required to support experimental activities can be identified in Figure 3. Primarily, the platform is being designed to allow the experi-mental evaluation of new research results and solutions in realistic settings. By con-ception, it will also make possible the involvement of real end-users in the first ser-vice design phases, applying user-driven innovation methodologies. Furthermore, it will be also used to provide real services to citizens. SmartSantander experimental facility is not envisaged as a closed, standalone system. Instead, it is being designed to become an open experimental research facility that can be easily expanded and feder-ated with other similar installations (i.e. through OneLab2 [34]).

A key aspect in SmartSantander project is the inclusion of a wide set of applica-tions. Application areas are being selected based on their high potential impact on the citizens as well as to exhibit the diversity, dynamics and scale that are essential in advanced protocol solutions, and will be able to be evaluated through the platform. Thus, the platform will be attractive for all involved stakeholders: industries, commu-nities of users, other entities that are willing to use the experimental facility for de-ploying and assessing new services and applications, and Internet researchers to vali-date their cutting-edge technologies (protocols, algorithms, radio interfaces, etc.).

Several use cases are currently under detailed analysis for their experimental de-ployment taking into account relevant criteria from local and regional authorities. An illustrative list of these use cases is:

- Monitoring several traffic-related events in the city such as: dynamic occupation, mapping and control of limited parking zones, places for disabled people, load/unload areas restricted to industrial; as well as traffic management services: creation of corridors for emergency vehicles, *ecoways* enablement proposing alter-native routes for vehicles based on pollution monitoring in different city zones.
- Tracking and monitoring of people with disabilities, especially mental disorders (Down syndrome, Alzheimer's disease, dementia, etc.) or heart disease that may require constant monitoring by their families or physicians.
- Alert services that, orchestrating several services such as such eHealth, environ-mental monitoring, traffic control and communication services, will inform and/or alert citizens of different critical situations (i.e. urgent medical attention, city ser-vices recommendations, etc.)
- Tourism information in different parts of the city through mobile devices using visual and interactive experiences and in different languages.

- Video monitoring for traffic areas, beach areas and specific events in public places, such as airports, hotels, train stations, concerts and sport stadiums.
- Smart metering monitoring in buildings and houses for electric energy, water and gas control consumption, including real-time information to the citizens on their own consumption and environmental impact.

Based on these, and future, use cases a main goal in SmartSantander project will be to identify a detailed set of functional (required capabilities) and non-functional (required constraints) requirements for a urban-scale ICT platform. A major challenge will be to leverage on state-of-the-art experimental, research and service oriented initiatives on both IoT and IoS areas as WISEBED [25], SENSEI [8] and the USN IoT Platform (presented in Section 3) including Web 2.0 and Telco 2.0 design principles. Additionally, the requirements elicitation process in SmartSantander will also consider the following viewpoints: the FIRE testbed user, the service provider, the service consumers (citizens), the SmartSantander facility administrators, and individual testbed administrators. The SmartSantander initial architectural model specifies the subsystems (collectively, the SmartSantander middleware) that provide the functionality described by these requirements and is expected to accommodate additional requirements coming up from the different smart city services (use cases). A set of basic subsystems can be identified: i) Access Control and IOT Node Security subsystem, ii) Experiment Support Subsystem, iii) the Facility Management Support Subsystem, and iv) the Application Support Sub-system. The architectural reference model also specifies, for each sub-system, required component deployments on the IoT nodes, and component interactions and information models used to fulfill the subsystem's functionality.

In summary, it can be said that main benefits underlying the SmartSantander partnership is to fuel the use of its urban-scale experimentation facility among the scientific community, end users and service providers. This will not only reduce the technical and societal barriers that prevent the IoT concept to become an everyday reality, but also will attract the widest interest and demonstrate the usefulness of the Smart-Santander platform.

Finally, it must be noticed that financial aspects are of the utmost importance considering the investment required to deploy city scale testbeds. For this reason, apart from serving to its research purposes, it is essential, when planning a Smart City open innovation platform, to introduce requirements allowing the support of real life services simultaneously. This will be very useful to open new business opportunities and, at least and not less important, provide the means to guarantee its day by day maintenance.

5   Conclusions

Future Internet potential, through IoT and IoS, for creating new real-life applications and services is huge in the smart city context. First time success of large IoT deployments is seriously jeopardized by the lack of testbeds of the required scale, and suitable for the validation of recent research results. Many existing testbeds just offer experimentation and testing limited to small domain-specific environments or application specific deployments. While those may suffice as proof-of-concepts, they do not allow conclusive experimentation with the developed technologies and architectural models, evaluation of their performance at an adequate scale under realistic operational conditions and, validation of their viability as candidate solutions for real life IoT scenarios.

At present, some practical implementations of advanced USN platforms [22] have been successfully demonstrated in real deployments for smart metering services, smart places scenarios, and environmental monitoring systems. Ongoing activities are extending its scope to broader M2M scenarios, and large scale deployments for experimental smart urban spaces. The described implementation has shown a big potential to create a fan of new services, providing the key components required to intertwining IoT and IoS worlds. Referred IoT USN platform is currently being evolved with the addition of new capabilities, and integrated within other components being previously developed by the EU projects SENSEI [8] and WISEBED [33] to implement a city scale infrastructure for IoT technologies experimentation within the SmartSantander project. In this project, a large infrastructure of about 20,000 IoT devices is addressed. Currently, the deployment of the first 2,000 sensors in the urban environment is been carried. Non-technical aspects are also of a big importance. The cardinality of the different stakeholders involved in the smart city business is so big that many non-technical constraints must be considered (users, public administrations, vendors, etc.). In this sense, what may be evident from a purely technique perspective it is not so clear when politics and business meet technology. Besides, and although market claims its readiness for supporting a vast range of sensing capabilities as well as the corresponding end-user services, the real situation is quite far different. Nowadays, there are no field experiences across the world allowing assessing, in the short term, the behavior of massive wireless sensor deployments.

Acknowledgements. Although only a few names appear on this paper, this work would not have been possible without the contribution and encouragement of many people, particularly all the enthusiastic team of the SmartSantander project, partially funded by the EC under contract number FP7-ICT-257992.

References

1. Smart Cities, Ranking of European medium-sized cities, http://www.smart-cities.eu/
2. The ICT behind cities of the future, http://www.nokiasiemensnetworks.com/news-events/publications/unite-magazine-february-2010/the-ict-behind-cities-of-the-future
3. Simonov, M.: Future Internet applications relevant for smart cities, an ICT application area example: smart & proactive energy management, Open Innovation by FI-enabled services, Brussels, 15 January (2010)

4. Position Paper: Research Challenges for the Core Platform for the Future Internet. In: M. Boniface, M. Surridge, C.U (Eds.) http://ec.europa.eu/information_society/activities/foi/library/docs/fippp-research-challenges-for-core-platform-issue-1-1.pdf
5. Sundmaeker, H., Guillemin, P., Friess, P., Woelfflé, S. (eds.): Vision and Challenges for Realising the Internet of Things, CERP-IoT, March 2010. European Commission, Brussels (2010)
6. Future Internet Assembly 2009, Stockholm, Sweden (November 2009), http://ec.europa.eu/information_society/activities/foi/library/docs/fi-stockholm-report-v2.pdf
7. The European Network of Living Labs, http://www.openlivinglabs.eu/
8. SENSEI – Integrating the Physical with the Digital World of the Network of the Future. State of the Art – Sensor Frameworks and Future Internet (D3.1). Technical report (2008)
9. Belissent, J.: Getting Clever About Smart Cities: New Opportunities Require New Business Models, 2 November 2010. Forrester Research (2010)
10. EC FI-PPP: http://ec.europa.eu/information_society/activities/foi/lead/fippp/index_en.htm
11. Towards a Future Internet Public Private Partnership, Usage Areas Workshop, Brussels, 3 March (2010), http://ec.europa.eu/information_society/activities/foi/events/fippp3/fi-ppp-workshop-report-final.pdf
12. Real World Internet (RWI) Session, FIA meeting, Prague (May 2009), http://rwi.future-internet.eu/index.php/RWISession_Prague
13. COM: A public-private partnership on the Future Internet. Brussels, 28 October (2009), http://ec.europa.eu/information_society/activities/foi/library/docs/fi-communication_en.pdf
14. DG INFSO Task Force on the Future Internet Content.Draft Report of the Task Force on InterdisciplinaryResearch Activities applicable to the Future Internet, Version 4.1 of 13.07.2009 (2009), http://forum.future-internet.eu
15. NESSI Strategic Research Agenda, http://www.nessi-europe.com/files/ResearchPapers/NESSI_SRA_VOL_3.pdf
16. Gluhak, A., Bauer, M., Montagut, F., Stirbu, V., Johansson, M., Bernat-Vercher, J., Presser, M.: Towards an architecture for a Real World Internet. In: Tselentis, G., et al. (eds.) Towards the Future Internet, IOS Press, Amsterdam (2009)
17. Fisher, S.: Towards an Open Federation Alliance. The WISEBED Consortium. Lulea, July 2nd, 2009.22. In: Balazinska, M., et al. (eds.) Data Management in the Worldwide Sensor Web. IEEE PERVASIVE computing, April-June (2007)
18. Panlab Project, Pan European Laboratory Infrastructure Implementation, http://www.panlab.net/fire.html
19. Global service delivery platform (GSDP) for the future internet: What is it and how to use it for innovation?, http://services.future-internet.eu/images/d/d4/Report GSDPpanel-FISO-FIA-Madrid-draft%2Breqs.pdf
20. Future Internet Assembly, Meeting Report, Madrid, Spain, 9th–10th December (2008), http://ec.europa.eu/information_society/activities/foi/library/docs/madrid-conference-report-v1-1.pdf
21. ITU TSTAG: A preliminary study on the Ubiquitous Sensor Networks. TSAG-C 22-E (February 2007)
22. Bernat, J., Marín, S.: González, A., Sorribas, R., Villarrubia, L., Campoy, L., Hernández, L. Ubiquitous Sensor Networks in IMS: an Ambient Intelligence Telco Platform. ICT Mobile Summit, 10-12 June, Stockholm (2008)
23. Internet of Things Architecture project, http://www.iot-a.eu/public/front-page

24. IoT European Research Cluster, http://www.internet-of-things-research.eu/
25. White paper on the FI PPP definition (Jan. 2010), http://www.future-internet. eu/fileadmin/initiative_documents/Publications/White_Paper/EFII_ White_Paper_2010_Public.pdf
26. Botts, M., Percivall, G., Reed, C., Davidson, J.: "OGC Sensor Web Enablement: Overview and High Level Architecture", Open Geospatial Consortium Inc. White Paper Version 3 (2007)
27. OMA Service Environment Archive, http://www.openmobilealliance.org/ technical/release_program/ose_archive.aspx
28. Oulu Smart City, http://www.ubiprogram.fi/
29. Cambridge (MA) Smart City, http://www.citysense.net/
30. Friedrichshafen Smart City, http://www.telekom.com/dtag/cms/content/dt/en/395380
31. Sense Smart City project, http://sensesmartcity.org/
32. SmartSantander project, FP7-ICT-2010-257992, http://www.smartsantander.eu
33. WISEBED - Wireless Sensor Network Testbeds, http://www.wisebed.eu
34. OneLab2, ONELAB project, http://www.onelab.eu/
35. NGNI, http://www.fokus.fraunhofer.de/en/ngni/index.html

Author Index